ENERGY CONSERVATION STANDARDS

McGRAW-HILL SERIES IN MODERN STRUCTURES:
Systems and Management
Thomas C. Kavanagh   Consulting Editor

Baker, Kovalevsky, and Rish   *Structural Analysis of Shells*
Coombs and Palmer   *Construction Accounting and Financial Management*
Desai and Christian   *Numerical Methods in Geotechnical Engineering*
Dubin and Long   *Energy Conservation Standards*
Forsyth   *Unified Design of Reinforced Concrete Members*
Foster   *Construction Estimates from Take-Off to Bid*
Gill   *Systems Management Techniques for Builders and Contractors*
Johnson   *Deterioration, Maintenance and Repair of Structures*
Johnson and Kavanagh   *The Design of Foundations for Buildings*
Kavanagh, Muller, and O'Brien   *Construction Management*
Kreider and Kreith   *Solar Heating and Cooling*
Krishna   *Cable-Suspended Roofs*
O'Brien   *Value Analysis in Design and Construction*
Oppenheimer   *Directing Construction for Profit*
Parker   *Planning and Estimating Dam Construction*
Parker   *Planning and Estimating Underground Construction*
Pulver   *Construction Estimates and Costs*
Ramaswamy   *Design and Construction of Concrete Shell Roofs*
Schwartz   *Civil Engineering for the Plant Engineer*
Tschebotarioff   *Foundations, Retaining and Earth Structures*
Tsytovich   *The Mechanics of Frozen Ground*
Woodward, Gardner, and Greer   *Drilled Pier Foundations*
Yang   *Design of Functional Pavements*
Yu   *Cold-formed Steel Structures*

ENERGY CONSERVATION STANDARDS

For Building Design, Construction, and Operation

FRED S. DUBIN, M.E., P.E., M.Arch.

CHALMERS G. LONG, Jr., Architect

McGraw-Hill Book Company

New York   St. Louis   San Francisco   Auckland   Bogotá   Düsseldorf   Johannesburg
London   Madrid   Mexico   Montreal   New Delhi   Panama
Paris   São Paulo   Singapore   Sydney   Tokyo   Toronto

Library of Congress Cataloging in Publication Data

Dubin, Fred S

Energy conservation standards for building design,
construction, and operation.

Includes index.
1. Buildings—Energy conservation. I. Long,
Chalmers G., joint author. II. Title.
TJ163.5.B84D79   697   78-6189
ISBN 0-07-017883-6

1234567890   FGRFGR   7654321098

Portions of this book have been taken from the following documents
previously published by the U.S. Government Printing Office:

Energy Conservation Manual I
Energy Conservation Manual II
Guidelines for Office Buildings

This material, written by Dubin-Bloome Associates under contract to
the Federal Energy Administration and the General Services
Administration, has been extensively edited, revised, and combined
with new material to form the present volume.

The editors for this book were Jeremy Robinson and Joseph
Williams, the designer was Elliot Epstein, and the production
supervisor was Thomas G. Kowalczyk. It was set in Caledonia by
University Graphics, Inc.

Printed and bound by Fairfield Graphics.

CONTENTS

PREFACE

We speak again of the litany of American energy waste: Energy is in short supply. It is expensive. Buildings require too much of it to keep people warm or cool, to light their work, or to power their equipment.

This book is written for architects, engineers, and other people who want or need to reduce energy usage in their own buildings, or in the buildings for which they are responsible. It is a book which attempts to establish a comprehensive framework for understanding energy flow in buildings and which offers a detailed checklist for the design of energy management programs.

Chapters 1 through 5 give scope to the problem of energy management in new and existing buildings and illustrate the potential for conservation. Case studies are made of a new building, the Federal Office Building in Manchester, New Hampshire, and an existing building, the Chemistry Building of the Brookhaven National Laboratory, Long Island, New York.

The heart of the book, Chapters 6 through 12, describes energy flows in buildings. It is organized in three parts: (1) energy which is used to provide environmental control and to power lighting and equipment—*the building load;* (2) energy which is used simply to move air, water, or electricity from one point of the building to another to satisfy the building load—*the distribution load;* (3) energy required to convert raw fuels or electricity into forms useful for heating or cooling with boilers or furnaces or refrigeration equipment—*the conversion load.* The chapter subheadings make up a plain-language checklist of energy conservation options.

The closing chapters (13 through 15) provide information on central control systems, economic analysis, and alternative energy sources.

This book is an outgrowth of an earlier set of manuals—*Guidelines for Saving Energy in Existing Buildings, ECM 1,* and *ECM 2*—which were written for the Federal Energy Administration by Dubin-Bloome Associates in 1975. *ECM 1* and *ECM 2* were published to help building owners, architects, and engineers implement and quantify energy management programs in existing buildings to conserve energy. This book mainly follows that orientation to existing buildings in example, language, and assumption, but principles of good energy management in existing buildings clearly apply with equal force to the design of new buildings. Additional principles relating to options only possible in new building—site adaptation, building planning, and the design of the building shell—are a part of Chapter 2, New Buildings. This material is taken from *Energy Conservation Guidelines for New Buildings,* prepared for the General Services Administration by Dubin-Bloome Associates, The American Institute of Architects Research Corporation, and Heery and Heery, Architects. Costs which are given throughout this text for first approximation of energy savings represent 1975 New York City costs and have not been escalated to account for any inflation since then.

Buildings are shaped by so many forces, but finally they are shaped by the expectations of our people. We look forward to a time when the primary demand by society of its architects and engineers will be for buildings that are not just energy-efficient in their equipment but are closely attuned to the sun and the winds of their environments, and that are wholly energy-responsive in their designs.

Fred S. Dubin, *M.E., P.E., M.Arch*
Chalmers G. Long, Jr., *Architect*

ACKNOWLEDGMENTS

Portions of the information contained in this book were developed by Dubin-Bloome Associates, P.C., Consulting Engineers and Planners under contract to the U.S. General Services Administration and the Federal Energy Administration under the direction of Fred S. Dubin. Many members of the firm participated in the studies and made valuable contributions to the project. Their efforts and those of the following individuals and agencies are gratefully acknowledged:

Selwyn Bloome, P.E., Partner—*Dubin-Bloome Associates, P.C.*
Robert Sparkes, P.E., Associate—*Dubin-Bloome Associates, P.C.*
Lou Mutch, P.E., Associate—*Dubin-Bloome Associates, P.C.*
Karl Christensen, Associate—*Dubin-Bloome Associates, P.C.*
Lawrence Newman, P.E., Associate—*Dubin-Bloome Associates, P.C.*
Thomas Ware and David Hattis—*Building Technology, Inc.*
Charles McGlure—*Charles J. R. McGlure, P.E. Associates*
William Adams—*Hammer, Siler George Associates*
Raymond Firmin—*G.A. Hanscomb Partnership*
George Heery, Architect—*Heery & Heery, Architects*
Walter Meisen—*General Services Administration*
Dr. Maxine Savitz and Steve Cavros—*Department of Energy*
David Rosoff—*Housing and Urban Development*
John Eberhard and Margo Villecco—*A.I.A. Research Corporation*

We are especially thankful to the following individuals for their efforts in helping produce the final manuscript:

Deirdre McCrystal and Art Sturz—*graphics*
Yvonne McDougal—*production manager*
Sarah Dubin-Vaughn—*editing*
Bernice Singh-Beharry and Marge Haffer—*typing*

ENERGY
CONSERVATION
STANDARDS

CHAPTER ONE

AN INTRODUCTION: ENERGY UTILIZATION IN BUILDINGS

1.1 ENERGY BOUNDARIES

Energy usage in buildings can be classed in many ways, by several dimensions. A first, useful classification is by energy purchased, that is, the building energy consumption in gallons of oil, pounds of coal, cubic feet of gas, pounds of steam, or kilowatthours of electricity. This is the classification by which monthly utility bills are paid, dollars per fuel unit. If all fuels are reduced to a common energy equivalent such as the British thermal unit (Table 1.1), their total is the final number by which building consumption can be measured. This number divided by the gross building area yields the energy usage in Btu per square foot, a most valuable unit for the comparative measure of building performance. Utility bills for energy are based on the measure of the energy sources that come through the building envelope, that is, the building boundary.

Coal, oil, and gas are raw fuels commonly converted to heat by on-site combustion. The efficiency of this conversion will be considered in detail in Chap. 12 as a major element of building energy management. But electricity is a secondary source, and, typically, raw-fuel conversion through combustion for the production of electricity takes place off-site. The losses of conversion are hidden from view, but these losses are high—65 to 70% of the raw fuel burned—and should not be discounted. For a more complete understanding of the costs for building energy sources, the building boundary can be extended to include the central electric generating plant.

The numbers and graphs in this book show energy usage within the building boundary (or building-complex boundary) regardless of source, generally reduced to the Btu common denominator. In several examples, however, the larger boundary that includes the power plant is used in computing raw-source requirements for the building in Btu per square foot by assuming 30% efficiency for electric production and distribution.

In a few similar cases where steam is generated off-site, raw-source fuel requirements are computed with an efficiency for steam manufacture and distribution of 70%.

Beyond the power plant or central steam generating plant there is yet another boundary that in a larger frame of reference must bear consideration with respect to national policy and concern. Ultimately, the value of an energy source is measured by drawing an inclusive envelope around losses in mining, drilling, pumping, refining, and distribution to the building consumer. How much energy is used to produce energy, and how much energy is required to deliver energy, will become increasingly critical as the nation develops ever-scarcer domestic resources.

1.2 ENERGY USE BY BUILDING SYSTEMS

Once inside the building boundary, energy is consumed for various functions of comfort, utility, and convenience and allocated by common title to heating, cooling, lighting, power for equipment, and heating for domestic water. These functions are arranged by common title below in order of national energy consumption; heating is the largest share, and domestic hot water the least.

- Heating

- Lighting

- Cooling

- Power for equipment

- Domestic hot water

The order of magnitude for these five variables will differ in any individual building with climate, with the function of the building, and with extremes of building systems design, construction, and operation.

In climatic zones with mild winters (below 2500 degree-days, see Fig.

TABLE 1.1   ENERGY EQUIVALENTS

No. 2 oil	138,000 Btu/gal
No. 6 oil	146,000 Btu/gal
Natural gas	1,000 Btu/ft³
Manufactured gas	800 Btu/ft³
Coal	26,000,000 Btu/short ton
Steam	900 Btu/lb
Propane gas	21,500 Btu/lb
Electricity	3,413 Btu/kWh

4.1), the seasonal cooling load may be larger than the seasonal heating load and may consume more energy (depending upon the respective efficiencies of the heating and cooling systems). In office buildings, schools, and retail stores in such mild zones, the electrical load for lighting, which is largely independent of climate, may exceed that for either heating or cooling. Buildings used for only a few hours per week, such as auditoriums and religious buildings, however, may consume more energy for heating unless indoor temperatures are set back during unoccupied periods and boiler or furnace efficiencies are kept high.

In cold climates, 6000 degree-days annually and above, heating in office buildings and schools usually consumes the most energy per year, with lighting and then cooling next. For retail stores in this zone, the most likely order of energy use is lighting-heating-cooling, or lighting-cooling-heating. Generally, in this zone heating consumes the most energy for religious buildings or other buildings used for only a few hours per week, with lighting and cooling following in order.

In mid-climates, 2500 to 6000 degree-days annually, the ordering of categories depends largely upon the type of mechanical and electrical systems and the characteristics of the building structure in which they are installed. The energy required for industrial buildings exclusive of process loads is generally similar to that required for commercial buildings in all zones.

The amount of energy required for domestic hot water is significant in hospitals, housing, and athletic and cooking facilities in schools and colleges. In housing, the amount of energy to heat water is frequently second only to space heating in the North and to air conditioning in the South. In hospitals anywhere, the amount of energy to heat water may exceed the amount of energy required for lighting.

Institutional buildings which include meeting rooms, offices, and school facilities are most likely to consume energy in the same pattern as office buildings in the same geographic location—but in smaller quantities per square foot of floor area.

In retail stores with high levels of general illumination and display lighting or with large numbers of commercial refrigeration units, electricity consumes the greatest amount of energy.

The matrix shown in Table 1.2 rates systems by building types and climates in the general order of annual energy usage, with 1 the greatest and 5 the least. Any particular building must be analyzed individually, however, to determine the pattern of actual annual usage.

The data for Fig. 1.1, which were produced by Dubin-Bloome Associates and the National Bureau of Standards with computer analysis, illustrate the relative magnitude of energy consumption for an office building located in a cold climate (Manchester, New Hampshire) and for the identical building if located in a warm climate (Orlando, Florida). Heating is the single largest user of energy for the office building in New Hampshire, and the amount used for cooling is relatively small. In Florida the energy used for heating has decreased, and cooling requires the greater amount.

The actual annual energy consumed by a building is a complex of operating procedures, equipment performance, and functional requirements, each of which is variable within broad extremes. The following example, a simulated building modeled to represent those frequently encountered, is included to give some feeling for the contributions of the several variables and for the changes that are typically made with an energy management program.

TABLE 1.2   COMPARATIVE ENERGY USE BY SYSTEM*

	Heating and Ventilating	Cooling and Ventilation	Lighting	Power and Process	Domestic Hot Water
Schools					
A	4	3	1	5	2
B	1	4	2	5	3
C	1	4	2	5	3
Colleges					
A	5	2	1	4	3
B	1	3	2	5	4
C	1	5	2	4	3
Office buildings					
A	3	1	2	4	5
B	1	3	2	4	5
C	1	3	2	4	5
Stores					
A	3	1	2	4	5
B	2	3	1	4	5
C	1	3	2	4	5
Religious buildings					
A	3	2	1	4	5
B	1	3	2	4	5
C	1	3	2	4	5
Hospitals					
A	4	1	2	5	3
B	1	3	4	5	2
C	1	5	3	4	2

*Climatic Zone A: Fewer than 2500 degree-days.
Climatic Zone B: 2500 to 5500 degree-days.
Climatic Zone C: 5500 to 9500 degree-days.

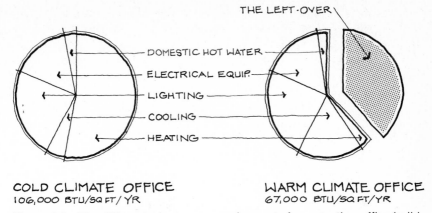

COLD CLIMATE OFFICE
106,000 BTU/SQ FT/YR

WARM CLIMATE OFFICE
67,000 BTU/SQ FT/YR

Figure 1.1 The difference in energy requirements for operating office buildings in warm and cold climates.

Example: An Office Building, Chicago, Illinois

The gross area per floor is 100 ft × 100 ft = 10,000 ft²; for 10 stories, the total gross area is 100,000 ft². The floor-to-floor height is 10 ft, and there is a 33% window/wall ratio.

Conditions before Operational Changes

Clear single glazing, U = 1.1. Negligible interior shading assumed during heating season. Shading by venetian blinds during cooling season; shading coefficient (SC) = .67. For wall, U = 0.3 (excluding glazed areas); for roof, U = 0.2. Occupied 40 hours per week. Outdoor ventilation air = 19,800 cfm at 30 cubic feet per minute (cfm) per person × 660 occupants. Average infiltration = 1/2 air change per hour. Interior heat gains: light = 4.0 W/ft², office equipment = 0.5 W/ft², fans = 1.0 W/ft²; total = 5.5 W/ft². No cooling during unoccupied periods. Indoor temperature is 75°F during the heating season, and 75°F during the cooling season, with 50% RH. Domestic hot-water flow rate = 2 gpd per person at 140°F. See Tables 1.3 to 1.5 for summaries of energy usage and savings.

TABLE 1.3 SITE ENERGY USED BEFORE CONSERVATION

Purpose	Btu/(ft²)(yr)
Heating (oil)	106,400
Cooling (electricity)	8,200
Lighting	31,400
Power	10,600
Domestic hot water (oil)	4,000
Total	160,000

A second example, Fig. 1.2, is included with the same purpose of showing both system requirements and the possibilities for reduction. In this example the interrelationship of two variables is graphically expressed. By reducing the electrical requirement for lighting, a portion of the cooling requirement (to remove the heat gain from the lights) is reduced as well.

1.3 ENERGY USE BY LOAD IDENTIFICATION

The classification of heating, cooling, lighting, power, and domestic hot-water systems is readily understood and useful for comparison purposes. But a somewhat different classification is equally useful, perhaps more so, in actually managing energy flows within buildings. This classification labels energy consumed as loads, that is, building loads, distribution loads, and primary conversion loads. This classification is the framework on which Chaps. 6 to 12 on energy management and conservation are based.

BUILDING LOADS Building loads are energy requirements for the environmental control of temperature, humidity, and ventilation and for building lighting and other equipment, such as elevators, computers, and coffee pots. Where the term *building load* is used in this book, it is a specification of the energy, in Btu or kilowatthours, required to maintain desired indoor space conditions and to operate building equipment, independent of losses in the distribution system and in energy conversion. In other words, only the end-use requirements are considered.

The magnitude of the building load is dependent on the location of the building and climate, the degree of environmental control maintained, the number of occupants and the period of occupancy, the thermal performance of the building structure, and the use made of the building. In general terms related immediately to goals of energy conservation, building loads will be reduced if: (1) the temperature and relative humidity indoors are maintained at lower levels in the winter and at higher levels in the summer; (2) heat loss,

TABLE 1.4   RAW-SOURCE ENERGY USED BEFORE CONSERVATION

Purpose	$Btu/(ft^2)(yr)$
Heating	106,400
Cooling*	27,340
Lighting*	104,660
Power*	35,330
Domestic hot water	4,000
Total	277,730

*Figures for raw-source energy reflect the energy conversion for electricity.

TABLE 1.5 SUMMARY OF ENERGY SAVINGS POTENTIAL

	Annual Site Energy	Btu/(ft²)(yr)	Percent Reduction in this Segment of Energy	Percent Reduction Total Energy
HEATING				
12° night setback	27,000 gal No. 2 oil	38,000	35	24.0
7° day setback	1,000 gal No. 2 oil	1,400	1	0.9
Increase boiler efficiency by 10%	6,600 gal No. 2 oil	9,200	9	6.0
Reduce outside air during occupied periods to 8 cfm per person	1,800 gal No. 2 oil	2,500	2	1.5
Caulk windows	12,000 gal No. 2 oil	16,700	16	10.0
Add storm windows	26,000 gal No. 2 oil	36,400	34	23.0
Add night barrier ($U = .1$)	29,150 gal No. 2 oil	40,000	38	25.0
Selected combinations:				
Increase boiler efficiency 10% plus 12°F night setback	31,000 gal No. 2 oil	43,000	40	27.0
Increase boiler efficiency 10% plus 7°F day setback	9,000 gal No. 2 oil	12,600	12	8.0
Increase boiler efficiency 10% plus caulk windows	17,800 gal No. 2 oil	25,000	23	15.5
COOLING				
Economizer cycle	47,700 kWh	1,600	20	1.0
Reduce lighting to 2.0 W/ft²	31,500 kWh	1,100	13	0.7
LIGHTING				
Reduce lighting to 3.0 W/ft²	226,000 kWh	7,700	25	5.0
Reduce lighting to 2.0 W/ft²	450,000 kWh	15,300	49	9.5
HOT WATER				
Lower water temperature to 100°F	1,070 gal No. 2 oil	1,500	37	1.0
Reduce flow rate to 1 gpd per person	920 gal No. 2 oil	1,250	32	0.8
Combined 100°F temperature plus reduced flow rate	1,990 gal No. 2 oil	2,800	70	2.0

NOTE: The savings for multiple changes are not directly additive.

heat gain, and infiltration through the building envelope are decreased; if outdoor air ventilation rates are reduced; (3) domestic hot-water temperature and quantity are reduced; (4) the level of illumination by electric lighting is lowered; and (5) the number of hours of operation of elevators, business machines, and cooking equipment is reduced. A further image of the nature of the building load is presented with the outline in Table 1.6, which makes up a checklist of energy and use conservation potential.

DISTRIBUTION LOADS   It takes energy to move energy. Distribution loads (often characterized as *parasitic loads*) are inevitable in the best of cooling, heating, and electrical distribution systems. Energy is required to operate fans and to operate pumps for hot, chilled, or condenser water. In addition, some load is expected in the form of fluid leakage or unwanted heat transfer in the heating and cooling ducting and piping systems. In electrical systems, there will be some loss through transformers and in the resistance of conductors and switchgear.

As with building loads, Table 1.6 indicates the scope of distribution loads in checklist form. The distribution load, relative to the building load which is served, is highly variable and dependent in large part on the skill of the designer in systems performance optimization. In certain existing high-velocity air systems designed in the 1960s for high-rise office buildings, before this decade of precious energy, it was not uncommon for pump and fan energy requirements to equal the heating and cooling building load requirements.

CONVERSION LOADS   Energy to satisfy the heating and cooling load is moved typically with some distribution system from prime conversion hardware: boiler, furnace, or refrigeration equipment. The gross energy requirement for the heating and cooling building load and the distribution load is

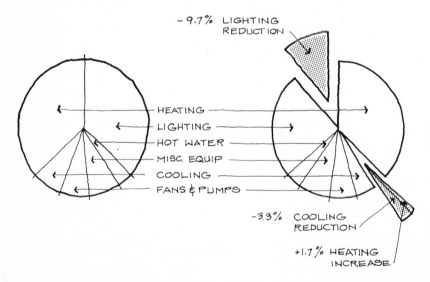

Figure 1.2 A typical reduction in building energy usage following a reduction in lighting levels.

TABLE 1.6 A CONSERVATION CHECKLIST

HEATING (see Chap. 6)

The building heating load: Introduction
Set back indoor temperatures during the heating season for unoccupied periods
Reduce indoor temperatures during occupied periods
Avoid radiation effects to cold surfaces
Reduce levels of relative humidity
Shut down ventilation system during unoccupied hours
Reduce ventilation rates during occupied periods
Reduce rate of infiltration
Increase the solar heat gain into the building
Install controls for space temperature and humidification
Control building ventilation
Use separate makeup air supply for exhaust hoods to reduce outdoor air ventilation
Reduce heat losses through windows
Reduce heat losses through windows with thermal barriers
Reduce heat losses through walls
Reduce heat losses through roofs
Reduce heat losses through floors
Reduce infiltration with building alterations

COOLING (see Chap. 7)

Heat gain and heat loss: Introduction
Increase levels of indoor temperature and relative humidity
Reduce ventilation
Reduce infiltration
Reduce solar heat gains through windows
Reduce heat gains through walls
Reduce heat gains through floors and roofs
Reduce internal heat gains from lights and equipment
Use outdoor air for cooling
Use evaporative cooling or desiccant dehumidification

DOMESTIC WATER HEATING (see Chap. 8)

The nature of the load
Reduce domestic hot-water temperatures
Reduce domestic hot-water quantity
Reduce domestic hot-water system losses
Improve domestic hot-water system performance with equipment modifications

LIGHTING (see Chap. 9)

The demand for light: Introduction
Reduce illumination levels
Improve lighting system operations
Use daylight
Improve space conditions for lighting
Improve lamp and fixture efficacy

TABLE 1.6   A CONSERVATION CHECKLIST (*Continued*)

POWER (see Chap. 10)

The nature of the loads
Reduce energy requirements for elevators and escalators
Reduce energy consumption for equipment and machines
Reduce energy requirements for commercial refrigeration
Reduce peak loads
Reduce transformer losses
Improve the efficiency of motors
Correct power factor

HEATING, VENTILATING, AND AIR-CONDITIONING SYSTEMS (see Chap. 11)

The distribution load for heating and cooling: Introduction
Reduce resistance to flow in air distribution systems
Reduce the volume of flow in air distribution systems
Reduce thermal losses in air distribution systems
Reduce resistance to flow in piping systems
Reduce volume of flow in water distribution systems
Reduce thermal losses in water and steam distribution systems
Direct hot-water and steam-heating systems
Single-duct, single-zone systems
Terminal reheat systems
Multizone systems
Dual-duct systems
Variable-volume air systems
Induction systems
Fan coils and unit ventilators
Heat pumps and air conditioners
Exhaust systems

BOILERS AND REFRIGERATION (see Chap. 12)

Primary conversion loads: Introduction
Improve boiler or furnace efficiency
Install flue gas analyzer
Isolate off-line boilers
Replace existing boilers with modular boilers
Increase boiler and furnace efficiency with preheating and air atomization
Reduce blowdown losses
Improve refrigeration efficiency
Improve compressor-evaporator performance
Increase condenser performance
Improve efficiency with equipment modifications

dependent on the efficiency of the primary conversion equipment. Any improvement in the efficiency of combustion, or in the efficiency of refrigeration heat pumping, will equate directly to overall energy savings. The elements of concern are also listed in Table 1.6.

The organization of building energy requirements into elements of building loads, distribution loads, and conversion loads is of interest, in passing, from another vantage point. The architect, traditionally, has been most concerned, and most expert, in the design and qualification of the building load: the shell of the building with respect to climatic force and fenestration, the lighting of the building, and the organization of the building's parts. The engineer, traditionally, has been most concerned, and most expert, in the design and selection of the distribution and conversion loads: the equipment for heating, cooling, and powering of the building. Not infrequently the engineer of the past has simply quantified the building load without involvement in the building design, and seldom has the architect known the difference between a compression and an absorption chiller.

Such traditions are no longer acceptable for the energy-conscious design of new structures or for the design of energy management programs for existing structures. It is most important that the engineer and architect work closely from the inception of the project and that both extend their expertise and responsibility to all elements of building loads, distribution loads, and conversion loads.

Energy conservation is most logically derived by first reducing building loads, then reducing distribution loads to match building loads (with further load reduction by equipment modification), then finally improving the efficiency of primary conversion equipment (at the same time that this equipment is being adjusted to the reduced total of building and distribution loads).

1.4 ANNUAL ENERGY AND PEAK ENERGY AND COMPUTATION

In the office building of the first example in this chapter (see Tables 1.3 to 1.5), the annual energy requirement before conservation was 160,000 Btu/$(ft^2)(yr)$. For the office building used as an example in Fig. 1.2, the annual energy requirement before conservation measures was 116,000 Btu/$(ft^2)(yr)$, but the Manchester federal office building (New Hampshire) was designed and finally built to an expectation budget of only 55,000 Btu/$(ft^2)(yr)$. A contemporary office building in Houston, Texas, was actually measured at an annual figure of 660,000 Btu/$(ft^2)(yr)$. Energy consumption in buildings varies widely, even for the same building type.

Very little can now be stated as to norms of energy usage by buildings in the United States. Important research is underway to give meaning to energy budgets and energy allowances for buildings, but these data are not yet available. At best we can say, for instance, that stores consume energy at a rate within a particular range and that religious buildings are at somewhat less than the lower end of this range, that hospitals generally use more energy per square foot than stores and that housing uses less, and that schools use less energy per square foot than office buildings. Realistic goals are on the order of 55,000 Btu/(ft^2)(yr) for new office buildings, 75,000 Btu/(ft^2)(yr) for existing office buildings and stores, 60,000 Btu/(ft^2)(yr) for existing schools, and 35,000 Btu/(ft^2)(yr) for existing religious structures. We have now been able to design office buildings which use approximately 35,000 to 40,000 Btu/(ft^2)(yr).

One must remember that the annual energy consumption in any particular building is not a fixed quantity dependent on the design of the building and its systems alone. Two huge variables—the mode of operation and the quality of maintenance—are extremely important and can easily push consumption up, or reduce it, by factors of 50% or more of some base requirement. This is well documented in the chapters that follow and points up the fact that ultimately it is people that use energy, not buildings.

Annual energy requirements determine the fuel bill, whereas peak-energy requirement defines the scale and the first cost of the equipment that serves the building. The two measures are closely interrelated, and, with increasing concern for life-cycle costs rather than capital costs alone, many decisions with respect to building design and system selection require economic modeling.

Such methods are given in some detail in Chap. 14. It is interesting to note at this point, however, that many, if not most, of the measures related to energy-conscious design and the reduction of annual energy usage will reduce peak-loading requirements and capital costs as well. (With the electrical system, peak demand is a part of the billing; reduction of peak demand saves dollars directly.) It is untrue that a building owner must necessarily pay more now to save later. Good design of new buildings should enable reduction in both capital *and* energy costs.

Both peak-load and annual energy requirements for new and old buildings alike are amenable to calculation, though annual energy usage is subject to greater error in new buildings because of the difficulty in forecasting the two big variables mentioned above: the mode of operations and the quality of maintenance. Peak-load estimates are common to routine design procedures for the purpose of sizing equipment. Annual usage estimates are less common but are increasingly required for design decisions with respect to life-cycle costing and in the future will be required by many jurisdictions as a measure of compliance with energy conservation codes.

Peak-load calculations can be computed manually with a given set of steady-state conditions (usually maximum and minimum expected conditions outdoors and occupied conditions indoors, including lighting levels and equipment operations). While these calculations give a reasonable indication of the size of equipment required to meet maximum loads (invariably conservative), they give no indication of energy requirements to meet the myriad part-load conditions that actually occur in an average year. A rough analysis of yearly energy requirements can be obtained by using heating degree-days and cooling degree-hours, but such an analysis totally ignores the building and system reactions to constantly changing conditions, such as internal heat gain, solar radiation, and wind effects.

Closer approximations of yearly energy requirements can be obtained by making a series of calculations for different outdoor conditions during both occupied and unoccupied times and correcting for the length of time these conditions are expected to exist. The accuracy of the results improves with the number of different conditions selected and the number of separate calculations, but this can develop into a long, tedious, and costly process if done by hand.

Computers are extremely good at making a series of repetitive calculations, and programs have been developed independently by many organizations to perform this task. Because these programs have been developed independently, each one has characteristics and functions which are unique. One program, for instance, may be very good at calculating heating and cooling loads but rather coarse in its approach to analyzing HVAC system reaction to these loads, while another may carefully analyze interreaction between HVAC systems but base its building-load calculations on extrapolations of one typical day or month.

1.5 COMPUTER MODELING

Calculation programs fall into two categories: those used for solving load design problems (design programs) and those used for figuring energy usage (energy programs). There is some overlap between categories as some of the energy programs have subroutines which can be used independently for design problems.

Table 1.7 lists design programs and Table 1.8 lists energy programs that are currently available. These lists are assembled from data from various sources and are not necessarily complete; neither are they an indication of equal quality and performance. When selecting programs for a particular application, prospective users must obtain detailed information from the program authors and then make their own evaluations based on their requirements, in-house computer capabilities, and projected costs.

DESIGN PROGRAMS Design programs have been developed to allow rapid solution of particular design problems such as pressure drop in piping and ductwork systems, and illumination levels obtained from a given lighting fixture layout. The value of these programs is that they allow more alternative schemes to be considered in the initial design stages than would

TABLE 1.7 COMPUTER DESIGN PROGRAMS

Type	Name	Author
Building form		
Form generation	Form Generation	University of Southern California
Shading analysis	SHADOW	University of Texas
	SUNNY	University of Texas
Solar gain analysis	SUNSET	Dubin-Mindell-Bloome Associates
Building exterior envelope		
Glass comparison	Glass Comparison	Libbey-Owens-Ford
Building interior planning		
Optimization	ARK-2	Perry, Dean & Stewart
	B.O.P.	Skidmore, Owings & Merrill
Lighting		
Conventional	Lighting II	APEC, Consulting Engineers Council
	Lighting	Dalton, Dalton, Little & Newport
	Interior Lighting Analysis & Design	Giffels Associates, Inc.
ESI	Lighting Program	Isaac Goodbar
	Lighting Program	Illumination Computing Service
	Lighting Program	Ian Lewen
	Lumen II	Smith, Hinchman & Grylls
Daylighting	Daylighting	Libbey-Owens-Ford
Power		
Distribution network	Electrical Feeder II	APEC, Consulting Engineers Council
	Electrical Feeder Sizing	Dalton, Dalton, Little & Newport
	Three-Phase Fault Analysis	Giffels Associates, Inc.
Demand study	Electrical Demand Load Study	Giffels Associates, Inc.
HVAC		
Equipment selection	HCC-III (Mini-Deck)	APEC, Consulting Engineers Council
	Equipment Selection	Carrier Air Conditioning Co.
	Equipment Selection	Trane Co.

TABLE 1.7 COMPUTER DESIGN PROGRAMS (*Continued*)

Type	Name	Author
HVAC		
Duct design	Duct Program	APEC, Consulting Engineers Council
	Several	Dalton, Dalton, Little & Newport
	Several	Giffels Associates, Inc.
Air-handling unit design	Fan Static Calculations	Giffels Associates, Inc.
Domestic water		
Piping design	Piping Program	APEC, Consulting Engineers Council
	Several	Dalton, Dalton, Little & Newport
	Several	Giffels Associates, Inc.
Vertical transportation		
Elevator design	Elevator Design	Dover Corporation
	Elevator Design	Otis Elevator
Operation and maintenance		
Automated control systems		Honeywell, Inc.
		Johnson Control Service
		Powers Regulator Company
		Robertshaw Controls
Solar energy systems		
Solar collector odels	Flat Plate Collector	Honeywell, Inc.
	Parabolic Trough Collector	Honeywell, Inc.
	Flat Plate Collector	Westinghouse

be possible with hand calculations and thus increase the choices, and evaluations, available to the designer.

Some programs allow the input of desired results and as output give the parameters of design. For example, some lighting programs will design a fixture layout based on inputs of room size, interreflectances, and desired lighting profile.

Design programs for equipment selection must be used with discretion as some programs only recognize certain makes of equipment and will not necessarily allow selective comparisons to be made.

ENERGY PROGRAMS   Energy programs have been developed to calculate the yearly energy requirements for buildings and systems combined, for

TABLE 1.8 COMPUTER ENERGY PROGRAMS

Name	Author
COMMERCIAL PROGRAMS	
ECUBE	American Gas Association
HCC-111	APEC, Consulting Engineers Council
Energy Analysis	Caudill Rowlett Scott
AXCESS	Electric Energy Association
Glass Comparison	Libbey-Owens-Ford
Energy Program	MEDSI
Energy Analysis	Meriwether & Associates
Building Cost Analysis	PPG Industries
TRACE	Trane Company
Energy Program	Westinghouse Corp.
HACE	WTA Computer Services, Inc.
RESEARCH PROGRAMS (Negotiable)	
CADS	University of California at Los Angeles
SIMSHAC	Colorado State University
FINAL	Dalton, Dalton, Little & Newport
HVAC Load	Giffels Associates, Inc.
Energy Program	Honeywell, Inc.
NBSLD (Honeywell)	Honeywell, Inc.
Energy Program	University of Michigan
NBSLD	National Bureau of Standards
B.E.A.P.	Pennsylvania State University
Post Office Program	
DEROB	University of Texas
TRANSYS	University of Wisconsin
IN-HOUSE PROGRAMS (Proprietary)	
Energy Program	General Electric Company
Residential and Small Commercial	Honeywell, Inc.
Energy Program	IBM

buildings only, or for systems only. Smaller programs or parts of larger programs are also available to calculate yearly energy requirements for discrete portions of a building or system considered in isolation.

Energy programs first calculate the building and/or system loads for a given set of conditions and then simulate the operation of the building and/or systems to determine their reaction to continuously changing conditions and to calculate yearly loads.

Energy programs calculate loads in many different ways. Some require U values, internal loads, and equipment loads to be calculated by hand, and the results are used as input. Other more complex programs will not accept a U value input but require the physical characteristics of each building surface to be fully described in scientific units. Most programs have the facility for printing out the building and/or system loads before the energy analysis phase is started, but some programs will not divulge, for instance, the calculated U value, and the load print-outs must be carefully screened for input errors at this stage.

Once a valid load is developed, the energy analysis phase of the program can be initiated. This part of the program simulates the building and systems reaction to changing weather conditions, occupancy, internal loads, equipment loads, part-load efficiencies, etc. The energy analysis is achieved by simulating the building and systems as mathematical models in the computer and recalculating loads on a time basis. Less-complex programs may calculate loads for only two conditions on one day in each month and extrapolate for yearly energy loads, while more-complex programs calculate loads for each of the 8760 h in a year and integrate these for yearly energy loads. Calculations on an hourly basis allow the dynamic flow of energy to be simulated by accounting for the thermal lag of building materials and the effect of thermal storage on loads. This effect can be important, resulting in reduced loads and possible changes of operating techniques.

PROGRAM SELECTION The choice of a computer program must be made on the basis of the prospective user's needs, expertise, and in-house capabilities. The following questionnaire will help establish these.

What degree of sophistication is required? If the problem is to evaluate various system designs, then the appropriate design program will suffice, and the choice should be limited strictly to programs dealing with the particular type of system. If the problem is to evaluate energy use of alternative building shapes and construction, then use an energy program biased toward buildings should be used. Conversely, if the problem is to evaluate various system reactions in a given building, then use an energy program biased toward systems.

How much money is available for computer calculations? The costs of computer time and data compilation vary widely from program to program and are not necessarily related to program complexity. Obtain estimates of run time from the program author.

What are the prospective user's in-house capabilities? Some programs are available for use on in-house computers only; others can be used either in-house or by remote batch or time sharing. Still others are limited to use on the author's computer operated by the author's personnel. Information on the application and availability of a particular program can be obtained from program authors.

CHAPTER TWO

NEW BUILDINGS: ENERGY-CONSCIOUS DESIGN

2.1 OPTIONS

The best way to save energy in new buildings is simply to forgo building. If we could forgo new construction, we could save much energy: energy required to manufacture the materials of the building, energy to build the building, and energy to provide some modicum of thermal and visual comfort to the occupants of the building once it is in use. But if new buildings are needed and are to be built (as we can afford them), it is our opportunity to grasp a new challenge: to make our buildings in all respects energy-conscious.

The later working chapters of this book attempt a comprehensive checklist of energy conservation options applicable to both new and existing buildings. Chapters 6 to 12 follow the ordering of building loads, distribution loads, and conversion loads. In this chapter additional considerations of building planning, site development, and design of the building envelope are discussed, all of which affect directly any building's performance but are options typically available only to new buildings.

Buildings proceed commonly from some functional requirement, be it commerce, residence, or industry. And buildings, once ordered, are shaped by many additional forces of site, budget, technology, and situation. It is difficult to unravel these forces for deliberate appraisal of their interactive nature or for their contribution to energy flows. What merit, for instance, is there in consideration of building high rather than low, of fronting south rather than north, or of building underground rather than on stilts?

The one variable that most informs such questions is climate. The United States ranges in climate from locations with no heating degree-days (Honolulu) to locations of 10,000 heating degree-days (Duluth) to locations with as much as 14,000 degree-days (Fairbanks, Alaska). The continent is blessed with desert and swamp, and dry grassland and moist forest. Each building site has a particular climatic pattern of air temperatures, winds, humidities, and sun which varies through each day and through the year. Each building site is placed within a regional climate but sees a particular microclimate that is further shaped by nearby mountains or skyscraper, by paved urban areas, or by the atmospheric discharges of cities.

In the United States, and worldwide for that matter, it is not uncommon to see identical buildings in locations separated by 20° of latitude or by 5000 ft of altitude, or situated on the coast and in the middle of the plains 1000 miles away. At the very least, in consideration of planning and building form, energy-conscious design must give force to a new regionalism, with buildings that are responsive to the environmental forces surrounding them.

Climatologists have worked out any number of classification schemes, but a very simple one of particular merit is given in Table 2.1. This designation

TABLE 2.1   CLIMATIC CLASSIFICATION*

WINTER

A. Long heating season	With sun and wind (A1)
	With sun without wind (A2)
	Without sun and wind (A3)
	Without sun with wind (A4)
B. Moderate heating season	With sun and wind (B1)
	With sun without wind (B2)
	Without sun and wind (B3)
	Without sun with wind (B4)
C. Short heating season	With sun and wind (C1)
	With sun without wind (C2)
	Without sun and wind (C3)
	Without sun with wind (C4)

SUMMER

D. Long cooling season	Dry (D1)
	Humid (D2)
E. Moderate cooling season	Dry (E1)
	Humid (E2)
F. Short cooling season	

*Definitions:

A. Long heating season: 6000 degree-days or more

B. Moderate heating season: 6000–4000 degree-days

C. Short heating season: 4000 or less degree-days
  1. With sun: 60% of daylight time or more
  2. With wind: 9 mi/h or more

D. Long cooling season: More than 1500 h at 80°F

E. Moderate cooling season: 1500–600 h at 80°F

F. Short cooling season: Less than 600 h at 80°F
  1. Dry: 60% RH or less
  2. Humid: 60% RH or more

of climates was developed by Dubin-Bloome Associates for the General Services Administration to give priority to the various options offered in *Energy Conservation Design Guidelines for Office Buildings*. The options from this publication applicable to building planning keyed to these types of climate are given in Table 2.2, for site development in Table 2.3, and for the building envelope in Table 2.4. Much of the material that follows is taken from this publication.

2.2 CLIMATE

The sun is all important. By its presence or its absence, it affects virtually every portion of a building's design, from siting and orientation to envelope and glazing, from HVAC and lighting systems to operating and maintenance policies.

The yearly energy consumption of two buildings in different geographic areas with the same number of heating degree-days could vary in energy required for heating by as much as 30% or more, depending upon the average cloud cover and available sunshine. North walls for two such buildings may be constructed of the same materials, with the same mass and color, but the other three facades should be designed for a particular sun with fenestration for heating in the winter and with appropriate solar controls in the summer to reduce heat gain.

Monthly solar loads must be accurately calculated to determine the sun's effect upon each building facade. The solar irradiation for each hour of the day may be an important factor in determining building shape, configuration, and orientation. Solar controls, such as internal or external shading devices for glazed areas, or cooling ponds or sprays for roof areas, can manipulate sunlight to achieve the maximum advantage and energy savings from it. Sunlight may also be used by solar collectors for space heating and cooling, for domestic water heating, and eventually for power generation as well.

The amount and duration of solar radiation can determine the use of color for walls and roof—dark for absorption on north walls in cold climates, light for reflection from roofs in warm climates.

The wind is as important to consider as the sun. Wind can wash away the insulating still-air film that surrounds a building and thus increase heating and cooling loads. Wind can carry solar heat away from a building and evaporate moisture on wet surfaces, thus cooling the skin to temperatures lower than the ambient air.

In the Northern Hemisphere, typically, the north and west sides of a building are most subject to cold winds. The most vulnerable portions of buildings (entrances and glazed areas) should be oriented away from these cold winds to avoid infiltration through cracks and openings. If an entrance must be on the north or the west, it should be shielded. The intensity and duration of outdoor temperature conditions determine the amount of energy required to overcome infiltration loads for any given quantity of air leakage in the building, and a combination of high winds and low temperatures in winter amplifies the need for wind protection.

Snowfall affects building energy consumption in a number of ways. In areas where snowfall is heavy, snow cover on roofs actually serves as insulation. The reflection of sunlight from snow on the ground and from adjacent snow-covered buildings increases the value of natural illumination systems and the effectiveness and efficiency of flat-plate solar collectors.

Though a given geographic area is characterized by a general set of climatic conditions, the specific site for a building can alter a particular climatic effect. The topography of the site, and existing or proposed structures adjacent to the building site, can increase or decrease wind velocities from any direction and modify the direction and intensity of sunlight which will touch the building envelope. The type of trees that can be grown in the area is dependent upon the climate. Almost anywhere, however, trees can be grown to provide either shade or windbreak.

2.3 BUILDING PLANNING AND CONFIGURATION

The orientation of each of the facades and of the roof has a unique effect upon energy consumption depending upon climate, site, and geographic location. A sloping roof facing south, for instance, will be subjected to more solar irradiation in any climate in the United States than one sloping north or even east or west. Similarly, cold winds are generally most severe on north and west facades. Each surface is subjected to different environmental influences, even for two buildings located in the same latitude in the United States.

Where the site permits choice in building orientation, the relative benefit (or disadvantage) of each environmental influence on total yearly energy consumption must be analyzed for the many alternatives available for envelope materials and construction. Orientation has little effect on ventilating loads or systems, but it can, however, influence the location of intakes or exhaust outlets. Orientation has little or no effect on the requirements of solid waste or domestic hot-water systems, but it may affect the power required for lighting systems affecting the available light from outdoors for natural illumination.

CONFIGURATION The configuration of a building, and the orientation and envelope, determine in large part the amount of energy used. Where energy can be saved by using natural illumination, the building perimeter should be increased and its interior space proportionately decreased, resulting in various building forms, such as multiple courtyards, atriums, light wells, finger buildings, and low buildings with skylights. Natural lighting saves energy for locations with exterior conditions close to interior conditions, where heat losses or gains through extensive glazing are low. However, in extremely cold climates, more energy is conserved by using artificial lighting systems with fewer windows, and a reduction of the perimeter exposure should be considered. A configuration that resists unwanted heat gains or losses results in an energy-conserving building.

A spherical or round building has less surface, hence less heat gain or loss than any other shape for an equal amount of total floor space. A square building has less surface than a rectangular one of equal area per floor and thus experiences less heat gain or loss. However, the number of stories modifies this relationship for the building as a whole. See Fig. 2.1.

A dome roof can permit warm air to rise and collect at the top, leaving the floor areas cooler. Pyramids, zigzag exterior wall configurations, rhomboid-shaped buildings, and other forms can all be used to control the influence of climate on consumption.

A tall building has a proportionately smaller roof and is less affected by

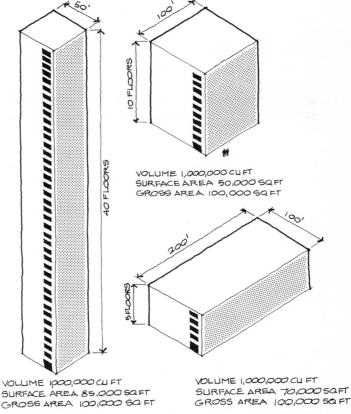

Figure 2.1 Variations in building surface areas with differing building forms.

solar gains on that surface. On the other hand, tall buildings are generally subjected to greater wind velocities, a condition which increases infiltration and heat losses. And tall buildings are less likely to be shaded or protected from winds by surrounding buildings and trees. They require more mechanical support systems, including elevators, and longer exhaust-duct systems, and the stack induction action in tall buildings increases infiltration.

Low buildings have greater roof area in proportion to wall area, and hence special attention must be given to the roof's thermal characteristics.

Designing with stilts or with small first-floor areas and large overhanging upper floors increases heat loss and heat gain owing to the extra exposed surfaces. While this may be of slight advantage all year in the southern regions, or anywhere in the summertime, it presents a serious increased heat loss in colder climates. Locating parking garages on intermediate levels similarly increases energy consumption by the additional exposed surfaces.

THE BUILDING PLAN The building plan responds to the functional program, but a critical review of a functional program that includes energy considerations often reveals opportunities to change the plan for energy conservation.

The building plan can have a major effect on the energy required to maintain comfort conditions in both summer and winter. In northern climates, corridors, equipment spaces, toilet rooms, and other service areas which do not require close temperature control can act as buffer spaces—against the cold in northern climates when located on a north wall, or against excessive thermal loads in southern climates when located on east, west, or south walls.

Grouping spaces that have similar environmental control needs can reduce the extent and complexity of the mechanical systems and permit heating, cooling, ventilation, and lighting to be concentrated in areas that have maximum requirements. It is not necessary to provide the same degree of environmental comfort throughout the entire floor area. Areas which have high internal heat gains are best located adjacent to exterior surfaces in cold climates to permit dissipation of excess heat without mechanical assistance. And increasing the percentage of usable areas to gross area can result in smaller buildings with corresponding conservation of energy.

The ceiling plenum, the floor thickness, and the floor-to-ceiling height together constitute the floor-to-floor height. The floor thickness is usually determined by construction and structural considerations. The floor-to-floor dimension most directly affects energy conservation in that it affects the area of the exterior building skin exposed to the weather. It also increases the amount of raw materials (and therefore energy consumed) which go into the construction of the building, especially the skin. When the height of the building is increased, more energy is used to raise materials, services, and people to greater heights during construction and occupancy. On the other hand, the dimension of the ceiling plenum can greatly affect the efficiency of the mechanical equipment. For example, floor-to-floor heights in commercial office buildings are typically 12 to 12½ ft. If this were to be increased to 13 ft (through an increase in the size of the ceiling plenum), significant savings could result from redesign of the mechanical system. Ducts could be larger, allowing greater volumes of air to be moved with smaller pressure drops permitting reduced fan horsepower. Duct aspect ratios could also be reduced to approach square or round sections, resulting in significant savings in the sheet metal and insulation in construction and reduced heat transmission and friction losses through the duct. Larger ceiling plenums may also permit the use of more effective energy conservation systems, such as variable volume and heat-of-light.

Grouping toilet rooms on a floor and above each other on multiple floors simplifies the exhaust systems, and the plumbing systems for such groupings generally require less energy for their distribution. Grouping also simplifies the installation of heat recovery devices or air purification devices, such as charcoal filters for exhaust air recirculation back into the building.

When kitchens and cafeterias are installed, locating them on the same level and adjacent to each other permits the use of exhaust air from the cafeteria to be transferred directly and efficiently to kitchen hoods for makeup air, reducing the energy required for that purpose.

Reducing the number of exits and entries into a building lowers the energy requirements in a number of ways: infiltration is reduced, more efficient utilization of space is possible, less security lighting is required, and less heat loss and heat gain are experienced.

The building plan can have a great effect upon the amount of energy required for the air and water distribution systems for heating and cooling. Long duct and pipe runs with many elbows and fittings offer more resistance to fluid flow and require more energy for pumps and fans, unless the ducts and pipes are increased in size and cost. A plan which can reduce the extent of the distribution systems leads to operating economy and energy conservation. Some planning options are given in Table 2.2.

2.4 SITE DEVELOPMENT

The site may influence the amount and duration of light available for natural illumination. In climates where natural illumination can provide a *net* gain in energy conservation, the location of the building on the site in relationship to adjacent structures and natural features (trees, hills) should be optimized.

Sites for rectangular buildings may have a length-to-depth relationship which will permit or hinder the ability of the building configuration to take maximum advantage of the sun and wind. The site also may influence the direction of the building axis so that the most adverse effects of sun and wind are felt. In the latter case, specific envelope design tactics are required to

TABLE 2.2 PLANNING OPTIONS

	Priority*			N/A†
	1st	2d	3d	
1. Group services rooms as a buffer and locate at the north wall to reduce heat loss or the south wall to reduce heat gain, whichever is the greatest yearly energy user.	A4 D1	A1 D2	A3 E1	C2 F
2. Use corridors as heat transfer buffers and locate against external walls.	A4 D1	A1 D2	A3 E1	C2 F
3. Locate rooms with high process heat gain (computer rooms) against outside surfaces that have the highest exposure loss.	A4	A1	A3	C2
4. Landscaped open planning allows excess heat from interior spaces to transfer to perimeter spaces which have a heat loss.				
5. Rooms can be grouped in such a manner that the same ventilating air can be used more than once, by operating in cascade through spaces in decreasing order of priority, i.e., office-corridor-toilet.				
6. Reduced ceiling heights reduce the exposed surface area and the enclosed volume: They also increase illumination effectiveness.				
7. Increased density of occupants (less gross floor area per person) reduces the overall size of the building and yearly energy consumption per capita.				
8. Spaces of similar function located adjacent to each other on the same floor reduce the use of elevators.				
9. Offices frequented by the general public located on the ground floor reduce elevator use.				
10. Equipment rooms located on the roof reduce unwanted heat gain and heat loss through the surface. They can also allow more direct duct and pipe runs reducing power requirements.	A4 D1	A3 D2	B4 E1	C2
11. Windows planned to make beneficial use of winter sunshine should be positioned to allow occupants the opportunity of moving out of the direct sun radiation.				
12. Deep ceiling voids allow the use of larger duct sizes with low pressure drop and reduce HVAC requirements.				

TABLE 2.2 PLANNING OPTIONS (*Continued*)

	Priority*			
	1st	*2d*	*3d*	*N/A†*
13. Processes that have temperature and humidity requirements different from normal physiological needs should be grouped together and served by one common system.				
14. Open planning allows more effective use of lighting fixtures. The reduced area of partitioned walls decreases the light absorption.				
15. Judicious use of reflective surfaces such as sloping white ceilings can enhance the effect of natural lighting and increase the yearly energy saved.				

*See Table 2.1 for meaning of abbreviations under Priority.

†N/A = not applicable

offset the hostile climate. A building with a north-south major axis receives more solar irradiation than one with an east-west major axis, anywhere in the country. Some site options are found in Table 2.3.

The following energy concerns are greatly influenced by the building shape and orientation which, in turn, may be influenced by a particular site:

- Trade-offs between heat loss and heat gain on each exposure with particular regard to wind and sun.

- Utilization of natural daylight versus heat loss and heat gain and glare problems peculiar to each exposure.

- Beneficial effects of direct solar radiation for heating versus adverse effects during the cooling season.

2.5 THE BUILDING ENVELOPE

The building envelope, consisting of walls, windows, doors, roofs, and floor surfaces, is subjected to varying influences of climate in each of its orientations. The qualities of the building envelope directly influence the heating and cooling peak requirements of the HVAC system and are major determinants of the yearly energy required for maintaining the building thermal environment. (But the envelope has little influence on energy loads due to ventilation, domestic hot water, or solid waste.) Table 2.4 lists some building envelope options.

TABLE 2.3 SITE OPTIONS

	Priority°			
	1st	2d	3d	N/A†
1. Use deciduous trees for their summer sun shading effects and windbreak for buildings up to three stores.	A1 D1	A2 D2	A4 D1	C4 F
2. Use conifer trees for summer and winter sun shading and windbreaks.	C4 D1	C1 D2	C2 E1	A2 F
3. Cover exterior walls and/or roof with earth and planting to reduce heat transmission and solar gain.	A1 D1	A2 D2	A4 E1	C4 F
4. Shade walls and paved areas adjacent to building to reduce indoor/outdoor temperature differential.	C2 D1	C1 D2	C3 D1	A2 F
5. Reduce paved areas and use grass or other vegetation to reduce outdoor temperature build-up.	C2 D1	C1 D2	C3 E1	A2 F
6. Use ponds, water fountains, to reduce ambient outdoor air temperature around building.	C2 D1	C1 E1	C3 D2	A4 F
7. Collect rainwater for use in building.				
8. Locate building on site to induce airflow effects for natural ventilation and cooling.	C2 F	C1 E1	C3 E2	A4 D2
9. Locate buildings to minimize wind effects on exterior surfaces.	A4 F	A1 E2	B4 E1	C2 D1
10. Select site with high air quality (least contaminated) to enhance natural ventilation.	C2 F	C1 E1	C3 E2	A4 D2
11. Select a site which has year-round ambient wet- and dry-bulb temperatures close to and somewhat lower than those desired within the occupied spaces.				
12. Select a site that has topographical features and adjacent structures that provide windbreaks.	A4 F	A1 E2	B4 E1	C3 D2
13. Select a site that has topographical features and adjacent structures that provide desirable shading.	C2 D2	C1 D1	B2 E2	A1 F
14. Select site that allows optimum orientation and configuration to minimize yearly energy consumption.				
15. Select site to reduce specular heat reflections from water.	C2 D2	C1 D1	B2 E2	A4 F
16. Utilize sloping site to partially bury building or use earth berms to reduce heat transmission and solar radiation.	A4 D1	A1 D2	A3 E1	C2 F
17. Select site that allows occupants to use public transport systems.				

*See Table 2.1 for meaning of abbreviations under Priority.
†N/A = not applicable

TABLE 2.4 BUILDING ENVELOPE OPTIONS

	Priority[°]			
	1st	*2d*	*3d*	*N/A*
1. Construct building with minimum exposed surface area to minimize heat transmission for a given enclosed volume.	A4 D1	A1 D2	A3 E1	C2 F
2. Select building configuration to give minimum north wall to reduce heat losses.	A4	A1	A3	C2
3. Select building configuration to give minimum south wall to reduce cooling load.	D1	D2	E1	F
4. Utilize building configuration and wall arrangement (horizontal and vertical sloping walls) to provide self-shading and windbreaks.	A4 D1	A1 D2	B4 E1	C3
5. Locate insulation for walls and roofs and floors over garages, at the exterior surface.	A4 D1	A3 D2	A1 E1	C2 F
6. Construct exterior walls, roof, and floors with high thermal mass with a goal of 100 lb/ft³.	A4 D1	A1 D2	A3 E1	C3 F
7. Select insulation to give a composite U factor from 0.06 when outdoor winter design temperatures are less than 10°F to 0.15 when outdoor design conditions are above 40°F.				
8. Select U factors from 0.06 where vol-air temperatures are above 144°F up to a U volume of 0.3 with sol-air temperatures below 85°F.				
9. Provide vapor barrier on the interior surface of exterior walls and roof of sufficient impermeability to condensation.				
10. Use concrete "slab-on-grade" for ground floors.	A4 D1	A1 D2	A3 E1	C2 F
11. Avoid cracks and joints in building construction to reduce infiltration.	A4 D2	A1 E2	A3 D1	
12. Avoid thermal bridges through the exterior surfaces.	A4 D2	A1 D1	A3 E2	C3 F
13. Provide textured finish to external surfaces to increase external film coefficiency.	A4	A1	B4	C2
14. Provide solar control for the walls and roof in the same areas where similar solar control is desirable for glazing.	D2	D1	E2	A
15. Consider length and width aspects for rectangular buildings as well as other geometric forms in relationship to building height and interior and exterior floor areas to optimize energy conservation.	A4 D1	A1 D2	A3 E1	C2 F

29

TABLE 2.4   BUILDING ENVELOPE OPTIONS (*Continued*)

	Priority°			
	1st	*2d*	*3d*	*N/A*
16. To minimize heat gain in summer due to solar radiation, finish walls and roofs with a light-colored surface having a high emissivity.	D1	D2	E1	
17. To increase heat gain due to solar radiation on walls and roofs, use a dark-colored finish having a high absorptivity.	A1	A2	A4	C2
18. Reduce heat transmissions through roof by one or more of the following items:				
a. Insulation	A4	A1	A3	C3
	D1	D2	E1	F
b. Reflective surfaces	C2	C1	C3	A4
	D1	D2	E1	
c. Roof spray	D1	E1	F	
d. Roof pond	D1	E1	F	
e. Sod and planning	A4	A1	A3	C2
	D1	D2	E1	F
f. Equipment and equipment rooms located on the roof.	A4	A1	A3	
	D1	D2	E1	
g. Provide double roof and ventilate space between.	D1	D2	E1	F
19. Increase roof heat gain when reduction of heat loss in winter exceeds heat gain increase in summer:				
a. Use dark-colored surfaces	A2	A1	B2	B1
b. Avoid shadows	A2	A1	B2	B1
20. Insulate slab on grade with both vertical and horizontal perimeter insulation under slab.	A	B	C	
21. Reduce infiltration quantities by one or more of the following measures:				
a. Reduce building height				
b. Use impermeable exterior surface materials.	A4	A1	A3	C4
c. Reduce crackage area around doors, windows, etc., to a minimum.	D2	E2	D1	F
d. Provide all external doors with weather stripping.				
e. Where operable windows are used, provide them with sealing gaskets and cam latches.				
f. Locate building entrances on downwind side and provide windbreak.				
g. Provide all entrances with vestibules; where vestibules are not used, provide revolving doors.	A4	A1	A3	C4

TABLE 2.4   BUILDING ENVELOPE OPTIONS (*Continued*)

	Priority°			
	1st	*2d*	*3d*	*N/A*
h. Provide vestibules with self-closing weather-stripped doors to isolate them from the stairwells and elevator shafts.	D2	E2	D1	F
i. Seal all vertical shafts.				
j. Locate ventilation louvers on downwind side of building and provide windbreaks.				
k. Provide break at intermediate points of elevator shafts and stairwells for tall buildings.				
22. Provide wind protection by using fins, recesses, etc., for any exposed surface having a *U* value greater than 0.5.	A4	A1	B4	C2
23. Do not heat parking garages.				
24. Consider the amount of energy required for the protection of materials and their transport on a life-cycle energy usage.				
25. Consider the use of the insulation type which can be most efficiently applied to optimize the thermal resistance of the wall or roof; for example, some types of insulation are difficult to install without voids or shrinkage.				
26. Protect insulation from moisture originating outdoors, since volume decreases when wet. Use insulation with low water absorption and one which dries out quickly and regains its original thermal performance after being wet.				
27. Where sloping roofs are used, face them to the south for greatest heat gain benefit in the wintertime.	A1	A2	B1	C4
28. To reduce heat loss from windows, consider one or more of the following:				
a. Use minimum ratio of window area to wall area.				
b. Use double glazing.				
c. Use triple glazing.				
d. Use double reflective glazing.	A4	B4	C4	
e. Use minimum percentage of the double glazing on the north wall.	A1	B1	C1	
f. Manipulate east and west walls so that windows face south.	A2	B2	C2	
g. Allow direct sun on windows November through March.	A3	B3	C3	

TABLE 2.4 BUILDING ENVELOPE OPTIONS (*Continued*)

	Priority°			
	1st	*2d*	*3d*	*N/A*
h. Avoid window frames that form a thermal bridge.				
i. Use operable thermal shutters which decrease the composite *U* value to 0.1.				
29. To reduce heat gains through windows, consider the following:				
a. Use minimum ratio of window area to wall area.				
b. Use double glazing.				
c. Use triple glazing.	D1	E1	F	
d. Use double reflective glazing.	D2	E2	F	
e. Use minimum percentage of double glazing on the south wall.				
f. Shade windows from direct sun April through October.				
30. To take advantage of natural daylight within the building and reduce electrical energy consumption, consider the following:				
a. Increase window size but do not exceed the point where yearly energy consumption, due to heat gains and losses, exceeds the saving made by using natural light.				
b. Locate windows high in wall to increase reflection from ceiling, but reduce glare effect on occupants.	C2	B2	A2	
c. Control glare with translucent drapes operated by photo cells.	C1	B1	A1	
d. Provide exterior shades that eliminate direct sunlight, but reflect light into occupied spaces.	C3	B3	A3	
e. Slope vertical wall surfaces so that windows are self-shading and walls below act as light reflectors.	C4	B4	A4	
f. Use clear glazing. Reflective or heat-absorbing films reduce the quantity of natural light transmitted through the window.	F	E	D	
31. To allow the use of natural light in cold zones where heat losses are high energy users, consider operable thermal barriers.	A4	A1	B4	C3
32. Use permanently sealed windows to reduce infiltration in climatic zones where this is a large energy user.	A1 / D1	A4 / D2	B1 / E1	C3 / F
33. Where codes or regulations require operable	A1	A4	B1	C3

TABLE 2.4   BUILDING ENVELOPE OPTIONS (*Continued*)

	Priority°			
	1st	*2d*	*3d*	*N/A*
windows and infiltration is undesirable, use windows that close against a sealing gasket.	D1	D2	E1	F
34. In climatic zones where outdoor air conditions are suitable for natural ventilation for a major part of the year, provide operable windows.	C2 F	C3 E1	C1 E2	A4 D2
35. In climate zones where outdoor air conditions are close to desired indoor conditions for a major portion of the year, consider the following:				
a. Adjust building orientation and configuration to take advantage of prevailing winds.				
b. Use operable windows to control ingress and egress of air through the building.				
c. Adjust the configuration of the building to allow natural cross ventilation through occupied spaces.	F	E1	E2	D2
d. Utilize stack effect in vertical shafts, stairwells, etc., to promote natural airflow through the building.				

See Table 2.1 for meaning of abbreviations (A1, A2, etc.) under Priority.

The thermal properties of the envelope are determined by the combination of wall mass, thermal resistance, insulation location, exterior surface color and texture, and the type and location of glazing. In both summer and winter the effect of each is dependent upon the mode of operation of the heating and cooling systems. The building wall must retard heat flow in and out, control air passage in and out, and accept thermal differentials with minimum harm or deterioration.

Walls of large mass have high thermal inertia, which modifies the effect of heat transmission by delay and by damping. Mass also reduces the short cycling of oil burner and hence increases combustion efficiency. A wall of high thermal inertia subjected to solar radiation for 1 h will absorb heat at its outside surface but transfer it to the interior over a time period as long as 12 h. Conversely, a wall having the same U value but low thermal inertia will transfer the heat more quickly, in perhaps 30 min. The value of adjusting the time lag depends upon general climate conditions, diurnal swings in temperature, and occupancy modes. In areas subject to long, cold winters or long, hot summers, with extreme peak temperatures and large diurnal swings, large thermal inertia has a high priority for energy conservation. Priority is low in areas subject to mild winters and summers with small diurnal swing typical of humid climates.

On exterior surfaces, light colors decrease solar heat gain; dark colors increase solar heat gain. In most cases, a dark-colored north wall and light-colored east and west walls will be the most energy-conserving setup. In hot climates light colors and high reflectivity are best. (But the color of the wall has relatively little effect on energy consumption when used on exterior walls of low U values and high thermal mass.)

Burying the building or parts of the building by use of earth berms minimizes solar gain and air infiltration caused by the effect of the wind and minimizes heat loss and heat gain due to air temperatures. Berms are most effective in areas of high winds, extreme solar gains, and extreme hot or cold temperatures.

Wall textures or vines can shade and can maintain a still-air film on building surfaces to reduce heat loss and heat gain. In addition, if the wall insulation is on the outside of the wall, the mass of the wall will provide thermal storage and will dampen the effects of diurnal weather variations and indoor occupied-unoccupied temperature cycles. Where diurnal variations are greatest, heavy construction (up to 100 lb/ft² or more) can be very effective in reducing energy consumption. The location of insulation becomes less important with construction of low mass (less than 25 lb/ft²); this is particularly true with curtain walls. When construction methods, such as stud wall framing and sandwich panels, dictate the location of the insulation, other methods of conserving energy equal to that amount saved by mass can be considered.

Insulation should be located on the outside of the structure to reduce air leakage through construction joints and to reduce heat loss by eliminating the effect of cold bridging from through-the-wall concrete or steel. The insulation should be protected from moisture; its insulation value decreases sharply when wet or damp. Insulation with low water permeability, such as Styrofoam plastic foam or urethane, and which dries out quickly and returns to original thermal performance is valuable.

Windows have a major effect on energy use due to transmission, solar gain, and air infiltration. Heat transmission is much greater through glass than through most opaque walls. U values for walls can be reduced to 0.04 or less, but single glass has a U value of about 1.15, double glass of 0.55 to 0.69, and triple glass of 0.35 to 0.47.

Windows are frequently provided in *gross* excess of any requirement for natural light, ventilation, or view. Large glass areas can cause discomfort for persons who must sit in front of them, owing to the sun's heat, radiation to the cold surfaces, and cold downdrafts and glare.

Elimination of all windows would exclude natural light and vision and would create somewhat incompletely understood psychological problems, even though windowless buildings would reduce the problems of solar heat gain in the summer, air infiltration, and heat loss. The percentage of window to opaque wall can be reduced if the window shape, placement in the wall, type of glazing, and use of shading devices are designed with an awareness of their combined impact on energy consumption and user needs.

The shape of a window can be important, even where the window area remains constant. Reducing the area of a window does not reduce the amount of natural light it admits in proportion to the reduction in size. The treatment of windows is particular to the site; a high percentage of sunshine in a cool climate may require design measures similar to those in hot zones, and a cloudy, rainy microclimate in a hot zone may alleviate the need for solar control typical for that zone. In some climatic zones, buildings with large internal heat gains caused by building function may suggest a larger window area or less insulation to permit heat dissipation to the outside. Reflective and heat-absorbing glass intercepts up to 80% of the radiant energy of the sun, which is very helpful for cooling in summer, but results in a loss of useful heat in winter and a loss of natural lighting.

An important relationship exists between natural light and window size. Larger windows provide more natural illumination but lose heat when the temperature is lower outdoors. An insulated thermal barrier installed over windows at night and weekends can easily reduce this heat loss or gain. The farther south a building site is located, the greater the energy-saving potential in utilizing natural light, so long as direct penetration of sunlight is controlled. The use of thermal barriers will extend this geographic limitation.

Solar control to reduce heat gain in the summer is most effective when located on the exterior of the building and is particularly effective when movable. Fixed solar fins and overhangs which eliminate direct solar penetration in the summertime, however, also block out some of the solar rays in the late spring and early fall, when rays could be useful for heating. Solar control is most effective when designed specifically for each facade, since time and duration of solar radiation vary with the sun's altitude and azimuth. Horizontal shading is most effective on southern exposures, but if not extended far enough beyond the windows, it will permit solar impingement at certain times of the day. On the east and west walls, a combination of vertical and horizontal sun baffles is required. The building configuration itself can be adjusted to give solar protection. In hot climates, it is effective to shade walls and roofs as well as windows. A double roof, or equipment located on the roof, can act as a solar shield as well as a thermal barrier during cold weather. In windy areas, the solar screens can be made to serve the double purpose of windbreaks.

Operable windows permit the use of natural ventilation but unless properly equipped with weather stripping, gasketing, and tight locking devices may increase infiltration loads. At times natural ventilation is not desirable to the extent that energy requirements exceed the savings due to natural ventilation—another trade-off which must be analyzed. Natural ventilation is delightful when outdoor conditions are such that the air is clean enough, pure enough, and dry enough to be enjoyed. However, natural ventilation cannot penetrate deeply into a building, and configuration and plan may have to be adjusted to allow its use. The number of hours in a year in which natural ventilation can be effectively used must be analyzed vis-à-vis the possible increased infiltration, heat loss, and heat gain for the hours when natural ventilation is not useful.

CHAPTER THREE

NEW-BUILDING CASE STUDY: THE MANCHESTER FEDERAL BUILDING

3.1 THE PROJECT

This case study is a condensation of *Designing an Energy Efficient Building* by Nicholas Isaak and Andrew Isaak, the architects for the building. This federal office building in Manchester, New Hampshire, consists of seven office stories abovegrade plus a mechanical penthouse and two levels of unheated parking structure belowgrade. The total building gross area is 176,000 ft², and the typical office floor measures 100 by 130 ft.

The building, for which the design phase had just begun, was designated an energy conservation demonstration project in December 1972 by the General Services Administration. The resources of the National Bureau of Standards, and Dubin-Bloome Associates as independent energy conservation consultant and designers of the solar energy system for heating and cooling, were added to the design team of Isaak and Issak, Architects, and their consultants Rose, Goldberg and Associates, structural engineers, and Richard D. Kimball Company, mechanical engineers.

Dubin-Bloome's task was to analyze all the factors that influence building energy consumption and recommend design criteria for a building that would use 20% less energy than a building designed to current standard practice. By use of computer programs, including those of the National Bureau of Standards's NBSLD, Meriwether & Associates' Energy Analysis, and others, mathematical buildings were modeled in computer memory and subjected to changing conditions, hour by hour, throughout a typical year using tapes of actual weather data. The final recommended design model showed a predicted savings over standard practice of more than 60% and an annual energy budget of only 55,000 Btu/(ft²)(yr). These recommendations were made to the architects in April 1973 and are largely incorporated in the now-finished building.

3.2 THE COMPARATIVE BUILDING MODEL

The standard practice building which was modeled for comparison purposes was based in large part on early design sketches that had been made by Isaak and Isaak prior to the building's designation as a demonstration project. The original concept was for six stories, 21,000 ft² gross area per floor, and a 2:1 length-to-width ratio with the long sides facing east and west. For modeling purposes, this conventional standard practice office building was given U values of 0.20 for the roof, 0.30 for the walls, and 1.13 for the single-pane windows, with inside shading to obtain a shading factor of 0.50, and a glass area 50% of the exposed room wall area, or approximately one-third of the total exterior wall.

Lighting was assumed to be on from 8 A.M. to 6 P.M. 5 days a week at 3.5 W/ft² for 75% of the floor area and at 1 W/ft² for the remaining 25% of floor area. Office equipment was assumed to use ½ W/ft² for 75% of the floor area. On the basis of an occupancy of 600 people weekdays from 8 A.M. to 6 P.M., the total volume for outdoor ventilated air was set at 15,000 cfm. Infiltration air was assumed to effect a constant air change of one half the total volume per hour, or 8925 cfm. All these calculations were based on indoor conditions from October through May of 70°F and 30% relative humidity from 8 A.M. to 6 P.M., during weekdays; 60°F and 42% relative humidity, 6 P.M. to 8 A.M. during week nights and 24 hours a day on weekends and holidays; and from June through September, 75°F and 50% relative humidity, 8 A.M. to 7 P.M. during weekdays. See Fig. 3.1.

3.3 THE ARCHITECTURAL DESIGN

SITE   The small size of the site, coupled with restrictions imposed by local zoning and fire safety regulations, left no room to manipulate the orientation of the building. The site shape is rectangular, with a directly north-south long axis, and the facades parallel the property lines.

A large parking garage located below grade occupies almost the entire area of the site, thereby denying the use of planting as a primary energy conservation measure. Planting is limited to trees and shrubbery around the site perimeter. Covering the north and west walls with ivy was considered at one time but was abandoned because of possible maintenance problems involving the plant's tendency to transmit moisture to the wall masonry.

BUILDING SHAPE   The change in the building's shape is one of the most noticeable departures from non-energy conservation design. The original narrow rectangular building with a long facade facing directly into the western sun was transformed into a more nearly cubical mass. Since the restricted site dimensions obviate a south-facing rectangular building, which

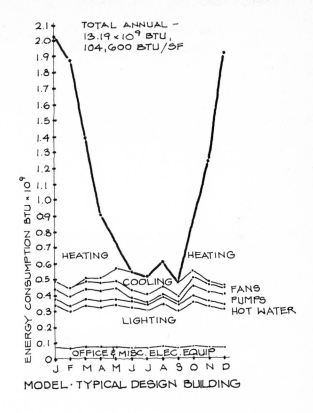

ENERGY CONSUMPTION BTU × 10⁹

2.1
2.0
1.9
1.8
1.7
1.6
1.5
1.4
1.3
1.2
1.1
1.0
0.9
0.8
0.7
0.6
0.5
0.4
0.3
0.2
0.1
0

TOTAL ANNUAL –
13.19 × 10⁹ BTU,
104,600 BTU/SF

HEATING   HEATING

COOLING   FANS
                       PUMPS
                       HOT WATER

LIGHTING

OFFICE & MISC. ELEC. EQUIP.

J F M A M J J A S O N D

MODEL · TYPICAL DESIGN BUILDING

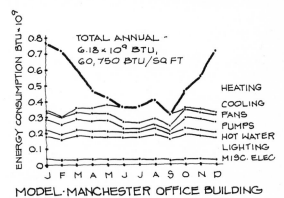

ENERGY CONSUMPTION BTU × 10⁹

0.8
0.7
0.6
0.5
0.4
0.3
0.2
0.1
0

TOTAL ANNUAL –
6.18 × 10⁹ BTU,
60,750 BTU/SQ FT

HEATING
COOLING
FANS
PUMPS
HOT WATER
LIGHTING
MISC. ELEC

J F M A M J J A S O N D

MODEL · MANCHESTER OFFICE BUILDING

Figure 3.1 A comparison of the energy requirements of a typical office building with the projected energy requirements for the Manchester Federal Office Building.

would have had some merit in terms of energy conservation, the calculations were made for width-to-length ratios of 1:1, 1.25:1, 1.5:1, and 2:1, and the 1:1 was found the most energy-efficient. An exception was made for the first floor, which extends beyond the general outline of the upper floors in order to provide sufficient space to house agencies that are expected to have substantial public contact or involve a great number of handicapped persons. See Fig. 3.2.

FENESTRATION The original Dubin-Bloome recommendation reduced heat loss by severely limiting fenestration to approximately 5% of exterior wall surface exposed to interior space (the area between floor and ceiling). The final design allowed 12% fenestration on the east, west, and south facades (a 2 ft × 5 ft window every 10 ft) and none on the cold north face. See Fig. 3.3.

In the selection of the windows, emphasis was given to tightness against air infiltration and resistance to thermal conductivity. Horizontal pivot-type

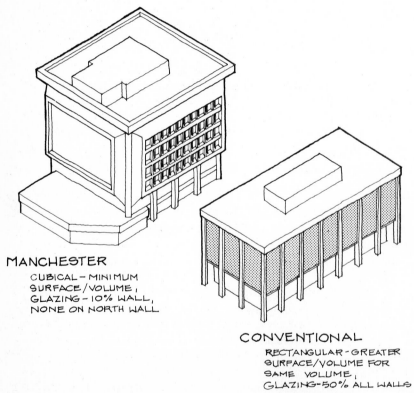

MANCHESTER
CUBICAL – MINIMUM
SURFACE/VOLUME,
GLAZING – 10% WALL,
NONE ON NORTH WALL

CONVENTIONAL
RECTANGULAR – GREATER
SURFACE/VOLUME FOR
SAME VOLUME,
GLAZING – 50% ALL WALLS

Figure 3.2 A comparison of the configuration of the Manchester Federal Office Building with the configuration of a conventional office design.

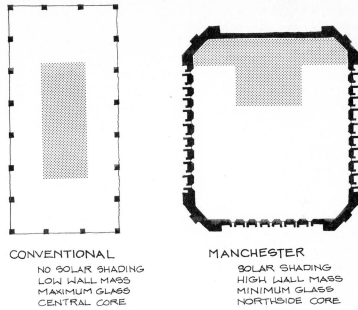

CONVENTIONAL

NO SOLAR SHADING
LOW WALL MASS
MAXIMUM GLASS
CENTRAL CORE

MANCHESTER

SOLAR SHADING
HIGH WALL MASS
MINIMUM GLASS
NORTHSIDE CORE

Figure 3.3 A comparison of the plan of the Manchester Federal Office Building with the plan of a conventional office design.

windows were finally selected to allow for easy access for cleaning from the inside. They are double glazed with venetian blinds built into the space between the panes and are equipped with a compression-type seal with a cam action dogging device; the frames are broken with a thermal barrier. The complete assembly has a U value of 0.55.

INSULATION AND WALL DESIGN Other basic changes in the building's design occurred in the construction of the external shell. The exterior was designed to attain a U value of 0.06 for the walls, roof, and the floor slab immediately above garage level.

In addition, the conservation recommendations suggested that the exterior walls, roof, and floors should have a mass of approximately 100 lb/ft² of surface to provide something of a thermal flywheel to damp overheated or underheated conditions. In the final design, some trade-offs in favor of lower initial cost were made by exempting interior floor slabs from these specifications and by reducing the mass to 80 lb/ft² for the remaining components. To achieve this weight, the exterior wall of the entire building was designed with a backup masonry wall consisting of 12-in-thick concrete block.

Exterior wall insulation is of two kinds, for different areas of the building. The first consists of 2½ to 3½ in of insulation applied to the backup wall by a combination of mechanical fasteners and adhesives, then protected by facing panels. The second consists of aluminum-clad insulating panels with a 3-in insulating core and factory-laminated aluminum sheets on both the exterior and interior faces. Both types are placed on the exterior of the wall mass. See Fig. 3.4.

SHADING DEVICES   Ideally, the purpose of shading devices is to control the amount of sunlight admitted through windows so that they will be totally shaded during the summer months and totally exposed to the sun during the winter. For the south facade this was accomplished with a 3-ft-deep eyebrow and 8-in-deep vertical fins on either side of each window. In the summer the vertical fins shade the early morning and late afternoon sun, and the overhead fin shades the high-altitude midday sun. In the winter the fixed shading devices do not interfere with the winter sun because of its low altitude and southerly azimuth at sunrise and sunset. The fins serve an additional purpose in reducing the cooling effect of cold winds blowing over the surface of the building.

The east and west facades are similarly fitted, except that the vertical fins on the south edges of the windows are approximately 8 in deep and the north-edge fins are nearly 3 ft deep. The north-edge fins exclude early

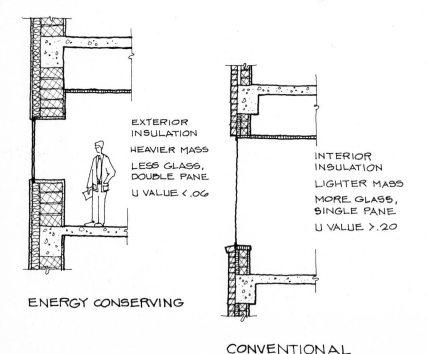

EXTERIOR
INSULATION

HEAVIER MASS

LESS GLASS,
DOUBLE PANE

U VALUE < .06

INTERIOR
INSULATION

LIGHTER MASS

MORE GLASS,
SINGLE PANE

U VALUE > .20

ENERGY CONSERVING

CONVENTIONAL

Figure 3.4 Typical wall sections for energy-conserving design and conventional design.

morning and late afternoon sun in the summer; the overall shading effectiveness is 80% on the east-west facades, 100% on the south facade. Shading performance during the winter months is 80% for the east and west facades and about 90% of ideal for the south facade.

SPECIAL GLAZING The second floor is designated a natural-lighting experimental area. Whereas other floors were designed with limited glazing to control heat loss, the enlarged window area on the second floor is designed as a testing area to determine whether increased natural lighting will lead to energy savings by reducing the requirements for artificial lighting.

In order to be effective, a system such as this must be used with a controlled system of artificial lighting so that when enough natural lighting is available, the artificial lighting is switched off. This can be achieved with manual switching, but in order to ensure the desired result, the three perimeter rows of lighting along each side of the building, with the exception of the windowless north side, have been connected to three independent photoelectric switches, so that whenever a sufficient amount of daylight is available, the three rows will be automatically switched off—one row at a time, from the extreme outer row to the inner row. The photoelectric cells are located in the ceiling, adjacent to the rows they control. They are equipped with a built-in time delay device so that lights will not switch on and off when outside light intensity levels change for a short period of time with passing clouds. The device selected for the Manchester project has a time delay of approximately 30 s.

The entire perimeter of the second floor is recessed so that the increased use of glass for this floor will not contribute to heat gains in the summer. This means that much potential heat gain in the winter has to be sacrificed, because much of the window area is shaded even from the low winter sun. This is an instance where the closest energy bookkeeping is necessary to effect a proper balance of conflicting goals.

ROOF The roof insulation, like the wall insulation, was placed on the exterior side of the roof. A 2-ft-high parapet was added around the roof perimeter to retain snowfall as an additional winter insulator. The roof U value is 0.053. Crushed white marble was used for the finished roof surface to reflect solar heat.

INTERIOR PLANNING Open planning, which allows 25% less power for lighting and easier heating or cooling air distribution, is used throughout the building. Necessary exceptions to open planning include the fourth floor, on which there are Armed Forces examination rooms, and some private offices and conference rooms. Enclosed rooms are grouped together by like function for economy of HVAC delivery. Rooms which require the greatest illumination (for typing, accounting, and drafting) were located next to windows, while conference rooms, closets, and storage rooms were located away from windows.

3.4 THE MECHANICAL SYSTEMS DESIGN

Because the building is experimental, different HVAC systems are used in different areas of the building, with instrumentation to monitor performance over time. All systems chosen were judged energy-conserving relative to others available, but diversity was needed for demonstration and documentation.

DESIGN CALCULATIONS AND CONDITIONS Heating conditions were calculated to satisfy 95% of the weather conditions during the occupied periods, rather than 100% or more as in conventional practice. In both cases, credit is taken for heat gain from people, light, and machines, but in the Manchester project positive means of heat recovery are also included.

The design conditions for high and low temperatures were substantially different from the conventional approach for the Manchester area. Normally, calculations would be set for parameters of -15 to $+73°F$ outside temperature, but this building was designed on a $+5$ to $+68°F$ basis.

Provision of outdoor air for office areas was predicated on a population of one person per 100 ft², except for specified high-density population areas such as lunchrooms and conference rooms. This assumed figure results in a total occupancy considerably greater than the actual occupancy contemplated by the GSA, but it is the minimum design occupancy load required under the building code of the City of Manchester, which the GSA honored wherever possible. The outdoor-air rate used is 6 cfm per person. This air, when combined with the recirculated air, provides a total air quantity in excess of 50 cfm per person. All the air systems are capable of delivering 100% fresh air, by simply altering damper settings, and are equipped with enthalpy controllers.

For the toilet exhaust system, 50% exhaust rather than the usual 100% exhaust is used; the other 50% is recirculated through activated-charcoal filters. The system is equipped to accommodate 100% exhaust if the need arises.

A recommendation to provide for 20% relative humidity during the winter was incorporated without any special provisions for humidification; calculations indicated that this condition will result automatically under the currently planned one person per 100 ft² occupancy with the reduced amount of outdoor air introduced into the building.

The final design permits temperatures in corridors, passageways, restrooms, and storage and equipment rooms to drop to 65°F in winter and to rise to 80°F in summer.

EQUIPMENT DESCRIPTION   The primary conversion equipment is a complex of four modular gas-fired boilers with electric ignition, a 60-ton reciprocal chiller, a 25-ton absorption chiller, and some 90 tons of heat pumps on the first three floors, all interconnected in a system that allows maximum interchange between overheated and underheated zones and storage of either hot or cold water in three 10,000-gal basement storage tanks.

The boilers, the two chillers, exhaust/return fan, and air supply fan for the upper four floors, plus most of the circulating pumps, are all located in the penthouse. The heat pumps and associated air handlers for the first three floors are located on those floors.

Hot-water heat exchangers are designed for a minimum temperature differential of 10°F water (kept low to make most efficient use of solar collection), a temperature differential of 17°F for chilled water, and a 23°F temperature differential between supply air and room air to reduce fan horsepower and total air supply requirements.

Closed-circuit coolers are used to provide part of the required ventilation for the garage, in addition to propeller-type fans. All ventilation systems in the garage area are provided with CO_2 monitoring devices.

The emergency generator is used as one of the prime movers for the air-conditioning system to power the reciprocal chiller, and heat recovered from the generator is used in the absorption machine. Since the generator normally operates cooling equipment, in the event of an emergency this load is automatically dumped, and the generator is switched over to the emergency circuit. This system is well matched to the project because the emergency power requirement nearly equals the power requirement for cooling. In addition, computerized demand limiters are incorporated as part of the control system to reduce peak loads for items such as air-conditioning compressors and elevators.

ENVIRONMENTAL CONTROL SYSTEMS   On floors 1 and 3, for the exterior zones (assumed to extend 15 ft deep inside the building perimeter) a unitary closed-loop water-to-air heat pump system is provided in each of the east, west, and south zones, complete with ductwork from each unit to its zone. To as great an extent as possible, ductwork is located not more than 15 ft from the perimeter wall. Variable-volume boxes are provided in branches from the main distribution supply duct from each unit as required for temperature control in subareas within each zone. Return air to each unit is taken from the ceiling plenum space. The interior zones for floors 1 to 3 are each handled by similar packaged heat pumps with variable-volume control as necessary for a typical area. These units are located in the north area cores for each floor.

All the outdoor air for the first, second, and third floors is taken down a shaft from the rooftop to the units servicing the interior zones. The outdoor-air intake is large enough to enable an economizer outdoor-air cycle and is dampered with an enthalpy controller sensitive to both temperature and humidity.

The second floor, designed for natural lighting with larger window areas, is provided with unitary heat pumps located beneath the windows in the perimeter zones on a 10-ft module.

All variable-volume systems are of the dump, or bypass, type so that there is full airflow across the direct expansion (DX) coils at all times.

Rejected heat from all heat pumps operating on the cooling mode during occupied periods is recovered and piped to one of three 10,000-gal hot-water storage tanks in the basement to provide the heat source for pumps operating on the heating mode at night, or at unoccupied periods when heating is required. These tanks serve as well as the storage system for solar energy collection. The storage tanks are fabricated of steel insulated with Gilsonite and are located in the garage area to provide access for the connecting piping.

For floors 6 and 7, separate fan-coil air-handling units with four pipes are provided at the exterior zones but are arranged for operation either as two-pipe or four-pipe systems. The fan-coil units draw air from the ceiling plenum to make use of heat of lights for heating during the occupied period. The interior zones are served by variable-volume boxes on the central supply system from the penthouse.

Floors 4 and 5 are provided with a separate air-handling unit on each floor for each of the perimeter zones, arranged for variable volume, with vaned inlet fan control. The units are located in the north zone and arranged to draw air from the ceiling plenum to make use of heat generated from lighting for heating during the occupied periods. Each unit is provided with a heating coil for use during warm-up and unoccupied periods. The variable-volume controls and diffusers are of the dual type which can handle warm air through one side and cold air supplied from the cold-air duct serving the interior area through the other. The interior cold-air duct is served from the penthouse central supply system. This central air system is provided with outdoor-air intake for an economizer cycle equipped with an enthalpy controller.

The central system serving floors 4 to 7 includes the emergency gas-engine generator for the elevator and other emergency electrical services. The generator is sized to produce the electricity required to operate the chiller, though the chiller is also furnished with a normal power source and a transfer switch which allows it to be served by purchased power in the event of maintenance or breakdown of the gas engine. Additional cooling is supplied by an absorption cooling unit powered with waste heat recovered from the gas engine or from the solar collector system. The absorption chiller is a 25-ton capacity unit, and the electric-driven chiller is approximately 60 tons. This latter unit is fitted with a double-bundle condenser and a double-bundle evaporator. The condenser water from both chillers is piped to one of the storage tanks located belowgrade and to a cooling tower located on the roof.

Heat can be stored in two of the tanks when there is excess energy to be saved. When the tank reaches capacity, or heat storage is not desired, the condenser water is cooled by the cooling tower. The waste heat from the engine, in addition to serving the absorption chiller, can heat domestic hot water at night and supply additional heat, when available, to the storage tanks.

Heat stored in tanks is available for night heating; it is supplemented by the heat rejected from the condenser for the electric-motor-driven chiller.

Heat requirements during occupied period are small and can be met by low-grade heat from the double-bundle condenser, while the chiller is working in the cooling mode.

The peak loads for the upper four floors can exceed the capacity of the electric-driven chiller and absorption chiller by about 15 tons of refrigeration. A third 10,000-gal capacity storage tank is provided belowgrade to store chilled water that can be set aside during a period of low demand and used during peak periods in the daytime. In this way, the chiller can be operated at maximum efficiency regardless of load.

The components and subsystems of the mechanical design, regarded individually, are not unique to this building, with the possible exception of the full-time use of the emergency generator. While it would be difficult to single out any item that would not have been used in conventional design, their collective use in the total system reflects the highest degree of care in planning a unique combination of available equipment and systems to achieve the primary goal of conserving energy. See Figs. 3.5 to 3.8.

3.5 THE LIGHTING SYSTEMS DESIGN

The Manchester project was selected to demonstrate by the use of various lighting methods the possibility of reducing raw wattage without the loss of visual performance. A basic criterion of 2 W/ft² average to obtain 70 fc in work areas and reduced levels in lounge, passageways, and equipment rooms was established.

The first floor will house agencies requiring extensive public contact, such as the Social Security Administration and the Veterans Administration. This area, with a ceiling height of 13 ft, is equipped with a uniform lighting system consisting of 3 ft × 3 ft "crosslamp" fixtures spaced approximately 10 ft on center in each direction in a checkerboard pattern, each fixture containing three 40-W lamps. High-intensity-discharge lighting fixtures are used in lobby areas throughout the building.

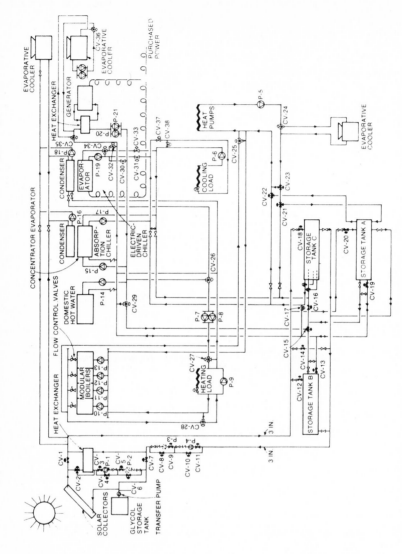

Figure 3.5 The Manchester Federal Office Building mechanical system. (*DeMichael, Don, "A GSA Energy Conservation Lab Grows in Manchester," Specifying Engineer, August 1975, Fig. 3.*)

48

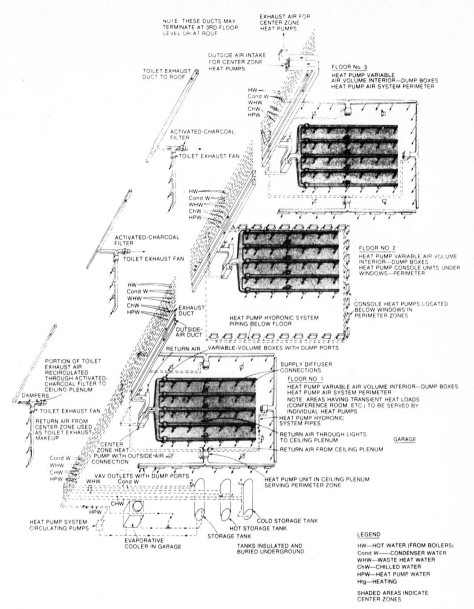

Figure 3.6   The Manchester Federal Office Building HVAC systems for floors one, two, and three. (*DeMichael, Don, "A GSA Energy Conservation Lab Grows in Manchester,"* Specifying Engineer, *August 1975, Fig. 4.*)

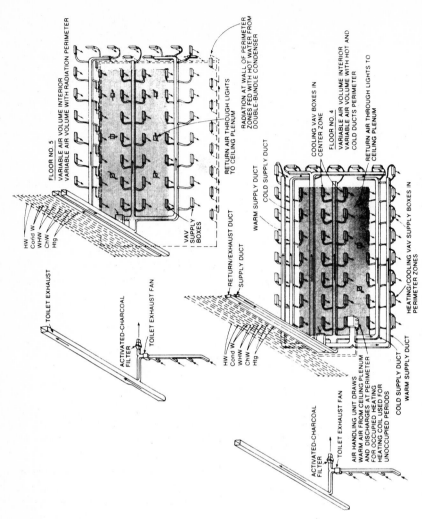

Figure 3.7 The Manchester Federal Office Building HVAC systems for floors four and five. (*DeMichael, Don, "A GSA Energy Conservation Lab Grows in Manchester,"* Specifying Engineer, *August 1975, Fig. 5.*)

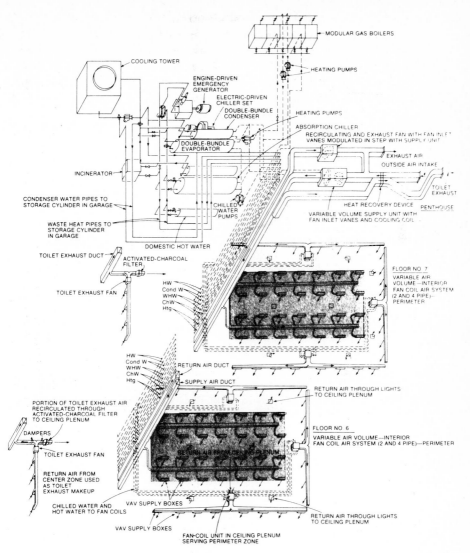

Figure 3.8 The Manchester Federal Office Building HVAC systems for floors six and seven. (*DeMichael, Don, "A GSA Energy Conservation Lab Grows in Manchester,"* Specifying Engineer, *August 1975, Fig. 6.*)

The second floor, with more window area on the south, east, and west exposures, is designated a natural lighting floor and will be compared with the sixth floor, which employs similar lighting fixtures but has less window area. This lighting system uses single-lamp 40-W fluorescent fixtures, mounted in coffered acoustical ceilings, having twin beam (bat wing) distribution lenses. Photocontrol switching of lighting fixtures by rows at 5-ft, 10-ft, and 15-ft increments from the respective zones' exterior wall is provided to make use of natural light when it is available.

The third floor demonstrates the use of polarizing lenses and will be compared with a floor having flat white acrylic lenses and with a floor having prismatic acrylic lenses. Lighting fixtures are the fluorescent crosslamp type with two 40-W lamps, which are interchangeable with those on floors 5 and 7. Interchangeable lenses are used to compare worker efficiency under different conditions.

The fourth floor is illuminated for task lighting by means of high-intensity-discharge lighting fixtures (150-W high-pressure sodium lamps) which are relocatable within adjacent four ceiling modules, depending on the task location.

The fifth floor tests flat matte-white acrylic diffusers in fluorescent crosslamp fixtures equipped with four 40-W lamps. The lenses are interchangeable with those on floors 3 and 7.

The lighting fixtures on the sixth floor are similar in layout and type to those on the second floor, but without the photocontrolled switching.

The seventh floor demonstrates prismatic refractive acrylic lenses in crosslamp fixtures equipped with four 40-W lamps. These lenses are interchangeable with those on floors 3 and 5.

At the basement and subbasement levels, the criterion is ½ fc at floor surface based on mercury-vapor street-lighting–type fixtures installed in ceiling-pan coffers. These are 100-W coated mercury lamps equipped with prismatic glass refractors. The unit produces an Illuminating Engineering Society (IES) type III (street lamp) distribution.

The emergency lighting system is designed to operate by the output of the gas-engine–driven generator which normally powers the chiller compressor unit, in case of a power failure.

3.6 THE SOLAR ENERGY SYSTEM

From the beginning of the design, provisions were made to accommodate a future solar energy system. Toward the end of the design phase, the GSA authorized the architects to include the solar energy system. They in turn retained Dubin-Bloome Associates to design the system.

The final design of the solar energy system includes approximately 4000 ft² of solar collector, mounted on the roof in four rows. The collector performance and space requirements are based on flat-plate collectors, facing south and arranged with a mechanism to provide adjustable tilt from 20 to 80°. The angle adjustment will be done seasonally to provide the optimum amount of collection.

The basic HVAC systems include coils in air-handling units which require only 100°F hot water to heat the building, three 10,000-gal insulated tanks (one to store hot or chilled water and two to store hot water only), and heat pumps for heating and cooling. The low-temperature coils permit operation with hot water at low temperatures, thus increasing the collector efficiency and enabling the systems to use low-grade waste heat from other sources.

In overcast weather, or during extremely low outdoor temperatures, the building system heat pumps will boost the hot-water temperature from the collectors when it is lower than 110°F.

In summer the collectors will provide higher-temperature hot water (a combination of higher outdoor ambient air temperature and the flatter collector tilt angle) for operation of the absorption refrigeration system for air conditioning. The three storage tanks, which were included in the basic HVAC design for energy conservation, provide the needed storage for hot water from the collectors to tide the system over three or four cloudy or sunless days.

To provide frost protection in winter, the collector system uses a 50-50 mix of ethylene glycol and water as the heat transfer fluid. In this mode a heat exchanger is used to transfer heat from the collectors to the building systems while preventing cross-contamination with the antifreeze solution. In summer the antifreeze solution is drained into a storage tank, and the heat exchanger is bypassed so that the collectors are connected directly to the building systems sharing the same water. This avoids the penalties of the approach temperature difference associated with the heat exchanger and allows water to be used at or near collection temperature. This is important as the absorption chiller output falls off rapidly at low generator temperatures.

Use of the multistorage tanks permits storage of hot water at different temperature levels without degradation, depending on the solar collector outlet temperature variations during the day. Two tanks are used for hot-water storage in the summer and three in the winter, the third being used to store chilled water generated at off-peak hours when lower nighttime condensing water temperatures can be obtained.

The solar system is designed strictly as an energy-saving supplementary system. At ideal weather conditions, the solar system can provide 100% of the heating and cooling requirements, but on a seasonal basis it will contribute about 30% of the heating and cooling load. The computerized control system, specified for the building, will also monitor and control the solar energy system for both on-line and experimental modes of operation.

It is believed that the system is the first in a commercial building to employ tilt-angle adjustment and the first to incorporate solar heating and cooling in combination with heat pumps, absorption refrigeration, and waste heat from a gas-engine–driven electric generator driving a central chiller. See Fig 3.9.

3.7 CONCLUSIONS

One important factor that controls the success or failure of any energy conservation project is the availability of technology in the marketplace. The Manchester project was envisioned as an example of energy efficiency achieved through the intelligent application of available technology.

It is essential that energy "bookkeeping" covering every facet of a building be done at a very early stage of design, so that the interdependent relationships between all components of a building are fully coordinated. This requires the assembling of a design team at the very start of a project's development so that each area of responsibility can be carefully weighed and balanced against its effect on other facets. The days when engineering consultants fitted their systems into the framework of design established by the architect should be looked upon as passé. What is needed now is the inclusion of energy conservation considerations as a major factor in the building program. This necessitates a comprehensive planning effort con-

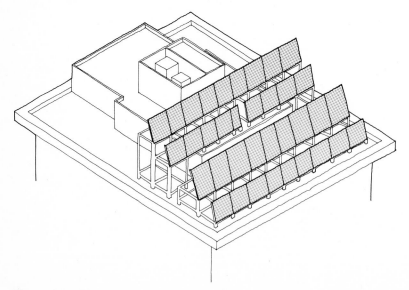

Figure 3.9 The flat-plate solar-collection configuration added to the Manchester Federal Office Building roof.

cerning all design factors before even a single line is drawn on a piece of tracing paper.

Since different systems are included for comparison, the Manchester project is not the least expensive energy-efficient building that can be built. It is a laboratory, albeit a functioning office building. With the proper use of existing technology, new buildings can be constructed that are energy-efficient at no additional first cost, and possible energy savings of 40 to 60% in the operation of well-designed buildings offer a great life-cycle-cost incentive for the pursuit of energy conservation throughout the construction industry.

CHAPTER FOUR

EXISTING BUILDINGS: AN ENERGY MANAGEMENT PROGRAM

4.1 THE NEED

New construction adds something on the order of 2% to the existing building inventory of the United States in any given year. This is an impressive investment. It is essential in this present day of energy concern that each of these new buildings be carefully put together with energy-conscious design, and this book attempts a framework of understanding for such design.

On the basis of the 2% figure for new construction, buildings already in existence will constitute a major portion of all buildings in use for many years to follow, and it is quite obvious that if energy is to be conserved in the building sector in meeting national goals, old buildings must receive first priority. This is recognized in the various chapters that follow by casting conservation and management suggestions in terms of existing buildings (though, of course, these principles apply to all buildings, old and new).

Old buildings are the buildings we live with and live in. With energy scarce and expensive, it is in the interest of owner, user, and administrator alike to examine carefully the way we heat, cool, and light our spaces, so that together we can discover where energy can better be conserved and managed. The pages below give procedure to this examination.

But with respect to energy and society there is a larger issue that should be mentioned in this introduction. It has been the nation's habit in our recent past to value but little the old and shabby buildings that form a large part of our cities and towns. Public renewal programs, and private renewal as well, have more frequently than not assumed new is automatically better than old, and the bulldozer more appropriate than conservation for such renewal programs. The energy quotients of tearing down and building anew have not been considered of import—energy was cheap and abundant. It is no longer so.

It is imperative that governmental policy, with respect to taxation, valuation, and depreciation, fully account for the true value of existing structures in terms of demolition and construction energy requirements. To this value must be added the enriching inflence of old structures on the quality of life in city and town; this imponderable is no small measure in the final worth of a building. Though in principle new buildings should be able to operate with lower operating energy budgets born of enlightened design than operating budgets required for older buildings, the difference in most cases cannot be used to justify demolition and rebuilding.

4.2 AN ENERGY MANAGEMENT PROGRAM

The purpose of an energy management program for existing buildings is (1) to minimize the energy consumed for heating, cooling, lighting, and other building services; (2) to use the particular type of fuel or electricity which is available at the lowest cost (and to switch from one fuel to another if the relative cost changes radically in the future); and (3) to carry out these objectives within the financial capability of the owner and within the emerging requirements of national energy and resources policy.

A management program will be given form and direction by objectives such as the following, which owners themselves must establish and order in priority:

- Reduce energy consumption without initial costs.

- Reduce energy consumption with capital investment.

- Reduce operating costs.

- Reduce the likelihood of shutdown due to reduced voltage or curtailment of power or fuel supplies.

- Extend the life of the mechanical and electrical equipment to reduce the frequency of equipment breakdown and replacement costs.

- Improve the performance of maintenance personnel to reduce operating costs and equipment failure.

- Reduce airborne pollution.

- Meet existing federal and state guidelines and forthcoming mandatory energy conservation standards.

- Enhance the rentability or salability of the building through lower operating costs.

- Enhance the internal environment by such measures as reducing disturbing glare from lighting and drafts from infiltration or cold wall surfaces.

Though some conservation options which are a part of energy-conscious design for new buildings are not available for existing building (options such as site selection, building orientation and configuration, height, and materials), energy flow in existing buildings can more readily be traced and then managed. The magnitude of the building load, the actual number of occupants, their activities and requirements, and actual energy usage can be measured. Existing mechanical and electrical systems can be examined for performance during full- and part-load conditions. With this data base, redesign and refitting of an existing building can be tailored to known requirements without need for safety margins necessary for new building design. In a final analysis, such safety margins so often included as contingency against unknown conditions contribute heavily to excessive energy use.

Conservation, as a part of an energy management program, is best stated as a target to be reached. In the near future definitive data from research will be available to set realistic energy budgets for each building type, size, and climatic location, but this information is not now accessible. In the interim, conservation goals can be set, however, in terms for existing buildings as percentage reductions from actual present energy consumption.

A realistic goal for energy reduction is 10 to 20% without any initial cost, another 10% with minimal first costs, and 10 to 20% more with an investment which will yield simple payback in 3 to 10 years by lower operating expense. Additional energy savings which at present are doubtful economic investments will become economically viable as fuel costs continue to escalate and will enable the yearly energy budget to be reduced even further.

4.3 THE ENERGY MANAGERS

Many people are involved in the use of energy in buildings and in its conservation. A well-considered and managed energy program must involve owners and administrators with policy decisions, professionals in consultation, and most especially building personnel who actually operate the building and the users of the building who by their support or lack of it can make a program work or make it worthless.

Energy management is a team effort. The role of each member is clear, but the efforts of each may occasionally overlap.

OWNER

- Select objectives and institute the program. Hire personnel and retain professional consultation.

- Collect and tabulate data on operating costs, fuel and energy use, and other pertinent subjects.

- Arrange access to the building for surveying and testing, balancing, adjusting, and retrofitting the mechanical and electrical systems and the structure.

- Control the conditions of building use. Pay those costs for services and goods which are not borne by the tenants.

ENGINEER AND ARCHITECT

- Analyze the existing conditions, survey the premises, identify the opportunities for conservation, and select systems for further analysis.
- Calculate loads and flow characteristics to determine the potential energy savings for each system or building component.
- Analyze utility rates and perform cost/benefit analyses for each system or group of systems (analyses may be simple payback, cash flow, present worth, discounted rate-of-return, or life-cycle costs) as dictated by the owner's fiscal needs.
- Prepare plans and specifications for operating procedures and/or construction programs.
- Prepare an operating manual for use by personnel retained by the owner.
- Train the permanent energy management coordinator and operating personnel to monitor, operate, and maintain the systems and structure in accordance with the developed energy management program.

UTILITY COMPANIES

- Provide information on existing and anticipated utility rates.
- Loan or rent meters to measure flow rates to test, balance, and adjust systems for optimal performance.

SERVICE AND MAINTENANCE ORGANIZATIONS

- Perform efficiency tests before and after modifications, under the direction of the consulting engineer, for combustion, boilers, chillers, compressors, cooling towers, air-cooled condensers, and other types of equipment (including elevators, escalators, motors, and automatic control systems).

OPERATING PERSONNEL OR OUTSIDE SERVICE PERSONNEL

- Operate, test, and service equipment.
- Direct the cleaning activities to maintain maximum efficiency of filters, louvers, heat transfer equipment, lighting fixtures, windows, etc.

- Direct activities to reestablish operating parameters in accordance with changes as they occur in occupancy and/or functional requirements of the building owner or tenants.

- Set up and maintain a data collection system to include a description of all systems, their operating characteristics, maintenance performed, and monthly fuel and energy usage.

- Develop specific operating programs and schedules, working from the operation manual provided by the engineer, to effectively monitor and control the operating systems.

TENANTS OR OCCUPANTS

- Help to implement specific conservation measures by shutting off lights, closing doors, turning off hot-water faucets, maintaining the specified thermostat and humidistat settings, etc.

COST ENGINEERS

- Estimate costs of labor and materials for preliminary analyses. (Final costs should be based on competitive bids or advice from reliable contractors.)

4.4 THE BUILDING OF THE PROGRAM

Step 1: Select the objective

Select the specific objectives which are consistent with the building owner's interests and resources. Then organize a conservation journal to record all findings, actions, calculations, and procedures. Include initial energy audits and profiles and arrange headings for entry of monthly utility bills with conversions to Btu for an accurate accounting of program effectiveness. Include headings for energy management capital costs and for expenses or savings related to labor and maintenance costs. A word of caution: Do not get bogged down in paper work.

Step 2: Establish climate profile

Identify the climatic conditions to which the building is exposed over extended periods of time with the aid of climatic maps, shown in Figs. 4.1 to 4.6, and enter each value on the appropriate line of a location and climatic zone profile form, shown in Table 4.1. The geographic locations should be plotted on each of the maps, and relevant conditions read from the contour lines. If the building lies between contour lines, the values can be interpo-

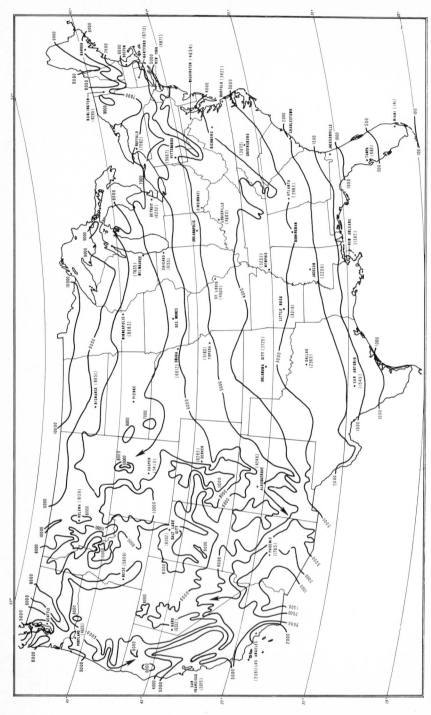

Figure 4.1 Annual mean heating degree-days below 65°F. [A heating degree-day is an average temperature 1° below a given temperature base, in this case 65°F. A day with an average temperature of 60°F thus has 5 degree-days (65–60°F). Degree-day totals are given on the map for all the days in the year with average temperatures below the base 65°F.] *(From Climatic Atlas of the United States, Department of Commerce, June 1968, p. 70.)*

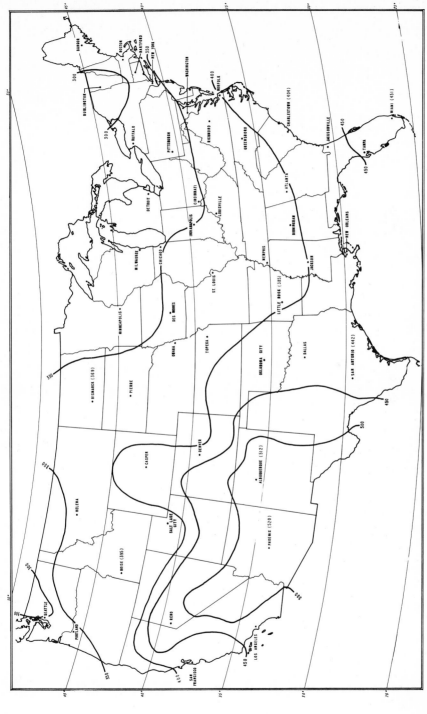

Figure 4.2 Annual mean daily solar radiation in langleys. One langley equals 3.6 Btu/(hr)(ft²). (*From Climatic Atlas of the United States, Department of Commerce, June 1968, p. 70.*)

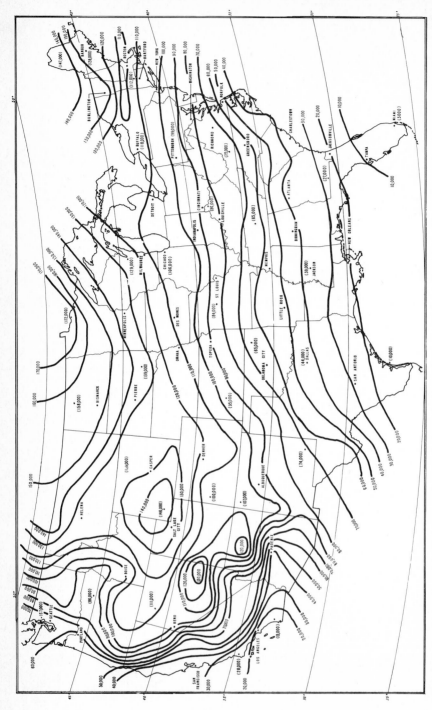

Figure 4.3 Annual wet-bulb degree-hours below 54°F WB and 68°F DB. [A WB degree-hour in this map is an average temperature 1°F WB below the given base conditions for one hour. One hour with an average WB temperature of 47°F (and below 68°F DB) thus has 7 WB degree-hours (54–47°F). The WB degree-hour totals are given for all hours, October through March, with average WB temperatures below the base 54°F WB and 68°F DB.] (*From AFM 88-8, U.S. Government Printing Office, June 15, 1967.*)

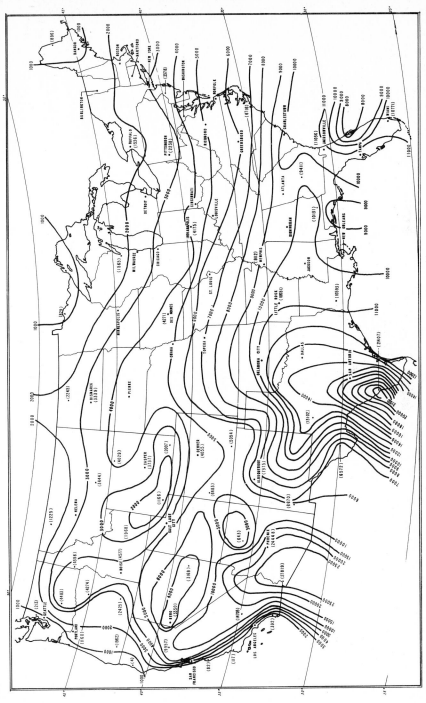

Figure 4.4 Annual dry-bulb degree-hours greater than 78°. [A DB degree-hour in this map is an average temperature 1° above 78°F DB for one hour. An hour with an average DB of 87°F thus has 9 DB degree-hours (87–78°F). The DB degree-hour totals are given for the working hours 0930 to 1730 for all the days of the year.] *(From AFM 88-8, U.S. Government Printing Office, June 15, 1967.)*

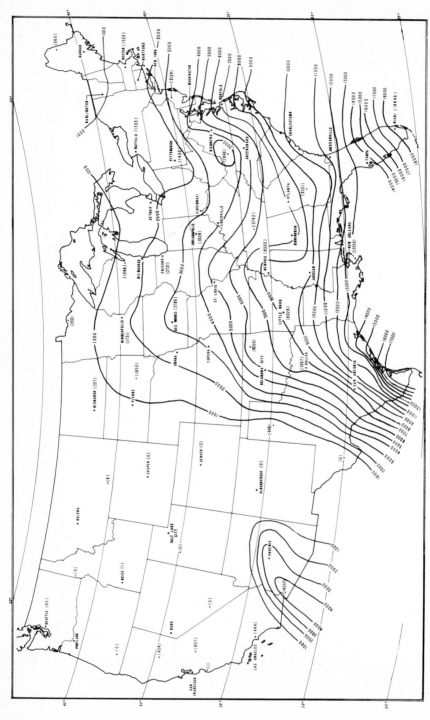

Figure 4.5 Annual wet-bulb degree-hours greater than 66°F. [A WB degree-hour in this map is an average temperature 1° above 66°F WB for one hour. An hour with an average WB of 74° thus has 8 WB degree-hours (74–66°F). The WB degree-hour totals are given for the working hours 0930 to 1730 for all days of the year.] (*From AFM 88-8, U.S. Government Printing Office, June 15, 1967.*)

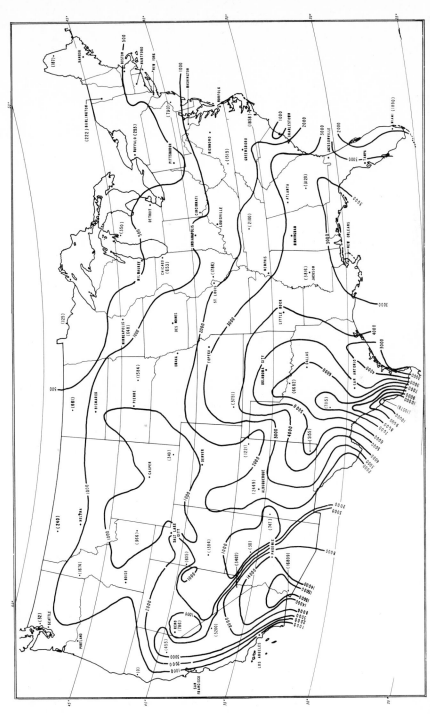

Figure 4.6 Annual dry-bulb degree-hours greater than 85°F. [A DB degree-hour in this map is an average temperature 1° above 85°F DB for one hour. An hour with an average DB of 87°F thus has 2 DB degree hours (87–85°F). The DB degree-hour totals are given for the working hours 0930 to 1730 for all the days of the year.] (*From APM 88-8, U.S. Government Printing Office, June 15, 1967.*)

TABLE 4.1   LOCATION AND CLIMATIC ZONE PROFILE FORM

	Column A Value	From
Latitude		Fig. 4.1
Heating degree-days		Fig. 4.1
Solar radiation, langleys		Fig. 4.2
Degree-hours less than 54°F WB when DB is less than 68°F		Fig. 4.3
Degree-hours greater than 78°F DB		Fig. 4.4
Degree-hours greater than 66°F WB		Fig. 4.5
Degree-hours greater than 85°F DB		Fig. 4.6

lated. The maps provide generalized information. More precise data for major cities can be obtained from the ASHRAE *Handbook of Fundamentals* or from local weather reporting stations.

The climatic maps used for Figs. 4.1 to 4.6 provide a general picture of conditions in any particular location; they cannot, however, indicate the variation of conditions experienced within local areas. Knowledge of the local climate should be used to modify the general conditions when more precise answers are required. For example, if one area is known to be 10°F colder in winter than the average of surrounding areas, the heating degree-days should be modified accordingly.

1. Estimate the annual heating degree-days (Fig. 4.1): Dry-bulb (DB) degree-days influence the annual energy consumption for building heating loads.

2. Estimate the annual mean daily solar radiation (Fig. 4.2): Solar radiation (measured in langleys or Btu) influences the annual energy consumption for the building cooling load by increasing heat gain by radiation through glazing and by conduction through opaque roofs and exterior wall surfaces. Solar heat gain reduces the building heating load in winter.

3. Estimate the annual wet-bulb (WB) degree-hours below 59°F WB when the DB degree-hours are below 68°F (Fig. 4.3): This value influences the annual energy consumption for building heating load to maintain relative humidity levels in the winter.

4. Estimate the annual DB degree-hours when DB temperatures are above 78°F (Fig. 4.4): This value influences the annual energy consumption for building cooling load due to heat gain by conduction through the building envelope, and for building cooling load required to lower the dry-bulb temperature of outdoor air (for ventilation and infiltration) to room conditions of 78°F.

5. Estimate the annual WB degree-hours when the WB temperature is above

66°F (Fig. 4.5): This value influences the annual energy consumption for building cooling load to cool and dehumidify outdoor air to room conditions of 78° DB and 55% RH.

6. Estimate the annual DB degree-hours above 85°F (Fig. 4.6): This value influences the annual energy consumption for building cooling load due to outdoor air when enthalpy conditions are low enough to permit full use of outdoor air, but existing cooling coils cannot maintain dry-bulb conditions with 100% outdoor air.

Step 3: Establish building construction, condition, and use profile

Examine available documents and conduct a thorough survey of the site and building to determine the actual dimensions and configuration of the building, the construction materials used, the physical condition of the structure, its compass orientation, and the amount of shading from adjacent buildings, plantings, or solar control devices. Determine which safety, ventilating, and lighting codes are applicable. Gather data on the complete lighting systems, the occupancy, type and duration of tasks in each specific area of the building, ventilation rates, probable infiltration rates, and appliances or processes which use energy and emit heat into the building (investigation has often disclosed discrepancies between actual conditions and those originally assumed for design). Table 4.2 is offered for data collection.

Use these data to calculate the peak heating, ventilating, and cooling loads at the extreme outdoor weather conditions (winter and summer) for the desired indoor temperature and humidity levels. (Consult the ASHRAE *Handbook of Fundamentals* for methods of performing calculations.) On the basis of design load calculations, estimate the annual heating and cooling loads with manual or computer procedures.

Step 4: Establish mechanical and electrical systems profile

Examine existing documents for information and make a thorough survey of the mechanical and electrical systems which consume fuel or electricity. Determine their operating characteristics, physical condition, control sequences, and materials of construction. See Table 4.3.

Step 5: Conduct an energy audit

Gather from monthly utility and fuel suppliers' bills the annual usage of energy in gallons of oil, cubic feet of gas, pounds of propane, tons of coal, and kilowatthours of electricity. Record the gross yearly quantity of fuel and power in profile Table 4.4, column A. Convert the units of fuel and electricity to equivalent Btu by multiplying the quantities of fuel and power by the appropriate conversion factor in column B. Record the product in column C.

TABLE 4.2   BUILDING TYPE, CONSTRUCTION, AND USE PROFILE
(Circle appropriate items and fill in blank)

Configuration and Construction:

Line No.

1. Primary building use:
2. Length, feet, and orientation _____ ft N. W. E. S.
3. Width, feet, and orientation _____ ft N. W. E. S.
4. Number of floors: _____
5. Height from floor to floor _____
6. Height from floor to ceiling _____
7. Floor area, gross ft²: Lines 2 × 3 × 4 = _____ ft²
8. Window glazing: single, double, clear, reflective
9. Window type: Fixed sash, double hung, casement
10. Window condition: Loose fitting, medium, tight
11. Windows: Number, area,
    orientation: North, no. _____ Gross area: _____
12. Windows: Number, area,
    orientation: West, no. _____ Gross area: _____
13. Windows: Number, area,
    orientation: East, no. _____ Gross area: _____
14. Windows: Number, area,
    orientation: South, no. _____ Gross area: _____
15. Door types and numbers:
    1, single; 2, vestibule; 3, revolving

                                      North, no. _____ type _____
16. Door types and numbers: East, no. _____ type _____
17. Door types and numbers: West, no. _____ type _____
18. Door types and numbers: South, no. _____ type _____
19. Gross wall area and
    orientation: North: Lines 2N × 3N × 4 × 5 = _____ ft²
20. Gross wall area and
    orientation: West: Lines 2W × 3W × 4 × 5 = _____ ft²
21. Gross wall area and
    orientation: East: Lines 2E × 3E × 4 × 5 = _____ ft²
22. Gross wall area and
    orientation: South: Lines 2S × 3S × 4 × 5 = _____ ft²
23. Net wall area and
    orientation: North: Lines 19-11 (area) _____ ft²
24. Net wall area and
    orientation: West: Lines 20-12 (area) _____ ft²
25. Net wall area and
    orientation: East: Lines 21-13 (area) _____ ft²
26. Net wall area and
    orientation: South: Lines 22-14 (area) _____ ft²
27. Exterior opaque wall construction (circle type):
    1, frame; 2, curtain wall; 3, solid masonry; 4, brick and masonry; 5, masonry cavity
28. Exterior opaque wall insulation: Material: _____
    Thickness: _____

TABLE 4.2 (*Continued*)

29. Roof construction (circle types): 1, masonry; 2, wood; 3, metal; 4, flat; 5, sloped; 6, pitched; 7, light; 8, dark
30. Roof insulation: Type: _____ Thickness: _____ U value: _____
31. Floor (circle type): 1, slab on grade; 2, over heated space; 3, over unheated space; 4, wood; 5, concrete; 6, other
32. 1) Number of working hours per week. 2) Number of occupants _____
    (a) For offices, employees and visitors; stores, employees and customers; religious buildings, occupants.
33. Number of custodial hours per week _____;
    after dark, summer _____; after dark, winter _____;
    Saturdays _____; Sundays _____;
34. Temperature and relative humidity (indoors):
    a. If heated, winter   Occupied hours ____°F_____ %RH
    b.   Unoccupied hours ____°F_____ %RH
    c. If air conditioned, summer   Occupied hours ____°F_____ %RH
    d.   Unoccupied hours ____°F_____ %RH
35. Ventilation (outside air):
    a. During occupied hours on/off, amount _____ cfm
    b. Per person: Line 35a ÷ line 32 (2) = _____ cfm
    c. During unoccupied hours on/off, amount _____ cfm

TABLE 4.3 ELECTRICAL AND MECHANICAL SYSTEMS PROFILE
(Circle appropriate item and fill in blank)

Line No.

1. Electric lighting system: Lighting fixtures in primary spaces such as office areas, halls of worship, store sales areas.
   (1) Incandescent; (2) Fluorescent; (3) Other _____ (4) Number of fluorescent fixtures _____; (5) Number of lamps per fixture _____; (6) Wattage per lamp _____; (7) Total wattage of all fluorescent fixtures _____; (8) Total wattage of all incandescent lamps _____; (9) Total wattage of incandescent and fluorescent lamps _____

2. Lighting fixtures in secondary spaces, such as corridors, toilet rooms, storage rooms.
   (1) Incandescent; (2) Fluorescent; (3) Other _____; (4) Number of fluorescent fixtures _____; (5) Number of lamps per fixture _____; (6) Wattage per lamp _____; (7) Total wattage of all fluorescent fixtures _____; (8) Total wattage of all incandescent lamps _____; (9) Total wattage of incandescent and fluorescent lamps _____

3. Total installed wattage: Lines 1 (9) + 2 (9) = _____

4. Average installed watts/ft²: Line 3÷7 of Table 4.2 =

5. Type lighting fixtures: (1) Pendant mounted; (2) Surface mounted; (3) Recessed; (4) Wall mounted; (5) Luminous ceiling; (6) Cove mounted; (7) Exterior lighting on walls; (8) Exterior lighting on standards

TABLE 4.3 *(Continued)*

6. Total wattage of exterior lighting for: (1) Security _____ (2) Parking lots and drive _____

7. Area of parking lots _____ length × _____ width = _____ ft²

8. Parking lot lighting in watts/ft², Lines 6 (2) ÷ 7 ft² = _____ W/ft²

9. Hours per week parking lot lighting is in operation _____

VERTICAL TRANSPORTATION

10. Escalators: Number _____ operation hours per day _____

11. Elevators: Number _____ Type: Gear, gearless, hydraulic operation, hours per day _____ Total connected HP _____

DOMESTIC HOT-WATER SYSTEMS

12. Method of generation and storage: Separate water heater: (1) Oil; (2) Gas; (3) Electric; (4) Coal; (5) Tankless heater on space-heating boiler; (6) Tank heater on space-heating boiler; (7) Storage tank size, if any _____ gal; (8) Tank insulation thickness _____ Type _____; (9) Aquastat setting _____ °F

13. Estimated annual usage:
    (1) Office buildings: Table 4.2, line 32(2) × 750 _____ gal/yr
    (2) Restaurants: Meals served/yr × 3 gal/meal = _____ gal/yr
    (3) Religious buildings: Table 4.2, line 32(2) × 50 gal/yr = _____ gal/yr (does not include special cooking facilities)
    (4) Stores: Table 4.2, line 32(2) × number of days = _____ gal/yr
    (5) Residential buildings: 7200 gal/yr per capita
    (6) Schools: 50 gal/week per capita
    (7) Hospitals: varies with type

HEATING AND AIR-CONDITIONING SYSTEMS

14. Boilers or furnace type for space heating (circle items):
    (1) Hot water; (2) Low-pressure steam; (3) High-pressure steam; (4) Fire tube; (5) Water tube; (6) Cast-iron; (7) Steel; (8) Gravity hot air; (9) Forced warm air

15. a) Boiler or furnace rating _____ Btu × 10³/h or _____ boiler HP
    b) Present measured peak load combustion efficiency _____ %

16. Compressors and chillers:
    (1) Number _____
    (2) Rating of each in tons of refrigeration _____
    (3) Total tons of refrigeration (1) × (2) = _____
    (4) If electric drive, total motor horsepower _____ HP
    (5) If absorption units, total peak steam consumption _____ HP

17. If central air-conditioning systems, indicate: (1) Cooling tower motor sizes, total _____ HP; (2) Air-cooled condenser motor sizes, total _____ HP; Condenser pumps, number _____; Total _____ HP.

TABLE 4.3 *(Continued)*

HEATING AND AIR-CONDITIONING SYSTEMS

18. If room air conditioners or through-the-wall units: Indicate (1) Total number _____ (2); Horsepower _____ /unit; (3) Total connected horsepower (1) × (2) = _____

19. If commercial refrigeration, indicate: (1) Number of cold cases or refrigerators: _____; (2) Number of condensing units: _____; (3) Total connected horsepower of condensing units: _____ HP

HVAC SYSTEMS

20. *All-air HVAC systems:* Check types, fill in blanks
   (1) Single zone _____ a) Number of air-handling units: _____
   b) Total horsepower: _____
   c) Total cfm/air-handling unit: _____
   (2) Terminal reheat _____ a) Number of air-handling units: _____
   b) Total horsepower: _____
   c) Static pressure: _____
   d) Number of reheat boxes: _____
   e) Type reheat coil: 1. hot water 2. electric 3. steam
   f) cfm/air-handling unit: _____
   (3) Variable volume _____ a) Number of air-handling units: _____
   b) Total horsepower: _____
   c) Dump-type system: _____
   d) Vaned inlet: _____
   e) cfm/air-handling unit: _____
   (4) Induction _____ a) Number of air-handling units: _____
   b) Total horsepower: _____
   c) Static pressure: _____
   d) Number of terminal units: _____
   e) cfm/air handling unit: _____
   (5) Dual duct _____ a) Number of air-handling units: _____
   b) Total horsepower: _____
   c) Static pressure: _____
   d) Number of terminal units: _____
   e) cfm/air-handling unit: _____
   (6) Multizone units _____ a) Number of air-handling units: _____
   b) Total horsepower: _____
   c) Static pressure: _____
   d) Number of terminal units: _____
   e) cfm/air-handling unit: _____
   (7) Forced warm-air furnaces, number: _____
   a) Total horsepower of blowers: _____
   b) cfm/furnace: _____

21. Water-air systems
   (1) Two-pipe fan coil _____ a) Number of units: _____
   b) Total connected horsepower: _____

73

TABLE 4.3 *(Continued)*

HVAC SYSTEMS

 (2) Four-pipe fan coil _____ a) Number of units: _____
             b) Total connected horsepower: _____
 (3) Unitary heat pumps _____ a) Number of units: _____
             b) Total connected horsepower: _____

22. Pumps
 (1) Chilled-water pumps: _____
           a) Number of units: _____
           b) Total connected horsepower: _____
 (2) Condenser water pumps: _____
           a) Number of units: _____
           b) Total connected horsepower: _____
 (3) Boiler feed pumps:_____
           a) Number of units: _____
           b) Total connected horsepower: _____
 (4) Hot-water pumps for space heating: _____
           a) Number of units: _____
           b) Total connected horsepower: _____
 (5) Recirculating pumps for domestic hot water:_____
           a) Number of units: _____
           b) Total connected horsepower: _____

23. (1) Outside air fans: _____
           a) Number of units: _____
           b) Total connected horsepower: _____
           c) cfm/fan unit: _____
 (2) Supply air fans *(check the number and total horsepower for all):*
   a) Number of backward-curved multivane fans _____ HP ____
   b) Number of forward-curved multivane fans _____ HP ____
   c) Number of axial fans _____ HP ____
   d) Number of propeller fans _____ HP ____
   e) cfm/fan unit _____
 (3) Exhaust air fans _____
   a) Number of backward-curved multivane fans _____ HP ____
   b) Number of forward-curved multivane fans _____ HP ____
   c) Number of axial fans _____ HP ____
   d) Number of propeller fans _____ HP ____
   e) cfm/fan _____
24. Check if installed:
 (1) Fin tube radiators _____ (5) Supply and return ducts _____
 (2) Cast-iron radiators_____ (6) Outside air dampers_____
 (3) Radiant heating coils _____ (7) Steam piping _____
 (4) Hot-water piping_____ (8) Exhaust duct work _____

TABLE 4.4 ENERGY USE AUDIT

	A	B Conversion Factor	C Thousands of Btu/yr

GROSS ANNUAL FUEL AND ENERGY CONSUMPTION

	A	B	C
1. Oil, gal	_____	× 138 (1) =	_____
	_____	× 146 (2) =	_____
2. Gas, ft³	_____	× 1.0 (3) =	_____
	_____	× 0.8 (4) =	_____
3. Coal, short tons	_____	× 26,000 =	_____
4. Steam, lb × 10³	_____	× 900 =	_____
5. Propane Gas, lb	_____	× 21.5 =	_____
6. Electricity, kWh	_____	× 3.413 =	_____
7. Total Btu × 10³/yr			_____

8. Btu × 10³/(yr)(ft²) gross floor area _____
(line 7 ÷ Table 4.2, Line 7)
Use for (1) No. 2 oil; (2) No. 6 oil; (3) natural gas; (4)
manufactured gas.

ANNUAL FUEL AND ENERGY CONSUMPTION FOR HEATING

	A	B	C
9. Oil, gal	_____	×138 (1) =	_____
	_____	× 146 (2) =	_____
10. Gas, ft³	_____	× 1.0 (3) =	_____
	_____	× 0.8 (4) =	_____
11. Coal, short tons	_____	× 26,000 =	_____
12. Steam, lb × 10³	_____	× 900 =	_____
13. Propane Gas, lb	_____	× 21.5 =	_____
14. Electricity, kWh	_____	× 3.413 =	_____
15. Total Btu × 10³			_____

16. Btu × 10³/(yr)(ft²) gross floor area _____
(line 15 ÷ Table 4.2, line 7)

ANNUAL FUEL AND ENERGY CONSUMPTION FOR DOMESTIC HOT WATER

	A	B	C
17. Oil, gal	_____	× 138 (1) =	_____
	_____	× 146 (2) =	_____
18. Gas, ft³	_____	× 1.0 (3) =	_____
	_____	× 0.8 (4) =	_____
19. Coal, short tons	_____	× 26,000 =	_____
20. Steam, lb × 10³	_____	× 900 =	_____
21. Propane gas, lb	_____	× 21.5 =	_____

TABLE 4.4   ENERGY USE AUDIT (*Continued*)

	A	*B* *Conversion* *Factor*	*C* *Thousands of* *Btu/yr*
ANNUAL FUEL AND ENERGY CONSUMPTION FOR DOMESTIC HOT WATER			
22. Electricity, kWh	_____	× 3.413 =	_____
23. Total Btu/h × 10^3			_____
24. Btu × 10^3/(yr)(ft²) gross floor area (line 23 ÷ Table 4.2, line 7)			_____

ANNUAL FUEL AND/OR ENERGY CONSUMPTION FOR COOLING (COMPRESSORS AND CHILLERS)

    (a) if absorption cooling

25. Oil, gal	_____	× 138 (1) =	_____
	_____	× 146 (2) =	_____
26. Gas, ft³	_____	× 1.0 (3) =	_____
	_____	× 0.8 (4) =	_____
27. Coal, short tons	_____	× 26,000 =	_____
28. Steam, lb × 10^3	_____	× 900 =	_____
29. Propane gas, lb	_____	× 21.5 =	_____
30. Total Btu/h × 10^3		=	_____

31.   Btu × 10^3/(yr)(ft²) gross floor area (line 30 ÷ Table 4.2, line 7)

    (b) If electric cooling

32. Electricity, kWh	_____	× 3.413 =	_____
33. Btu × 10^3/(h)(ft²) gross floor area (line 32 ÷ Table 4.2, line 7)			_____

ESTIMATED ANNUAL ENERGY CONSUMPTION FOR INTERIOR LIGHTING

34. Kilowatthours (Table 4.3, line 3 × Table 4.2, lines 32(1) + 33, × 52)	_____	× 3.413	_____
35. Btu × 10^3/(yr)(ft²) gross floor area (line 34, column C ÷ Table 4.2, line 7)			_____

ESTIMATED ANNUAL ELECTRICAL ENERGY CONSUMPTION FOR ALL MOTORS AND MACHINES IF BUILDING AND HOT WATER ARE NOT ELECTRICALLY HEATED

36. Total kilowatthours (line 6, column A)_____ less kilowatthours for lighting (line 34, column A) _____ =_____ kWh

37.* kWh/(yr)(ft²) gross floor area = _____ (line 36 ÷ Table 4.2, line 7)

38.* Btu × 10^3/(yr)(ft²) floor area = _____ (line 37 × 3.431)

*If building heat and hot water are electrically produced, deduct the kWh/(yr)(ft²) and Btu/(yr)(ft²) for heating and hot water (lines 37 and 38).

Total the annual Btu building requirement, and compute the annual energy usage in Btu per square foot.

Next, break down the average annual Btu consumption by system, using Table 4.4, sections 2 to 5, to record the data. The heating system utilization will vary with the duration and severity of winter from year to year; to obtain the most accurate results, five or more years of energy consumption should be averaged. If records are available for only the previous year, consumption for an average year can be obtained with the following equation:

$$\frac{\text{Fuel consumption for specific year}}{\text{Degree-days for specific year}} \times \text{average yearly degree-days}$$
$$= \text{average yearly fuel consumption}$$

Degree-days for a specific year can be obtained from the weather bureau or a local fuel supplier. Average yearly degree-days can be obtained in like manner, or from Fig. 4.1.

If oil, gas, or coal is the primary fuel and is used for both heating and domestic hot water, the usage should be broken down between the two. The space-heating load occurs in the winter, but the domestic hot-water load is continuous for the whole year at a rate that can be assumed constant. To determine the amount of the monthly fuel bill that can be attributed only to hot water, select one average summer month's consumption. This number times 12 is the yearly value for domestic hot-water heating alone.

If the building is heated by electricity and the total electrical usage is metered and billed in a lump sum, the bill will include energy for heating, lighting, and power. To arrive at the amount of electricity used for heating only, it is necessary to assess the quantity used for lighting and power and subtract this from the total billing. In small buildings, a quick assessment of the electricity usage for lighting can be made by counting the number of lighting fixtures and multiplying the wattage of each lamp and the average number of hours that these are switched on during the heating season. This will give the total number of watthours consumption that can be attributed to lighting. Similarly, a survey can be made of all electrical motors that are in use during the heating season and their nominal horsepower rating (multiplied by 0.800) to determine the approximate amount of electricity in kilowatthours used for each hour of running. (This formula assumes an efficiency of 93% for electric motors.) The number of kilowatthours should then be multiplied by the number of hours of operation during the heating season to determine the total kilowatthours that can be attributed to power. The sum of the kilowatthours assessed for lighting and power should then be subtracted from the total power consumed by the building for the heating season, to determine the amount used for heating. In large, complex buildings where simultaneous heating and cooling are likely to occur, energy flow analysis is more difficult but can be accomplished with the same procedure for the full 12 months.

To determine the amount of energy used for air conditioning, estimate the energy for fans and pumps as outlined above. For electric-driven refrigeration units, the usage of kilowatthours can be estimated by deducting the energy used for lighting and other motors from the June, July, August, and September electric utility bills.

Step 6: Measure systems efficiency

If systems can be made to operate at their maximum efficiencies at both full-load *and* partial-load conditions, then seasonal efficiency, too, will be maximum. If partial-load efficiencies drop off, then energy usage through the season will increase.

Measure first the actual performance for each system at full load with the following tests:

- Make a full-load combustion test, measuring CO_2, CO, or stack temperature and fuel rate for all burners and boilers.

- Measure the air and water consumption, brake horsepower, voltage, and amperage for all air systems handling more than 3000 cfm and water systems with flow rates exceeding 10 gpm. Refer to the ASHRAE *Systems Handbook*, Chap. 40, "Testing, Adjusting, and Balancing."

- Measure the pressure drop across each item of equipment such as air filters, strainers, heat exchangers, and coils to determine the system characteristic and identify options to reduce system resistances and power input to fans and pumps. The data from these tests will indicate system efficiencies and whether the flow rates and energy input are proper for the load.

- By calculation estimate the duration of each building load at specific percentage intervals of full load and the number of hours at these intervals that equipment should operate to meet the part and full conditions. Estimate the annual energy required for this load-following operation. Compare this figure with actual energy consumed for an indication of savings possible with system tuning or modification. (Any modification or adjustment, however, should of course come *after* all efforts have been completed to reduce the building loads.)

If equipment or expertise is not available for the measurement tasks above, these testing and balancing sources may be useful:

- Associated Air Balance Council (AABC): The certifying body of independent agencies.

- National Environmental Balancing Bureau (NEEB): Sponsored jointly by the Mechanical Contractors Association of America (MCAA) and the Sheet

Metal and Air-conditioning Contractors National Association (SMACNA) as the certifying body of the installing contractors' subsidiaries.

- Construction Specifications Institute (CSI): In the CSI Specification Series the guide specification Document 15050 entitled "Testing and Balancing of Environmental Systems."

Step 7: Select energy conservation measures for analyses

After careful study of the building and after consideration of all the various possibilities for energy conservation with respect to building, distribution, and conversion loads, and after consideration of alternative energy sources, make an initial selection of conservation options from Chaps. 6 to 12.

Step 8: Analyze selected energy conservation options

Calculate for each option the potential annual savings for each item singly or in combination with other options if they are related; analysis for lighting reductions, for instance, must include heat gain reductions as they affect cooling and heating. Calculate savings (in Btu or kilowatthours), then apply seasonal efficiencies for applicable systems to determine actual savings. The Btu quantities can then be converted to final quantities, and with kilowatt-hours can be given a dollar value.

For options requiring capital expenditure, the economic model, Chap. 14, can be used for a detailed financial feasibility study within guidelines established by the owner. Not infrequently, however, a building owner will entertain many good conservation possibilities requiring capital improvements, but will have money for only one or two. Rather than going through the full economic model for each item, a "simple payback" comparison will indicate the most favorable.

To make a simple payback comparison, first determine, for each opportunity, the prime cost of implementation (PC), the annual cost savings (AS), and the average useful life of the equipment (AUL). The annual cost savings are equal to the annual energy cost ± operating cost changes. Then apply the following formula to each opportunity.

$$\frac{PC}{AS} \times \frac{AUL_L}{AUL_A} = \text{payback factor}$$

where AUL_L = longest average useful life of those opportunities considered
$\quad\quad AUL_A$ = actual useful life of the opportunity

The most favorable opportunity will be the one which has the *lowest* payback factor.

Example:

Find the most favorable choice among three alternatives.

	PC	*AS*	*AUL*
1	$10,000	$2,000	21 years
2	$ 4,000	$1,500	3 years
3	$ 7,000	$2,250	11 years

$$1 = \frac{10{,}000}{2000} \times \frac{21}{21} = 5.0 \text{ payback factor}$$

$$2 = \frac{4000}{1500} \times \frac{21}{3} = 18.7 \text{ payback factor}$$

$$3 = \frac{7000}{2250} \times \frac{21}{11} = 5.9 \text{ payback factor}$$

The most favorable choice for investment is choice 1, and a full economic analysis of this opportunity should now be made using present worth (PW) for expenditures and revenues or savings which occur in different years.

$$PW = \frac{1}{(1 + \text{annual interest rate } I)^n}$$

without fuel escalation, or

$$PW = \left(\frac{1 + \text{annual escalation rate } E}{1 + \text{annual interest rate } I} \right)^n$$

with escalation, where n = number of years.

Step 9: Prepare report of energy conservation survey and analysis

Prepare a fully documented report with recommendations, costs, calculations, and benefits for the owner's consideration.

Step 10: Prepare construction documents

Prepare working drawings and specifications for all proposed construction work and assist the owner in taking competitive bids. Analyze the bids, draw up contract agreements with the supervision of the owner's attorney, supervise construction work, and certify payments to contractors.

Step 11: Postconstruction services

- Develop specific operating programs and schedules to monitor and control the operating systems, maintain environmental conditions, conserve the maximum amount of energy, reduce operating costs, and utilize manpower most effectively.

- Train operating personnel and supervise the educational program to acquaint the building occupants with new operating conditions and goals.

- Supervise the preparation of log books, which record all changes, energy use on a monthly basis (by subsystems), and costs, in order to authenticate the program and ensure continuity of effort. Monthly consumption, adjusted for degree-days, cooling hours, and operational changes, will indicate the effectiveness of the current program.

- Install suitable monitoring equipment to record the performance of modified equipment or procedures. Give consideration to the procedures that make up the balance of this list.

- Set up logs covering maintenance of equipment to identify hours of operation and responsibility for start and stop.

- Install elapsed-time meters to record hours of operation of prime movers.

- Install electric watthour meters to record usage of electric energy for specific operations or departments.

- Install fuel meters to record consumption of fuel for specific operational purposes.

- Install Btu meters, steam flow meters, or condensate meters to identify heat input or output of processes of building environmental apparatus.

- Install recording thermometers to verify performance of critical automatic temperature controls.

- Install recording ammeters to monitor relative load changes in electric prime movers and power distribution feeders.

- Develop reporting procedures and summaries to monitor the execution of the energy conservation program by the operating and maintenance personnel. Daily, weekly, and monthly logs of equipment use, meter readings, and performance evaluations should be summarized and reported to operating personnel. Continued comparison between performance goals and actual achievement is necessary to sustain interest and reward successful effort and skill at the level at which operating decisions are actually made.

  Each month, total start-stop equipment logs and elapsed-time meters to allow performance comparisons with the energy budget allocation.

  Total submeters on electricity, fuel, steam, condensate, chilled water, and hot-water usage monthly to allow performance comparison with the energy budget allocations.

- Prepare statements of energy use semiannually to compare the actual purchases of electricity and fuel with the energy budget and the adjusted record of energy purchased in the past. Graphic presentations may provide a continuing synopsis of the effects of energy conservation programs. Semiannual calculation of the cost effectiveness of energy conservation efforts should reflect the changes in unit costs of the energy purchased. It is necessary to identify the differences in energy units resulting from the conservation efforts and then apply current unit prices to determine the monetary value.

CHAPTER FIVE

EXISTING-BUILDING CASE STUDY: BROOKHAVEN NATIONAL LABORATORY

5.1 THE PROJECT

On Dec. 31, 1975, Dubin-Bloome Associates were retained by the Brookhaven National Laboratory (BNL) in Upton, New York, to study the chemistry building with goals of reducing annual requirements for electricity, steam, and water, of reducing peak electrical demand, and of reducing operating costs. This case study is from the summary of the DBA report of June 11, 1976. Though this building is certainly not a typical commercial-institutional building in energy profile, the study findings make a fair demonstration of the energy conservation potential for existing buildings.

The prestudy annual steam and electrical consumption was 558,000 Btu/ft^2 at the building boundary. The after-study potential for energy savings with less than 20-year simple-payback recommendations (payback equals initial cost divided by annual energy cost savings) was some 80% for steam and 30% for electricity. Consumption could be reduced to 170,000 Btu/ft^2 at the building boundary.

5.2 STUDY SCOPE AND METHODOLOGY

Existing conditions were determined by on-site inspections, interviews with BNL personnel, and a review of construction documents and metered data as follows:

- Annual electrical energy in kilowatthours and peak demand in kilowatts

- Maintained internal environmental conditions

- Building utilization and program

- Building shell configuration, materials, and construction

- Operating procedures and the physical characteristics and condition, control sequences, and hours of utilization of mechanical, electrical, and process systems, and maintenance practices
- Meteorological and climatic conditions at the BNL site

Climatic and meteorological profiles were prepared from the weather tapes for occupied and unoccupied periods including wet-bulb and dry-bulb temperatures, solar radiation, heating degree-days to base 65 and 50°F, and cooling dry-bulb degree-hours and wet-bulb degree-hours for an average year. Weather data for 1971, obtained from weather tapes provided by the Brookhaven Atmospheric Sciences Center, were used to construct temperature profiles. For preliminary analysis, weather data from La Guardia Airport were used.

Current annual energy consumption was determined using computer and hand calculations. For monthly steam requirements, the categories were these:

- Heating outdoor makeup air to balance laboratory hoods and general exhaust systems
- Transmission losses through the building envelope
- Humidification
- Domestic and process hot water
- Terminal reheat during the cooling season
- Piping and duct thermal losses

For monthly electricity consumption for chillers, the categories were these:

- Outside-air sensible load
- Outside-air latent load
- Internal heat gain due to occupants' lights and equipment
- Solar heat gain
- Transmission heat gain through the building envelope

Monthly electrical energy requirements were gathered for the entire building by the following categories:

- Exhaust air fans and blowers
- Supply air fans

- Chilled-water pumps

- Condenser water pumps

- Cooling-tower fans

- Condensate return pumps

- Hot-water circulating pumps

- Vacuum pumps and air compressors

- Computers, laboratory process equipment, office equipment, and other electrical equipment

The potential reduction in annual energy consumption and peak electrical demand was calculated for measures involving specific building shell components, mechanical and electrical systems components, measures using the tables and figures in *Energy Conservation Manual I* and *Energy Conservation Manual II,* hand calculations, and Ross F. Meriwether computer programs. The dollar values of energy savings were based on current energy costs.

Cost/benefit ratios were based on simple payback in accordance with instructions by BNL. The following formula was used:

$$P = \frac{IC}{OS}$$

where P = payback in years
   IC = incremental initial cost
   OS = yearly savings in energy costs

Incremental maintenance and operating labor costs were not included. (The useful life of all equipment and systems proposed was at least 20 years.) Also, in accordance with BNL criteria for this study, fuel cost escalation, inflation, and discount rate were not included. (For other building ownerships these items might be included.)

The energy conservation measures and their resultant effects which are summarized in Table 5.1 were calculated in a sequential manner; the savings in energy for any particular measure are based on the assumption that all previous measures which reduce the energy consumption from existing conditions are first implemented.

A number of other measures which were considered and analyzed proved to have payback periods in excess of 20 years at present fuel costs. It should be noted that payback periods will be reduced for all measures as fuel and electricity costs escalate.

The report also included recommendations for further consideration for which the savings of energy consumption could not be quantified without further study, including more efficient utilization of the building.

TABLE 5.1 ENERGY CONSERVATION MEASURES ANALYZED AND RESULTS

No.	Name	Annual Energy Savings	Annual Savings, $	Initial Cost, $	Payback Period, Years
IMMEDIATE PAYBACK					
HVAC 1	Lower room thermostat settings to 68°F or lower in winter.	$6,218 \times 10^6$ Btu	17,290	0	Immediate
L and P 1	Turn off main lights in basement and mezzanine mechanical rooms—use emergency lighting circuit only.	8×10^3 kWh	220	0	Immediate
L and P 5	Turn off all service chase, corridor, and lobby lighting after 10 P.M.	50×10^3 kWh	1,330	0	Immediate
L and P 6	Remove 50% of lamps in rooms 212, 211, 210, 115, 115B, 112, 111, and 111A.	29×10^3 kWh	347	0	Immediate
L and P 10	Switch off unneeded transformers.	33×10^3 kWh	800	0	Immediate
Subtotal:	120×10^3 kWh +	$6,218 \times 10^6$ Btu	19,987	0	
0- TO 1-YEAR PAYBACK					
HVAC 2	Install new humidistats in each zone to control steam humidifiers in each air-handling unit; set to maintain 30% RH or lower in winter; install room humidifiers as required.	$3,878 \times 10^6$ Btu $(-) 41 \times 10^3$ kWh $(-) 41 \times 10^3$ kWh	9,700	7,160	0.7
HVAC 3	Adjust and repair chillers to improve efficiency.	478×10^3 kWh	12,800	6,000	0.5
L and P 3	Turn off toilet room lights when the room is unoccupied; add pilot light switches.	8×10^3 kWh	214	160	0.8

		Energy			Payback (yr)
L and P 9	Choose replacement motors with high efficiency ratings.	199 × 10³ kWh	5,300	2,180	0.5
	Subtotal, 0- to 1-year payback: 644 × 10³ kWh + Subtotal, less than 1 year: 764 × 10³ kWh +	3,878 × 10⁶ Btu 10,096 × 10⁶ Btu	28,014 48,001	15,500 15,500	
1- to 3-YEAR PAYBACK					
HVAC 4	Install a new central control system to: (1) set back winter temperatures to 55° during unoccupied periods; (2) raise summer supply air temperatures and chiller water temperatures during periods of light loads; and (3) turn off reheat in summer during unoccupied hours in noncritical areas.	14,814 × 10⁶ Btu 220 × 10³ kWh	47,000	124,900	2.7
HVAC 7	Modify air handler for zone K-9 and its controls to allow shutoff during unoccupied periods.	317 × 10⁶ Btu 21 × 10³ kWh	1,444	4,000	2.8
HVAC 9	Use reject chiller heat for reheat in summer.	2,396 × 10⁶ Btu	6,660	14,200	2.1
DHW 2	Add additional insulation to 700 gal storage tank and booster heater.	231 × 10⁶ Btu	642	1,600	2.5
L and P 7	Relamp fluorescent fixtures with 34-W lamps.	28 × 10³ kWh	750	1,600	2.1
	Subtotal, 1- to 3-year payback: 269 × 10³ kWh + Subtotal, less than 3 years: 1033 × 10³ kWh +	17,758 × 10⁶ Btu 27,854 × 10⁶ Btu	56,496 104,497	146,300 161,800	
3- to 5-YEAR PAYBACK					
HVAC 5	For each laboratory, install an individual three-position exhaust fan switch and sup-	9,764 × 10⁶ Btu 565 × 10³ kWh	42,300	151,700	3.6

TABLE 5.1 ENERGY CONSERVATION MEASURES ANALYZED AND RESULTS (*Continued*)

No.	Name	Annual Energy Savings	Annual Savings, $	Initial Cost, Dollars	Payback Period, Years
3- to 5-YEAR PAYBACK					
HVAC 5	ply duct control dampers; add controls to vary air-handling unit supply air rate to match exhaust rate.				
HVAC 6	Add a return air system with economizer cycle for north wing.	$2,388 \times 10^6$ Btu 159×10^3 kWh	10,900	34,700	3.2
DHW 3	Use steam condensate to preheat hot water.	190×10^6 Btu	528	2,587	4.8
Subtotal, 3- to 5-year payback: 724×10^3 kWh +		$12,342 \times 10^6$ Btu	53,728	188,987	
Subtotal, less than 5 yr: $1,757 \times 10^3$ kWh +		$40,196 \times 10^6$ Btu	158,225	350,787	
5- to 10-YEAR PAYBACK					
HVAC 8	Add a runaround coil system to recover heat from exhaust air and preheat incoming outside air.	$4,952 \times 10^6$ Btu	15,200	111,230	7.3
HVAC 10	Reinsulate steam piping and heating-water piping; insulate condensate system in basement.	966×10^6 Btu	2,690	15,807	6.5
DHW 4	Add insulation to piping in basement and service chases.	117×10^6 Btu	325	1,948	6.0
L and P 2	Increase usage of natural light by relocating desks in perimeter areas; add new switches in library and task lights to laboratory offices where required.	18×10^3 kWh	480	4,500	9.4

Measure	Description	Energy			Payback
L and P 4	Install photocell controls to shut off lights in east, west, and main stairs during daylight hours.	5×10^3 kWh	140	1,200	8.64
	Subtotal, 5- to 10-year payback: 76×10^3 kWh +	6035×10^6 Btu	18,835	134,685	
	Subtotal, less than 10 years: $1,833 \times 10^3$ kWh +	$46,231 \times 10^6$ Btu	177,060	485,472	

10- to 20-YEAR PAYBACK

Measure	Description	Energy			Payback
BS 1	Add an additional pane of glass to all single-glazed windows.	132×10^6 Btu	367	5,720	16
BS 2	Add additional roof insulation during reroofing, except under fan loft.	740×10^6 Btu	2,057	41,360	20
L and P 8	Replace incandescent fixtures in rooms 306 and 306A with fluorescent fixtures.	3×10^3 kWh	79	1,325	16.8
	Subtotal, 10- to 20-year payback: 3×10^3 kWh +	872×10^6 Btu	2,503	48,405	
	Subtotal, less than 20 years: $1,836 \times 10^3$ kWh +	$47,103 \times 10^6$ Btu	179,563	533,877	

OTHER

Measure	Description	Energy			Payback
HVAC 11	Add a steam meter on existing high-pressure steam main supplying the building.			10,890	
DHW 1	Install a water meter to measure hot-water consumption.			2,190	
	Subtotal:			$13,080	
	Grand total:	$47,103 \times 10^6$ Btu $1,836 \times 10^3$ kWh	$179,563	$546,957	3.0

5.3 EXISTING CONDITIONS

ANNUAL ENERGY USE AND PEAK ELECTRICAL DEMAND

- 5,830,000 kWh electricity.

- Total cost for electric power at $0.0268/kWh: $156,000.

- Peak power electric demand: 1800 kW.

- Electricity cost attributable to peak demand: 11%.

- 57.2 billion Btu requiring 48.0 million lb steam at 125 lb/in² (1191 Btu/lb).

- Total cost for steam at $2.78/million Btu: $159,000.

- Total cost of energy $156,000 + $159,000: $315,000.

- Equivalent Btu per gross square feet including thermal and electrical energy: 558,000 Btu/ft² at building boundary and 847,000 Btu per square foot of raw source. (Raw source energy = thermal energy at boiler plant + electrical energy at building boundary + thermal energy at the power plant to produce the electricity used in the building.)

INTERNAL ENVIRONMENTAL CONDITIONS

- Winter (October 1 to April 30)
  72° DB, 40 to 45% RH during occupied and unoccupied periods
  200,000 cfm outdoor air for 103 hours per week
  135,000 cfm outdoor air for 65 hours per week
  (These are estimates of the number of hours exhaust and supply fans are operated at full and reduced capacity.)

- Summer (May 1 to September 30)
  72° DB, 40 to 45% RH during occupied and unoccupied periods
  200,000 cfm outdoor air for 103 hours per week
  135,000 cfm outdoor air for 65 hours per week

- Average illumination
  Offices and laboratories: 70 fc for 45 h per week
  Corridors and service chases: 25 fc for 168 hours per week

BUILDING UTILIZATION

- Design occupancy: 130 personnel.

- Average occupancy: 80 personnel.

- Normal hours of occupancy are 8:30 A.M. to 5:00 P.M. for 5 days per week.

BUILDING SHELL CONFIGURATION AND CONSTRUCTION The three-story building's major axis is east-west, with a small two-story wing jutting out at right angles to the north and a one-story wing to the south. Gross floor area is 139,400 ft².

Most windows, 8440 ft², are fixed, nonoperable, double glazed, with light-colored venetian blinds between the two panes; about 50% face north and 50% face south. There are approximately 2200 ft² of fixed single-glazed window; about 1200 ft² face south, 60 ft² east, 330 ft² west, and 530 ft² north. One skylight is located over the main stairwell.

Cavity masonry walls are not insulated and have a U value of 0.30. The panel construction walls are uninsulated and have a U value of 0.32. The concrete roof deck is covered with built-up roofing and has a U value of 0.23.

MECHANICAL SYSTEMS AND OPERATING PROCEDURES Steam is provided from the central boiler plant at 125 lb/in² and reduced inside the building to 50 and 5 lb/in².

There are nine central air-handling units on the basement level, each one serving a separate zone.

There is no return-air duct system in the entire building except for the conference room zone. Air-handling units draw in 100% outdoor air, year-round, and exhaust through laboratory hoods, room exhaust grilles, and toilet rooms. Each laboratory is equipped with a separate two-speed exhaust-air fan (controlled by the same switch as the zone supply fan) to handle all hoods in the room. Air supplied to the offices is transferred through open doorways or grilles into the adjacent corridor and from there through open doorways to adjacent laboratories. Approximately 70% of the makeup air for each laboratory is supplied directly into the laboratory, and 30% is drawn from the corridors.

During the heating mode, the preheating coils maintain a supply duct temperature of about 48°F and hot-water coils in the supply duct branch to each office and each laboratory are individually controlled by room thermostats. The individual hot-water coils serve as booster heating coils or reheat coils, depending upon climatic conditions.

Two 600-ton centrifugal chillers with vaned inlet capacity control provide chilled water to the cooling coils in each air-handling unit. The individual room booster heating coils reheat 50°F supply air as required to meet room conditions.

Domestic and process hot water is supplied by a storage-type steam water heater with a separate booster for lavatories, shops, glass washing, and process uses with a circulating pump operating continuously.

LIGHTING AND ELECTRICAL SYSTEMS AND OPERATING PROCE-DURES Purchased power is supplied to the building from the central substation at 13.8 kV and is reduced to 480 and 208/120 V by six transformers located in the basement. All supply fans, centrifugal exhaust fans, chillers, and pumps have 440-V motors.

Lighting is predominantly fluorescent with 2- or 4-lamp fixtures with 40-W tubes. All lighting fixtures in any one laboratory are controlled by a single switch. See Fig. 5.1.

5.4 SUMMARY OF CONCLUSIONS

FACTORS CONTRIBUTING TO EXCESSIVE ENERGY CONSUMPTION

A combination of major factors contributes to excessive energy consumption in the chemistry building. The basic functions performed in the building require that exhaust hoods operate for a considerable number of hours per year to expel noxious fumes and other contaminants from the building. The quantity of air which is expelled through laboratory hoods and room exhaust grills must be replaced with an equal volume of outdoor makeup air, requiring thermal energy in the winter to heat and humidify the air to room conditions. Humidification is required for occupant comfort and health (when outside air is heated, the relative humidity becomes very low if no moisture is added) and for quality control in computer rooms and where paper products are used.

In the summertime, outdoor makeup air must be cooled, dehumidified, and reheated, thus imposing a heavy thermal load on the steam system and a heavy power demand upon chillers and their auxiliaries. With the basic system design (a terminal reheat system) the outdoor air is overcooled for many spaces that have relatively low internal and solar heat gains, and then the too-cool air must be reheated with thermal energy. In addition, considerable power is consumed by supply and exhaust fans and, to a lesser extent, by pumps to move the air and water to meet the loads.

The quantity of exhaust air and corresponding makeup air is greater than that required to meet the building functional program. All laboratory spaces were designed to accommodate fume hoods, and the exhaust system for each laboratory, whether equipped with exhaust hoods or not, is operated in the same manner and capacity, even though the exhaust requirement for a laboratory without hoods is considerably lower. The exhaust requirements for the counting rooms are similar to those with hoods. Since the supply-air quantity into each laboratory is predicated upon the exhaust requirements and not upon the thermal environmental condition to be maintained, excess outdoor air is introduced into the building and wasted through the exhaust system for a considerable period of time when the actual requirements are either greatly reduced or nonexistent. To compound the problem, all exhaust fans and the supply fan in any one zone must be operated at full capacity, even though the actual exhaust requirement may be limited to one or two fume hoods, since in the existing installation all exhaust fans in any one zone containing many laboratories are controlled by a single switch. On-site observations showed that many air-handling systems were operating at full capacity (high fan speed during most occupied hours and to a large extent during unoccupied periods as well) when actual exhaust air requirements are only 25 to 50% of full-load conditions.

While 100% outside air is required in many parts of the building, those areas which could accommodate return air ducts with corresponding reduction in outside air have not been fully exploited. Specifically, the north wing could utilize a return-air system. In addition, there is no heat recovery system to capture energy from exhausted air and transfer it to makeup air to other uses.

The basic design of the HVAC distribution system and space planning

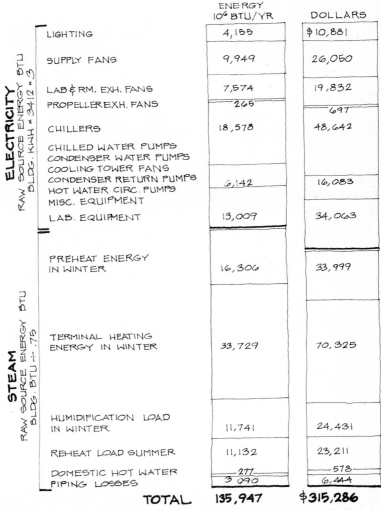

Figure 5.1 Existing raw-source energy usage and costs at the Brookhaven Chemistry Laboratory 1976.

contribute to excessive energy use. In most of the zones a common duct supplies conditioned air at the same temperature and humidity conditions to laboratories which are in the interior of the building and have no outside wall exposure, as well as to perimeter offices; in some cases, the same supply duct serves north-facing offices as well as south-facing offices. Each of these areas has widely differing thermal requirements owing to intermittent periods of solar heat gain, wind velocity, and internal loads. The duct layout and controls require that air be introduced into all zones at the temperature and humidity required for the most critical area within the zone (during the cooling season). All other areas then must compensate by adding more reheat energy than would be needed if separate ducts were to supply only areas having common environmental requirements (i.e., separate ducts with coils and controls for north, south, and interior zones).

In all but two of the air-handling zones, *there are scattered computer facilities in one, two, or three rooms within the zone.* The special temperature and humidity requirements for the computers at present dictate not only the environmental conditions of the rooms in which they are located, but also the environmental conditions for the entire zone. Thus excess energy is consumed to maintain conditions in a large area in order to meet the conditions required in as little as 1% of the total space.

Excessive energy is used for cooling since the system is designed and operated to provide chilled water at a constant temperature of 43°F to the coils of each air-handling unit. There are no control valves on the coils, and so the quantity of chilled water and the resultant cold-duct temperature runs wild. Most air-handling units deliver air at temperatures below that required to satisfy the cooling load of some or all spaces within each zone and thus require excessive amounts of reheat energy to maintain comfort conditions for the summer hours.

Chiller efficiency is penalized by the low chilled-water temperature which is maintained at all times, even though the ambient outdoor air conditions would permit operation for long periods of time at higher chilled-water temperatures when cooling loads are light (about 1¾% of the power requirements for chillers can be saved for every degree rise in chilled-water temperature). The setting of the condenser water controller at 85°F contributes to high energy consumption, since much of the year weather conditions permit lower condenser water temperatures which would reduce chiller power input.

While there are *light switches* in most individual rooms (as compared with some buildings where there is only one switch per floor), these *do not permit turning off selected fixtures in each room, each laboratory, or in corridors which are unused or where daylight is adequate.* In addition, the lamps and lighting fixtures in the north wing are essentially similar to those in the other wings, even though the higher ceiling levels in the north wing would permit the installation of more efficient lamps and fixtures, the type which can be used only in high-ceilinged areas.

Finally, *the space utilization and low occupant density contribute to excessive energy usage.* While the building was originally planned for about 130 personnel, an average of 80 use the building at any one time, and at night and weekends, there are only 5 or 10 occupants who are scattered throughout the entire building. Because no attempt was made to concentrate personnel, it is necessary to light, cool, and ventilate virtually the entire building even when only a portion of it is actually in use.

MAJOR OPPORTUNITIES TO CONSERVE ENERGY   Table 5.1 presents various recommendations by system and number and their payback periods. They are as follows:

1. Reduce the volume of outdoor makeup air to reduce energy for heating, humidification, cooling, and dehumidification (HVAC 5 to 7).

2. Lower thermostat settings in winter to reduce heating energy (HVAC 1, 4).

3. Reduce the humidification load in the winter (HVAC 2).

4. Reduce energy for reheat in the summer (HVAC 4, 7).

5. Reduce the energy required for chillers and auxiliaries (HVAC 3, 4).

6. Recover the energy lost through exhaust air to reduce heating and cooling loads (HVAC 8).

7. Reduce energy required to offset building conductive heat loss and heat gain (BS 1, 2).

8. Reduce energy consumption for domestic and process water heating (DHW 1 to 4).

9. Reduce heat losses from piping and ducts (HVAC 10).

10. Reduce energy consumption for lighting and power (L and P 1 to 3).

UNQUANTIFIED OPPORTUNITIES TO CONSERVE ENERGY

1. Relocate personnel into selected wings and shut down unused portions of the building.

2. Select all replacement and new laboratory equipment for lowest power requirements for the required function.

3. Concentrate existing and new computer equipment in only one or two air-handling unit zones to reduce the environmental requirements of the noncomputer areas.

4. Provide enclosures, with unit conditioning equipment, around existing computers.

5. Provide self-contained air-conditioning units on existing computer equipment and purchase any new computer equipment with integral cooling, humidification, and heating unit.

6. Provide a new steam meter and domestic hot-water meter to record usage and to provide data for monitoring energy usage in a continuing energy management program.

7. Replace motors on each air-handling unit with four-speed motors and switches when existing motors wear out.

8. Consider solar energy for preheating outdoor air when solar collectors are reduced in price to $5/ft^2 and each square foot can produce 220,000 Btu/yr at useful fluid temperature.

9. Select all controls with provisions and suitability for compatibility with a future facility-wide central control system to be developed under the 5-year plan.

10. Seal all building leaks which permit infiltration through the structure. Conduct a gas test for infiltration to locate cracks and leaks.

11. Provide a reflective shade below the skylight to control heat gain in summer.

12. Check all ductwork for leakage and seal joints where leaks are found.

13. Reduce impeller size of pumps where a permanent reduction in capacity results from reduction of loads.

14. If chillers are replaced, purchase new units with double bundle condensers to permit operation in heat pump mode.

15. Use portable air-cooled air-conditioning units for supplementary space cooling in selected areas where conditions permit operating the central equipment at higher chilled water and supply air temperatures for the majority of the spaces.

16. If a runaround coil system with new preheat coils is not installed, then provide an additional heating control valve on the two-bank steam heating coil assemblies of each air-handling unit to correct overheating.

17. When purchasing new furniture, consider furniture-mounted lighting systems to permit individual control of task lighting.

18. Provide water treatment for chilled-water circuits; maintain present condenser water loop treatment.

19. Check all room air conditioners, laboratory refrigerators, and experimen-

tal chillers for clogged condenser coils, dirty filters, and improper functioning of temperature controls.

20. Reduce peak electric demand. The major opportunity to reduce peak demands in this building occurs by implementation of the measures designed to save energy. Examples include relamping with 35-W fluorescent tubes and improving chiller efficiency. Since the building has a fairly steady peak load from 9:00 A.M. to 5:00 P.M., depending upon outdoor ambient temperatures, it is not feasible to shut off supply or exhaust fans or chillers to reduce peak loads without interrupting envi-

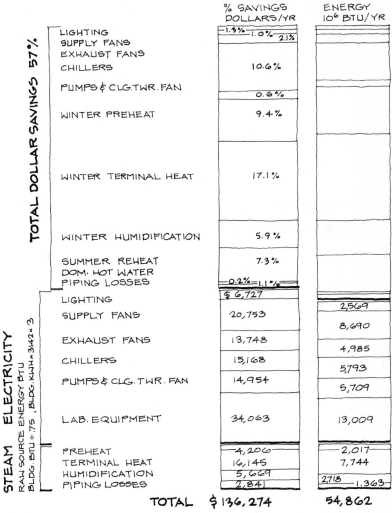

Figure 5.2 Projected raw-source energy requirements and costs with specified energy conservation measures.

ronmental conditions in the laboratories for extended periods of time. Some peak load reductions can be obtained, though, by scheduling experiments involving electrical equipment with a large power draw at night or during weekends in the summer (building peak load occurs in the summer).

21. Reduce water consumption. Since there are no meters for total domestic water or domestic hot water in this building, it is nearly impossible to accurately determine present water consumption or the potential savings. The study recommends that water meters for both total domestic and process hot water be installed to determine consumption. Water consumption can be reduced with replacement toilets that flush with 2.5 gal, installing new lavatories that have ½ gal/min flow rates, and by shutting off cooling water for experimental apparatus when equipment is in operation.

SAVINGS AND COST/BENEFIT ANALYSIS   More than 80% of the annual thermal energy and 30% of the annual electrical energy presently consumed can be saved by a comprehensive energy conservation program. This amounts to savings of $180,000 per year (1976 dollars) for an initial investment of approximately $600,000 including engineering fees, providing an overall payback of slightly over 3 years. A breakdown of the resultant energy consumption and the savings for this program is shown in Fig. 5.2.

CHAPTER SIX

HEATING

6.1 BUILDING HEATING LOAD: INTRODUCTION

Heating is, in much of the country, the major element of the building load. The subheadings below suggest ways of reducing it. Many of the recommendations for reducing heat loss from the building will reduce heat gain to the building as well, of some concern in the next chapter on cooling. These include measures to improve the thermal barrier qualities of the building skin and means to reduce excess outdoor-air quantities which infiltrate the building or are introduced as ventilation.

The most dramatic savings in heating are obtained quite simply: Reduce levels of air temperature and humidity and ventilation. Ways of achieving these savings are documented in the following paragraphs.

6.2 SET BACK INDOOR TEMPERATURES IN HEATING SEASON DURING UNOCCUPIED PERIODS

Energy expended to heat unoccupied buildings up to comfort conditions is wasted, and most buildings most of the time are indeed unoccupied. Energy is conserved by setting back the temperature level at these times. The energy saved will vary with the length of time and the number of degrees that temperatures are set back. The percentage of energy saved will be greater in warmer climates, but the gross energy saved will be greater in cold climates.

In areas where it is not necessary to maintain high temperatures during occupied periods, in corridors and lobbies, for instance, temperatures can be lowered further. Setback can be made by resetting thermostats manually (if automatic setback control has not been installed) or by adjusting controls to suggested temperatures (if clock, day-night, or other automatic reset controls are available). Climate, type of system, and building construction will determine the length of the startup period required to attain daytime temperature levels. Simple experimentation is necessary to find the optimal setback temperature and startup time for any particular building. If, in extremely cold weather, experience indicates that the heating system does not raise the temperature sufficiently by the time the building opens for the day, temperatures can be set back to a level higher than those recommended here for those periods of time only.

ACTION GUIDELINES

☐ 1. Shut off radiators or registers in vestibules and lobbies.

☐ 2. Reduce the hours of occupancy to the greatest extent possible during periods of severely cold weather.

☐ 3. Adjust automatic timers or add time clocks to automatically set back temperatures for night and weekend operation.

☐ 4. When buildings are used after hours for meetings, conferences, cleaning, or other scattered activities, reduce the number of spaces occupied and, to the extent possible, choose areas in the same section of the building. Reduce the temperature and turn off humidifiers in all other parts of the building.

☐ 5. When there is no danger of freezing, turn off radiators or supply registers in areas that do not have separate thermostats. Open them only when the building is occupied.

☐ 6. Check relevant codes requiring protection for plumbing, fire sprinkler systems, and standpipes before implementing extreme temperature reduction. Public utilities may set temperature requirements for their equipment rooms.

Use Fig. 6.1 to determine the probable setback energy savings for an average winter for any particular building with thermostats set at 68°F.* From Table 4.1, the climate profile, select the degree-days for the location and from Table 4.4, line 16, find the number of Btu per square foot per year now consumed for heating.

Enter the graph at the appropriate present heating energy consumption and degree-day axes, intersect with the proper setback line, and follow the example line to determine the savings in Btu per square foot per year. Multiply this value by the gross square foot floor area to give the total yearly savings in Btu that can be expected for the entire building.

Five locations with heating degree-day totals ranging from 1400 to 8400

*65°F is now the daytime setting requested by President Carter.

were analyzed regarding time-temperature distributions below 68°F DB. The percentage for temperature distributions for each of the 5°F ranges shown in the lower half of Fig. 6.1, with respect to the total number of degree-hours below 68°F, was determined 24 hours per day, 365 days per year. This percentage, plotted against annual degree-days and expanded to cover the entire range of degree-days, is shown as the lower half of Fig. 6.1.

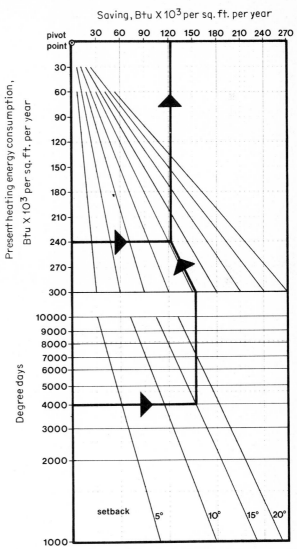

Figure 6.1 Heating energy saved by night setback. (*From AFM 88-8, U.S. Government Printing Office, June 15, 1967, and Climatic Atlas of the United States, U.S. Department of Commerce, June 1968.*)

The upper half of Fig. 6.1 represents the range of heating energy consumed per square foot for various buildings over the range of 1000 to 10,000 degree-days. Analysis of energy usage for heating of office buildings in several locations showed that approximately 10% of the total heating energy consumption can be assumed to occur during occupied hours. Therefore, savings by night setback are applicable to only 90% of the total heating energy consumption. The extreme right-hand line in the upper half represents this 90% of present consumption when projected vertically. The remainder of the upper half of this figure simply proportions the energy saved, on the basis of the point of entry from the lower section.

6.3 REDUCE INDOOR TEMPERATURES DURING OCCUPIED PERIODS

Maintaining lower indoor temperatures during occupied periods conserves energy, although savings are not as great as those for unoccupied hours which occur generally in the colder night. See Table 6.1 for recommended indoor temperatures during the heating season. The temperatures in less critical areas such as corridors and lobbies can be reduced even lower. The total energy conserved will be greatest in buildings which normally have longer periods of occupancy, stores as opposed to schools, for instance, and in buildings or sections of buildings which have little internal heat gain or solar radiation to help with the heating load.

ACTION GUIDELINES

☐ 1. Stores are commonly overheated and uncomfortable for transient patrons who are wearing winter clothing. Reduce temperatures to a level that is comfortable for heavily dressed patrons and encourage the staff to dress more warmly or provide local heaters for them.

☐ 2. In all buildings encourage occupants to wear heavier clothing so that they are comfortable at lower indoor temperatures.

☐ 3. Some buildings contain large heated areas such as storage spaces that are unoccupied or occupied by only one or two people. In such areas, reduce the temperature to a low level just sufficient to prevent damage to other systems (freezing sprinklers, etc.) and provide local radiant heaters for occupants.

☐ 4. Corridors and stairwells are unoccupied areas used only by people who are physically active in moving from one heated space to another. Provided that the temperature does not fall below 55°F, turn off heating in these areas. Keep all doors closed between unheated corridors and heated spaces.

TABLE 6.1   SUGGESTED INDOOR TEMPERATURES DURING HEATING SEASON

	Dry Bulb °F Occupied Hours (Maximum)	Dry Bulb °F Unoccupied Hours (Set-back)
OFFICE BUILDINGS, RESIDENCES, SCHOOLS		
Offices, school rooms, residential spaces	68	55
Corridors	62	52
Dead-storage closets	50	50
Cafeterias	68	50
Mechanical equipment rooms	55	50
Occupied storage areas, gymnasiums	55	50
Auditoriums	68	50
Computer rooms	65	As required
Lobbies	65	50
Doctors' offices	68	58
Toilet rooms	65	55
Garages	Do not heat	Do not heat
RETAIL STORES		
Department stores	65	55
Supermarkets	60	50
Drug stores	65	55
Meat markets	60	50
Apparel (except dressing rooms)	65	55
Jewelry, hardware, etc.	65	55
Warehouses	55	50
Docks and platforms	Do not heat	Do not heat

RELIGIOUS BUILDINGS*		24 h or Less	Greater than 24 h†
Meeting rooms	68	55	50
Halls of Worship	65	55	50
All other spaces	As noted for office buildings	50	40

*And other spaces used for only a few hours per week.

†When outdoor temperatures are above 40°F.

☐ 5. Spaces which are heated by adjacent areas or which receive solar heat through windows require less heating. If thermostats are unavailable in these areas, shut off radiators, registers, fan coil units, or any other terminal heating devices until the temperature levels suggested in Table 6.1 can be maintained.

☐ 6. Do not operate refrigeration systems or introduce outdoor air for the purpose of cooling in winter to reduce indoor temperature levels to the heating set point of 68°F. If gains exceed losses, allow space temperature to rise to 78°F.

☐ 7. When choosing new temperature levels, refer to applicable codes. Minimum temperature requirements are specified in the Occupational Safety and Health Act (OSHA) and other occupational regulations to ensure that working conditions are suitable for employees.

6.4 AVOID RADIATION EFFECTS TO COLD SURFACES

In cold climates the temperature of an interior surface of an exterior wall or window is considerably lower than room temperature, and people located near the surface radiate heat to it. Even if the room temperature is 70 or 75°F, occupants in these areas will feel cold, particularly if they are near windows, where the radiation effect is most severe. In order to keep warm, they often will request that room thermostats be set higher. Overheating of the interior of these particular rooms results, and heat loss and energy consumption increase accordingly. A few simple remedies will save energy and at the same time enhance the comfort of the occupants.

ACTION GUIDELINES

☐ 1. When the winter sun is not shining on the windows, draw drapes or close venetian blinds to reduce radiation of heat from occupants to cold surfaces. (The cost of adding venetian blinds ranges from $1.00 to $1.60/ ft².)

☐ 2. Encourage those working near the exterior walls and windows to wear heavier clothing in the winter.

☐ 3. Close windows tightly in the winter.

☐ 4. Have occupants sit together in interior spaces away from cold walls in sparsely occupied spaces.

☐ 5. Rearrange desks and task surfaces away from cold exterior surfaces. Put circulation and storage in cold locations.

6.5 REDUCE LEVELS OF RELATIVE HUMIDITY

Humidification systems vaporize water into the dry ventilating air to increase moisture content and achieve the desired relative humidity within the building. This humidification process requires a heat input of approximately 1000 Btu to vaporize each pound of water. The amount of moisture (water vapor) required to maintain any desired level of relative humidity is proportional to the amount and dryness of outdoor air which enters the building, less the moisture added to the air by the building occupants or moisture-producing processes.

Humidification systems, though not universally used, are often installed to maintain the comfort and health of occupants and to prevent drying and cracking of wood, furniture, and building contents. Where preservation of materials is not a problem, humidity may be lowered, and for this situation there is no reason to humidify when the building is unoccupied. Relative humidity during winter in cold climates drops to 5 or 10% in buildings without humidifiers. In the absence of sufficient evidence to support the contention that higher levels are more comfortable or promote health, it is suggested that 20% relative humidity be maintained in all spaces occupied more than 4 hours per day. If complaints of dryness and discomfort result, humidity levels can be raised in 5% increments until the appropriate level for each area of the building is determined. Relative humidity of 30% may be required to reduce static electricity and eliminate shocks. (Approximately twice as much energy is required to maintain 30% RH as to maintain 20%.)

ACTION GUIDELINES

☐ 1. Turn off all humidifiers at night and during unoccupied cycles.

☐ 2. Reduce the amount of infiltration and outdoor air ventilation to reduce humidification requirements.

☐ 3. Reduce or eliminate any introduction of moisture for humidification in corridors, storerooms, equipment rooms, lounges, and lobbies, and for buildings which by function or occupant density already have higher humidity levels, such as laundries, supermarkets, stores with high-density occupancy, religious buildings, cafeterias, and kitchens.

☐ 4. Use waste steam condensate for winter humidification.

☐ 5. Whenever condensation is running freely on the inside of window surfaces, shut off humidifiers.

☐ 6. When humidifiers are maintained to eliminate static electricity, shut them off if shocks from static electricity are not a problem.

☐ 7. If humidifiers are located in the return air or outdoor-air mixing box in the air-handling section of the system, relocate them to the hot-duct section.

☐ 8. In pan-type humidifiers adjust floats or controls to eliminate overflow onto hot furnace sections.

To determine the energy required for humidification, enter the lower section of Fig. 6.2 with annual wet-bulb degree-hours below 54°F WB and 68°F DB (see Table 4.1 and Fig. 4.3), intersect the indoor relative humidity level, then proceed vertically to intersect the hours of controlled humidity per week. Read horizontally the energy required to humidify 1000 cfm. Note that this quantification is for controlled outside air humidification and does not include humidification for uncontrolled infiltration. Infiltration for a building of 100,000 ft² of floor area might well amount to an additional 15,000 cfm.

Energy used is a function of the WB degree-hours below the base conditions, the RH maintained, and the number of hours of controlled humidity. Figure 6.2 expresses the energy used per 1000 cfm of outside air humdified.

An analysis of the total heat content of air in the range under consideration indicates an average total heat requirement of 0.522 Btu/lb for each degree WB change. By utilizing the specific heat of air, this can be further reduced

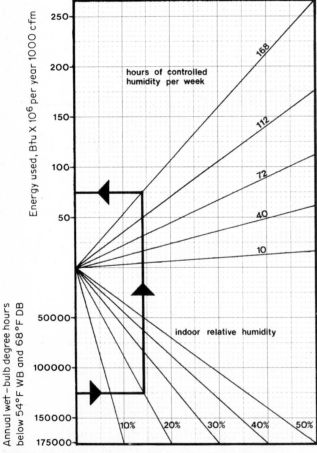

Figure 6.2 Energy required for winter humidification.

to 0.240 Btu/lb sensible heat and 0.282 Btu/lb latent heat. One thousand cubic feet per minute is equal to 4286 lb/h; since we are concerned with latent heat only, each degree Fahrenheit WB hour is equal to 4286 × 0.282, or 1208 Btu. Further investigation of the relationship between WB temperature, DB temperature, and total heat shows that latent heat varies directly with RH at constant DB temperature. The lower section of Fig. 6.2 shows this proportional relationship around the base of 40% RH. The upper section proportions the hours of system operation, with 168 hours per week being 100%.

6.6 SHUT DOWN VENTILATION SYSTEM DURING UNOCCUPIED HOURS

Ventilation is responsible for a large percentage of the building heating load. Cold outdoor air, introduced for ventilation, must be heated to indoor temperatures. The load which this imposes on the heating system is directly proportional to the temperature differential between indoor temperature and outdoor temperature and to the quantity of air introduced for ventilation. The yearly energy used to heat ventilation air is a function of heating degree-days.

Ventilation may be provided through an air intake duct to an air-handling unit, through outside-air intakes of fan coil, window, and through-the-wall units, through intake ducts to rooftop or package heating and cooling units, or by a separate outdoor-air fan. Examine the building system carefully to determine the way ventilation is being supplied and the control or damper devices that are available to reduce and shut off the supply of outdoor air.

Many buildings are ventilated at a rate far in excess of that necessary to maintain comfort, dilute odors, or meet code requirements. Ventilation systems are not infrequently operated 24 h/day, even when buildings are unoccupied or lightly occupied. These buildings contribute to a gross waste of heating energy.

ACTION GUIDELINES

☐ 1. If there are separate outdoor air supply fans, turn them off during unoccupied periods. Use an automatic timer if one is available, or switch the fans off manually. If the fans have automatic dampers that close when the fans shut down, make sure that they close tightly. Repair the dampers with felt edges if they are not airtight.

☐ 2. If air-handling units which supply conditioned air to interior spaces are equipped with outdoor-air inlet duct and damper, close the damper as described above. If no damper exists, install a felt-lined damper with a remote or local control device and timer control.

□ 3. If windows are used to provide ventilation, close them at night.

□ 4. If exhaust fan hoods serving kitchens, bakeries, cafeterias, and snack bars are interlocked with outside-air fans or dampers, shut down these exhaust systems when not needed.

□ 5. Close outdoor-air dampers for the first hours of occupancy when outdoor air must be heated or cooled.

□ 6. Provide ventilation in accordance with changing occupancy and not on a continuous basis. In many buildings heavy occupancy can be monitored or occurs at regular intervals, and so ventilation can be shut off for certain periods during the day. In a department store designed for peak loads of customers, ventilation can usually be shut off for those slack sales periods when only 10 or 20% of the peak occupancy occurs.

□ 7. If, during the heating season, the outdoor-air temperature in the morning is above desired room conditions, use it for heating by opening outdoor-air dampers.

□ 8. Generally the amount of outdoor air quantities for window units, through-the-wall units, and fan coil units is fixed. When these units are used to maintain nighttime or unoccupied cycle room temperatures, the outdoor air intakes are generally open. If dampers are available, close them when outdoor air is not needed. When infiltration is sufficient for daytime ventilation requirements, block off the outdoor-air damper completely for all or some of the units.

To quantify the energy required to heat outdoor air, enter the lower half of Fig. 6.3 with annual heating degree-days (from Table 4.1) and intersect the indoor temperature (from Table 4.2). Proceed vertically to the number of hours per week for shutdown of the ventilation system and read horizontally the annual energy saved for each 1000-cfm reduction of ventilation air.

Energy used is a function of the number of degree-days, indoor and outdoor temperature, and the number of hours that temperature is maintained. It is expressed as the energy used per 1000 cfm of air conditioned. The energy used per year was determined as follows:

$$\text{Btu/yr} = (1000 \text{ cfm}) (\text{degree days/yr}) (24 \text{ hours per day})^{(1.08)*}$$

Since degree-days are base 65°F, the other temperatures in the lower section of Fig. 6.3 are directly proportional to the 65°F line. The upper section proportions the hours of system operation, with 168 hours per week being 100%.

*1.08 is a factor which incorporates specific heat, specific volume, and time (0.24 × 0.075 × 60).

6.7 REDUCE VENTILATION RATES
DURING OCCUPIED PERIODS

Ventilation is required for replacement of expended oxygen, but this requirement is a very small proportion of the outdoor air typically introduced into most buildings. Odor dilution, especially smoke dilution, is the essential problem. No more cold outdoor air than is actually necessary for odor or smoke control should be brought into the building. Recommended ventilation standards are given in Table 6.2.

The building owner or operator can perform most of the procedures for shutting off ventilation at night or on weekends, but professional advice and help may be necessary to reduce the ventilation rate in all or portions of the building during occupied periods.

The actual ventilation rate must first be established by measuring the

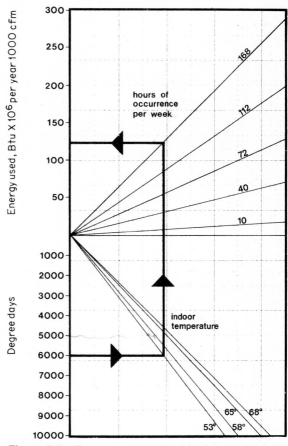

Figure 6.3 Energy required for heating outdoor air.

volume of air at the outdoor air intakes of the ventilation system. Local code requirements must be determined and compared with the measured ventilation rate to find the magnitude of possible savings. If the code requirements exceed the recommended ventilation standards listed in Table 6.2, a code variance request might be considered. Although code figures are often used as design criteria, with dampers and ducts installed accordingly, these figures do not necessarily reflect actual building requirements for the number of people occupying spaces for specific periods of time; they often exceed real requirements.

If ventilation is supplied to a building to provide makeup air for toilet, kitchen, and other exhaust-air quantities, the exhaust systems should be operated only when needed. Savings of fan horsepower for exhaust fans and intake fans, and heating energy to temper the supply air, can be substantial.

When two or more spaces with different requirements are served by one system, it is not always possible to control the ventilation rates separately. Not all spaces, however, need be supplied at the rate required for the most critical area. Refer to Table 6.2 and calculate the maximum amount of outdoor air by the sum of the individual room requirements. This procedure will generally provide satisfactory conditions for all spaces.

TABLE 6.2 RECOMMENDED VENTILATION STANDARDS

OFFICE BUILDINGS

Work spaces	5 cfm/person
Heavy smoking areas	15 cfm/person
Lounges	5 cfm/person
Cafeterias	5 cfm/person
Conference rooms	15 cfm/person
Doctors' offices	5 cfm/person
Toilet rooms	10 air changes/hour
Lobbies	0
Unoccupied spaces	0

RETAIL STORES

Trade areas	6 cfm/customer
Street level with heavy use (less than 5000 ft^2 with single or double outside door)	0
Unoccupied spaces	0

RELIGIOUS BUILDINGS

Halls of worship	5 cfm/person
Meeting rooms	10 cfm/person
Unoccupied spaces	0

ACTION GUIDELINES

☐ 1. Close outdoor-air damper, if provided, until desired volume reduction is achieved.

☐ 2. If damper is not provided, either fit one or blank off part of the outdoor-air intake.

☐ 3. Take particular care in setting outdoor-air quantities where window air conditioners, through-the-wall units, fan-coil units, and unit ventilators with fresh air intakes are used. Generally, the amount of outdoor air is fixed but can be reset.

☐ 4. Use an anemometer at the outdoor inlet or louver to measure the amount of outdoor air introduced into the ventilation system; the actual damper setting often differs from the damper indicator setting. A hinged volume damper blade does not proportion air directly with its openings; a blade with a 10% opening permits much more than 10% flow.

☐ 5. Cut off direct outdoor-air supply to toilet rooms and other "contaminated" or potentially odorous areas. Permit air from "clean" areas to migrate to the "dirty" areas through door grilles or undercut doors. This is effective if toilet areas have a mechanical exhaust system, or even if ventilated by open windows, especially those on the lee side of the building.

☐ 6. Use odor-absorbing materials in special areas rather than provide outdoor air for odor dilution.

☐ 7. Supply ventilation air to parking garages according to levels indicated by a CO_2-monitoring system.

☐ 8. Provide baffles so that the wind does not blow directly into an outdoor-air intake.

☐ 9. Operate exhaust systems intermittently throughout the day. Turn them off, if possible, when they are not needed, and operate them at times such as noon and coffee breaks, and during heavy cooking operations.

☐ 10. Shut off exhaust hoods in kitchens, snack bars, and cafeterias when cooking or baking operations are completed.

☐ 11. Many hoods exhaust much more air than necessary. Reduce the quantity of exhaust air from hoods in kitchens and similar operations by closing off a portion of the hood, by changing hood type from canopy to high velocity, or by slowing down exhaust fans to the point just necessary to satisfy exhaust requirements.

☐ 12. If outdoor temperature is below 45°F, do not operate the ventilation system any longer than needed. Noticeable odors are good indicators of the need to operate the fans or open outdoor-air dampers. Flushing out the building until odors disappear will usually be satisfactory.

☐ 13. Where separate ventilation systems are installed, group occupants by type of activity when possible to reduce the total ventilation rate.

☐ 14. Concentrate smoking areas together so that one ventilation system can serve them. Adjust outdoor air quantities to serve these areas, and reduce outdoor air quantities to all other systems.

☐ 15. Use window ventilation where possible. Doing so reduces the power required for mechanical exhaust systems and reduces the air intake which normally balances the exhaust. Shut off supply air system when window ventilation alone is adequate.

☐ 16. In high-ceilinged buildings used for short periods of intermittent operation, for example, religious buildings and auditoriums, outdoor air is frequently unnecessary. For longer periods of light occupancy, outdoor air is still unnecessary. A warehouse, for example, may have less than one person per 1000 ft^3 for less than 4 hours and has no need for outside air ventilation.

☐ 17. Check codes, where applicable, for specification of minimum exhaust air and for rules on recirculation or filtering of air.

Computation of the energy required for ventilation during occupied periods is the same as that for unoccupied periods. Enter the lower half of Fig. 6.3 with annual heating degree-days (from Table 4.1) and intersect the indoor temperature (from Table 4.2). Proceed vertically to the number of hours per week that the ventilation system will be in operation (Table 4.2) and read horizontally the annual energy required for 1000 cfm of ventilation.

Using the ventilation recommendations in Table 6.2 and the actual ventilation rate (from Table 4.2), compute the reduced ventilation in cubic feet per minute. With the reduced rate of ventilation and the energy required for 1000 cfm of ventilation, compute total energy savings due to reduced ventilation.

6.8 REDUCE RATE OF INFILTRATION

Outdoor air infiltrates a building through cracks and openings around windows and doors, through construction joints between individual panels in a panel wall construction, and through porous building materials of the exterior walls, roofs, and floors over unheated spaces. Infiltration increases with wind velocity and penetrates the windward side of the building, usually the north or west exposures in cold climates. However, in high winds, a negative pressure is often created on the lee side, which, if the north and west exposures are windowless and tight, may induce air into the building through open doors and passageways.

Stack effect is always a potential problem for vertical spaces such as service shafts, elevator shafts, and staircases. The density difference between the warm air in the shaft and the cold outdoor air induces air to leak into the bottom of the shaft and out of the top. In tall buildings stack action induces air leakage through exterior cracks and openings and can cause a serious heat loss.

Infiltration is also induced into the building to replace exhaust air unless mechanical inlet ventilation balances the exhaust.

Infiltration, which often accounts for a major portion of the heating load, cannot, like ventilation systems, be turned off at night or during weekends, although it may be decreased if exhaust fans are shut down. It can, however, be reduced at all times. Particularly effective measures include caulking cracks around window and door frames and weatherstripping windows and doors. Weatherstripping doors costs about $30 per single door or $50 per double door. To weatherstrip metal-framed doors with aluminum and rubber costs about twice as much. Weatherstripping costs for windows depend upon type and number, and range between $15 and $35 per window. To rake out old caulking and recaulk around window edges costs about $15 per window (or about $25 per 30 square feet of window). Figure 6.4 gives infiltration rates for various types of windows. In most climates the average wind velocity in the winter is 10 to 15 mi/h, and in some locations it is considerably higher. The local weather bureau can be consulted for relevant information, and individual experience can be very useful; wind patterns and velocities vary widely, even on the same street in the same city.

ACTION GUIDELINES

☐ 1. Operate the exhaust systems as outlined in Secs. 6.5 and 6.6.

☐ 2. Inspect the building's exterior and interior surfaces and caulk all cracks that allow outdoor air to penetrate the building's skin.

☐ 3. Caulk around all pipes, louvers, or other openings which penetrate the building skin.

☐ 4. Repair broken or cracked windows.

☐ 5. As a temporary measure cover windows with 4-mil plastic sheets and extend the covering over the frame. Hold plastic sheets in place with continuous nailing strips.

☐ 6. Weatherstrip exterior doors and windows in climates with more than 2000 degree-days.

☐ 7. Cover porous exterior wall and roof materials with epoxy resin. (A 4-mm coat of epoxy resin costs about 35¢/ft² for material and labor.)

☐ 8. Remove or cover window air conditioners with plastic covers in the winter (if not used for heating).

☐ 9. Install automatic closers on exterior doors. (Cost ranges from $60 to $90 per door depending on whether closers are installed by the building's maintenance staff or outside labor.)

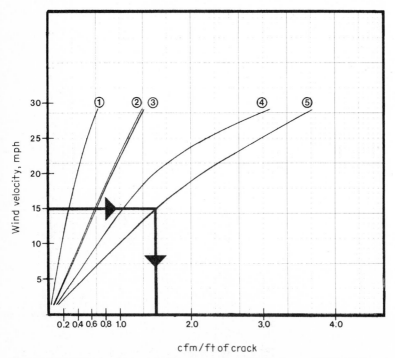

key to window infiltration chart			
(leakage between sash & frame)			
type	material	weatherstripped?	fit
① all	wood·	yes	avg.
hinged	metal	yes	avg.
② all	wood	no	avg.
hinged	metal	no	avg.
dbl. hung	steel	no	avg.
③ all	wood	yes	loose
dbl. hung	steel	yes	avg.
④ casement	steel	no	avg.
⑤ all	wood	no	loose
hinged			

Figure 6.4 Window-frame infiltration rates. (*Data from ASHRAE Handbook of Fundamentals, 1972, pp. 333, 337, and 338.*)

☐ 10. Reduce temperature in stairwells (but protect piping from freezing).

☐ 11. Seal elevator shafts at the top and bottom and ensure that the doors on the machine room penthouse are weatherstripped and closed.

☐ 12. Weatherstrip and close doors in the basement and roof equipment rooms where these are connected by a vertical shaft which serves the building.

To estimate window infiltration, note window type (Fig. 6.4) and find the wind velocity on the Fig. 6.4 graph, which gives the intake (in cubic feet per minute) per foot of crack. Add all window cracks on the windward side (and include the crack between sash). Multiply total windward cracks times infiltration per foot of crack to obtain total infiltration. The exercise can be repeated with assumptions of window refitting or weatherstripping to obtain a reduced infiltration rate. The energy required to heat infiltration air per 1000 cfm can be computed on an annual basis on Fig. 6.3 with entrants of heating degree-days (from Table 4.1), indoor temperature (from Table 4.2), and 168 hours per week (all hours) infiltration. From this heating requirement and the reduced infiltration rate an energy conserved figure can be computed. This energy savings, converted to fuel and dollars, can be compared with the cost of window refitting or weatherstripping to determine the cost advantage of this action.

6.9 INCREASE SOLAR HEAT GAIN INTO THE BUILDING

Although solar heat gain adds to the cooling load, it can be very helpful in reducing the heating load. Solar radiation impinging upon an opaque building envelope raises the exterior surface temperature and reduces conduction losses.

When solar radiation penetrates the building through windows and glass doors, it is particularly valuable, especially on south orientations in northern latitudes. The amount of sunlight that penetrates the windows depends upon the number of panes of glass, the area of the windows, the orientation of the windows, the type and cleanliness of the glass, the type of solar control device, the latitude of the building's location, and the percentage of sunshine at that location. Charts which address all the variables and permit calculations for exact benefits of increasing the solar heat gain in the winter are impossible within the scope of this manual, but a general rule is possible. In Minneapolis, there is approximately 25% more sunshine on the south glass and 15% more on the east and west glass than there is on the north face; in Florida, the amount of sunshine striking all facades is nearly equal. Between these two locations, owing to sun angles at different latitudes, the percentages of sunlight change linearly with latitude. The heat contributed by solar radiation through windows can save about ¾ gal of oil per square foot of south-facing glass in Minnesota and about ¼ gal of oil per year for east- or west-facing glass in Miami.

About 10% less sunlight penetrates double glazing than single glazing. However, double glazing reduces the heat load due to conduction, and the benefits from this more than offset the loss of solar radiation.

ACTION GUIDELINES

☐ 1. Clean windows to permit maximum sunlight transmission.

☐ 2. Operate drapes and blinds to permit sunlight (when available) to enter windows during the winter. Move desks or work stations out of the direct path of sunlight to avoid occupant discomfort.

☐ 3. A percentage of direct solar radiation is stored in the structure and furnishings where it will help to offset the heat load at night. Permit the space temperature to rise so that excess heat can be stored in the structure and be available for heating at night or cloudy periods. Even on cloudy days diffuse radiation is considerable; allow it to be transmitted into the occupied spaces. Treat skylights and display windows in the same manner as windows in occupied areas.

☐ 4. If windows are not fitted with blinds, drapes, or shutters, consider installing them to control the rate of heat flow into and out of the building.

☐ 5. During heavily clouded weather and at night, reduce the heat loss through the windows by drawing shades and drapes or closing shutters.

☐ 6. If direct sunlight or excessive window brightness causes glare, add a light translucent drape which cuts glare but permits solar heat to enter.

☐ 7. Readjust blinds during the day if overheating occurs.

☐ 8. Before installing shades or blinds, check the fire code for prohibitions against certain types of materials.

☐ 9. Trim all foliage shading the southern, eastern, or western faces of the building in winter. Reduce any evergreen foliage badly blocking the winter sun.

☐ 10. Where possible, remove shading devices and any other objects casting shadows on the building surfaces during winter.

☐ 11. Add thermal barriers in the form of shutters or screens to the inside of the windows to be closed when solar heat gain is not available.

6.10 INSTALL CONTROLS FOR SPACE TEMPERATURE AND HUMIDIFICATION

Sections 6.2, 6.3, and 6.5 of this chapter have documented the value of reducing indoor temperatures, setting back indoor temperatures at unoccupied times, and reducing relative humidity levels. Each of these actions is a system control function. Careful consideration should be given to the manual and automatic devices best suited to put into effect such control on a continuing basis.

ACTION GUIDELINES

☐ 1. Install a 7-day dual thermostat to operate the oil, gas, stoker, or electric heating elements. The thermostat should be set to maintain temperature levels during occupied periods and to reduce levels when the building is unoccupied at nights, weekends, and holidays.

☐ 2. Install 7-day controls to operate pumps for forced-circulation hot-water systems and automatic valves for hot-water or steam systems when boilers are operated by aquastat or pressure control.

☐ 3. Install a 7-day timer control to reset operating aquastat or pressure control for dual-level settings.

☐ 4. Install additional thermostats for individual zones where duct or piping systems permit control of individual zones. Operate controls on a 7-day cycle.

☐ 5. Provide room thermostats and dampers with automatic damper controls in supply air duct systems for additional zoning to permit further reduction of temperature levels in noncritical areas.

☐ 6. Where missing, install manually operated radiator valves or duct dampers to control or shut off the heat supply to noncritical building areas.

☐ 7. Install on-off switches to control the water supply to humidifiers to permit shutdown at night.

☐ 8. If a central humidifier serves the entire building, remove and relocate it and install duct-type or packaged room humidifiers to serve only those zones which require humidification.

☐ 9. Where room thermostats control both heating and cooling systems, exchange them for ones with a dead band level to prevent back-and-forth cycling between heating and cooling.

☐ 10. Provide locking devices on all room thermostats to prevent tampering.

☐ 11. Where supply-duct dampers are adjusted to reduce airflow into a space during the heating season, provide dampers to reduce return air from the space.

☐ 12. Relocate room thermostats to the most critical area, and rebalance the air or water system to reduce temperature and humidity levels in the other, less critical, areas.

☐ 13. Where window air-conditioner units or through-the-wall units provide heating as well as cooling, provide 7-day temperature-control thermostats to operate the heating elements.

6.11 CONTROL BUILDING VENTILATION

Section 6.7 of this chapter documented the value of reduced ventilation rates. This section suggests additional energy savings that can be achieved by additions to or modification of the various ventilation systems.

Many ventilation systems, particularly rooftop units, are initially provided with poor-quality outdoor-air dampers which do not allow accurate control and, even when fully closed, leak in such quantities as to exceed the minimum outdoor-air requirements. These dampers should be replaced with a good-quality opposed-blade damper with seals at the blade edges and ends. See Fig. 6.5. Such dampers can be installed for approximately $20 to $25 per 1000 cfm. If economizer cooling is used in the summer, the outdoor-air dampers must be sized to pass the full volume of the supply fan. If, however, economizer cooling is not contemplated and minimum quantities of outdoor air will be used year-round, then the outdoor-air intake louvers and dampers can often be drastically reduced in size, and the unused portion of the outdoor air inlet can be blanked off.

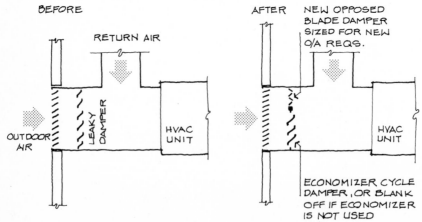

Figure 6.5 Damper modifications for better control of outdoor-air ventilation intakes.

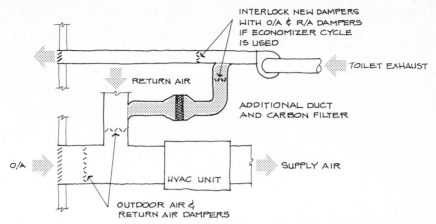

Figure 6.6 Carbon filtration for recirculation of toilet exhaust.

The quantity of air exhausted from a building often exceeds the minimum outdoor-air requirements and requires introduction of excess outdoor air to make up the difference and to maintain the pressure balance of the building. To balance the minimum outdoor-air intake and the total air exhaust from the building, the quantity of exhaust air may be reduced either by reducing the rate of exhaust or by retaining the same rate of exhaust but recirculating a proportion of the air back into the building.

Recirculated untreated exhaust air is often unacceptable because it smells bad. This air, however, can be treated by being passed through activated-carbon filters which will remove the odors. Candidates for such treatment are central exhaust systems from such spaces as toilet areas, dining rooms, and lounges.

To recirculate the air, ducting must be installed to connect from the discharge side of the exhaust system into the return air of the HVAC system. This ductwork should contain the activated-carbon filters. Recirculate that portion of the quantity of total exhaust air for the building which exceeds 90% of minimum outdoor-air requirements.

Activated-carbon filters remove odors from air by absorption, and after a period of time they become saturated and their performance falls off. At this point the filters must be removed and regenerated or replaced with freshly regenerated filters supplied by the manufacturer. Heat energy is required for regeneration but can be derived from a waste heat source.

The length of time the filters can remain in service depends on the circumstances of each situation; the manufacturer's advice should be sought on this point. The cost of installing the recirculating ductwork and filters will vary for each case, depending on the disposition of exhaust and HVAC systems and the quantity of air handled. To obtain an accurate assessment, designs should be prepared and quotations solicited from local contractors. For order of magnitude, the cost of installing activated carbon filters should be approximately $190 per 1000 cfm. See Fig. 6.6.

ACTION GUIDELINES

☐ 1. Repair leaking intake-air dampers or replace them with opposed-blade dampers with seals at the blade edges and ends.

☐ 2. Install automatic damper controls on outdoor-air intakes.

☐ 3. Provide a 7-day timer to operate automatic outdoor-air damper controls or to shut off all outdoor air for ventilation during unoccupied periods (for outdoor-air supply fans).

☐ 4. Add separate ventilation fans for zones requiring outdoor air.

☐ 5. Regulate interlocked exhaust-air–outdoor-air makeup units or dampers with 7-day controls.

☐ 6. Install charcoal filters or other devices to control the quality of supply air and reduce the amount of outdoor air required for ventilation.

To quantify savings in reduction in outdoor-air intake:

1. Determine the average hours per week during which the exhaust system is operated, then compute the reduction per week of outdoor air for ventilation.

2. Enter Fig. 6.3 at degree-days (from Table 4.1) and follow the direction of the example line to an intersection with indoor temperature (from Table 4.2) and hours of occupancy.

3. Read horizontally the yearly heating energy required per 1000 cfm.

4. Multiply this figure by the ventilation air reduction divided by 1000 to obtain the net yearly heating load reduction.

6.12 USE SEPARATE MAKEUP AIR SUPPLY FOR EXHAUST HOODS TO REDUCE OUTDOOR-AIR VENTILATION

Exhaust hoods in kitchens, hospitals, and laboratories and in process equipment areas pull large quantities of air along with smoke and fumes from a building. If the exhaust air up the hood is drawn from room air, the HVAC system must heat or cool outdoor air to make up this exhaust. This can be a considerable heating or cooling load; there is a better way.

It is not necessary for hood makeup air to be heated or cooled to the same degree as that required for occupant comfort. A separate makeup-air system comprising an outdoor-air fan, a tempering heating coil, and ductwork supply to the hood edges will save energy. This air need be heated only to 50 to 55°F in the winter and should not be cooled in the summer. The air, introduced close to and around the perimeter of the hood case, acts as a curtain and mixes with the exhaust fumes. In the winter any cooling effect on cooks or workers from the relatively low-temperature makeup air is offset by radiation from the equipment itself (except, of course, in laboratories working with cold processes).

The effectiveness of an exhaust hood in capturing heated air, fumes, smoke, or steam is a function of the air velocity at the edge of the hood. To maintain a satisfactory capture velocity, large open hoods require large volumes of air. With either room air or separate makeup air systems, face or capture velocities can be maintained or even increased while the total quantity of exhaust air is decreased by fitting baffles or a false hood inside an existing open hood. See Fig. 6.7.

ACTION GUIDELINES

☐ 1. For open exhaust hoods install baffles to allow reduction of the quantity of exhaust air without reducing the face velocity at the edge of the hood.

☐ 2. Install a makeup-air system to introduce outdoor air equal in quantity to that exhausted. Do not heat makeup air to more than 55°F, and introduce the air as close as possible to the hood in several positions around the perimeter to promote even airflow. Do not cool or dehumidify makeup air.

To determine savings with hood modifications:

1. Count the number of hours per week that the exhaust system is operated and compute the reduction per week in ventilation due to the separate makeup-air system.

2. Enter Fig. 6.3 at degree-days (from Table 4.1) and follow the direction of the example line to an intersect with indoor temperature (from Table 4.2) and hours of exhaust system operation.

3. Read horizontally the yearly heating energy required per 1000 cfm.

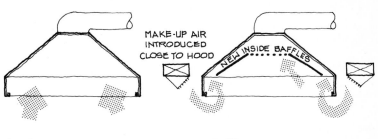

Figure 6.7 Exhaust-hood modifications to reduce conditioned room-air exhaust.

4. Multiply this figure by the ventilation reduction in cubic feet per minute divided by 1000 to obtain the net yearly load reduction.

5. Deduct from this the extra heating energy that must be expended to temper the air introduced by the makeup system and the energy required for the additional fan.

6.13 REDUCE HEAT LOSSES THROUGH WINDOWS

Heat loss from a building depends upon the temperature difference between indoors and out, the mode of operation of the heating system, and the mass, color, and insulating value of the exterior walls, roof, windows, and floors. Wind blowing on exterior surfaces increases heat loss, but solar radiation lowers heat loss by raising the temperature of the exterior surfaces. Heat loss varies greatly with the materials from which the building is constructed: A single layer of glass transmits 1.1 Btu/(h)(ft²) of surface per degree of temperature difference, double glass transmits about 0.55 Btu/h, an uninsulated frame wall about 0.3 Btu/h, a 12-in masonry and 4-in brick wall about 0.25 Btu/h, and insulated walls of various types and thicknesses transmit down to 0.027 Btu/(h)(ft²) per degree temperature difference.

Conversion of the windows from single glazing to double glazing will halve the conduction heat loss. Table 6.3 is the result of a computer program designed to study the reaction of both single- and double-glazed windows to various weather conditions, with runs made for 12 cities chosen to provide a typical cross-section of climates. Table 6.3 tabulates the heat loss in Btu per year for each square foot of glazing for the 12 selected locations. Figures 6.8 and 6.9 extrapolate loads from the calculations of Table 6.3 and allow prediction of annual heat loss based on inputs of heating degree-days, solar radiation, and hours of occupancy per week. These two graphs distinguish between latitudes less than 35°N and greater than 35°N.

Heat loss through windows is greatest in areas with high numbers of heating degree-days, but for any one geographic location this loss is modified by orientation. Heat loss is greatest through windows of north exposure and least in windows of south exposure, owing to the beneficial heating effect of the sun. This modifying effect of the sun is more marked in cold climates and in locations where winter sun altitudes are low but where intensity is fairly high. The greatest economic return from double glazing will be on facades which do not receive direct sunshine. Differences among various orientations are affected by geographic location and solar intensity, as indicated in Table 6.3.

Storm windows may be applied to either the outside or inside of an existing window to reduce heat loss by conduction and infiltration. If the existing windows and frames are of poor construction and allow a high rate of infiltration, the storm windows should be fitted on the outside, if possible, to reduce infiltration without incurring the extra cost of caulking the window.

TABLE 6.3 HEAT LOSS THROUGH SINGLE- AND DOUBLE-GLAZED WINDOWS

| | | | | Heat Loss, $Btu/(ft^2)(yr)$ | | | | | |
| | | | | North | | East and West | | South | |
City	Latitude	Solar Radiation, Langleys	Degree-Days	Single	Double	Single	Double	Single	Double
Minneapolis	45°N	325	8,382	187,362	94,419	161,707	84,936	140,428	74,865
Concord, N.H.	43°N	300	7,000	158,770	83,861	136,073	73,303	122,144	67,586
Denver	40°N	425	6,283	136,452	70,449	117,487	62,437	109,365	59,481
Chicago	42°N	350	6,155	147,252	75,196	126,838	65,810	110,035	58,632
St. Louis	39°N	375	4,900	109,915	56,054	94,205	49,355	84,399	45,398
New York	41°N	350	4,871	109,672	54,986	93,700	48,611	82,769	44,580
San Francisco	38°N	410	3,015	49,600	25,649	43,866	23,704	41,691	23,239
Atlanta	34°N	390	2,983	63,509	31,992	55,155	28,801	51,837	28,092
Los Angeles	34°N	470	2,061	21,059	11,532	19,487	10,954	19,485	10,989
Phoenix	33°N	520	1,765	25,951	14,381	22,381	12,885	22,488	12,810
Houston	30°N	430	1,600	33,599	17,939	30,744	17,053	30,200	16,861
Miami	26°N	451	141	1,404	742	1,345	742	1,345	742

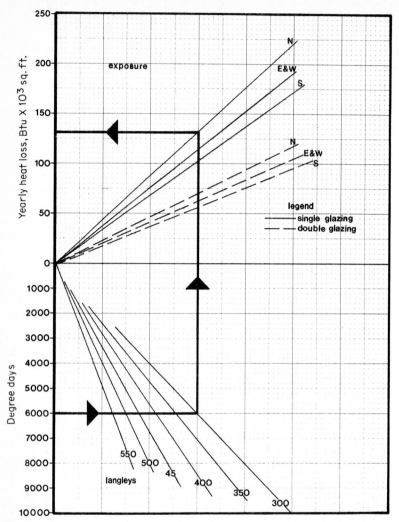

Figure 6.8 Winter heat loss through windows in latitudes 25 to 35°N.

The space between the storm window and the existing window should be vented with small openings to the outdoors and should be provided with drainage weepholes to prevent moisture buildup. If storm windows are added to existing windows, reflective or tinted glass should be considered to reduce solar gain and glare in summer.

Where natural lighting is desirable but appearance and visibility through the window are not important, a low-cost interim solution which will provide double glazing is the use of clear plastic film attached to a simple subframe. This will not provide a durable storm window but should be considered as a temporary means to save heating energy where money is not immediately available for a permanent installation. Materials will cost about 10¢/ft² and installation by unskilled labor will take about 30 min per window.

Existing single-glazed windows may be converted permanently to double-glazed windows by the addition of a new glazing frame to accept the additional pane of glass. The space between the glazing should be vented and drained to the outside, and provisions should be made for cleaning both sides of each sheet of glass.

Where the existing window frame is in good condition and the glazing system permits, a single sheet of glass may be replaced with a sealed double-glazed unit. Where the existing window frame is in poor condition and is scheduled for renovation or replacement, a double-glazed unit should be considered.

In all cases, whether storm windows are added or single windows are replaced by double glazing, the frame of the selected unit should not form a

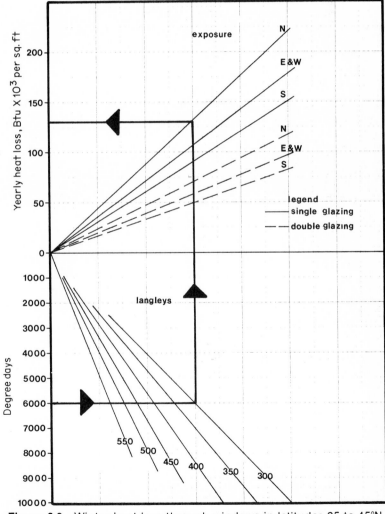

Figure 6.9 Winter heat loss through windows in latitudes 35 to 45°N.

heat bridge. Each frame should incorporate a thermal break between the inside and the outside surfaces. Such frames are commercially available. This is particularly important in metal frames, which readily conduct heat. A more desirable framing material is wood, which is a poor heat conductor.

In addition to the energy savings made by reducing heat losses, double glazing also reduces heat gain in summer, resulting in a slight reduction of cooling loads. Double glazing is never worthwhile, however, for cooling benefits alone.

Where existing single windows are very loose and allow large volumes of infiltration, the addition of double glazing will reduce infiltration and will save both heating and cooling energy. The saving of heating energy due to lower infiltration rates can be considerable in locations experiencing more than 2000 degree-days, and of cooling energy in locations experiencing more than 8000 wet-bulb degree-hours above 66°F WB, and should be evaluated in any economic study.

The cost of converting to double glazing varies with the methods used and with the type of existing glazing. For order of magnitude, assume costs of $7.50/ft² for additional single glass and $14.00/ft² for replacing existing frames and single glazing with sealed double-glazed units.

ACTION GUIDELINES

☐ 1. Where the heating season is 7500 degree-days or greater, use triple glazing. Remove existing single windows and frames and replace with triple glazing, or add double-glazed storm windows to existing single-glazed windows.

☐ 2. Where the heating season is 2500 degree-days or greater, install storm windows to reduce both conduction loss and infiltration. Where windows are tight, remove glazing from window frames and install double-glazed units. (Where the heating season is less than 1500 degree-days, double glazing cannot be justified by savings in heating energy alone but may be desirable to reduce outside noise.)

☐ 3. When replacing or adding glazed surfaces, use one layer of reflective glass or reflective coating to reduce solar heat gains if windows are in direct sunlight for more than 3 hours per day during the cooling season.

☐ 4. Where windows are subjected to high winds for a long duration of time in the winter and to sunlight in the summer, use prefabricated sun-screens on the exterior of the window to serve a dual purpose: to minimize the effect of heat loss due to wind and to reduce solar heat gain in the summer.

☐ 5. Install storm doors in all locations where the number of degree-days exceeds 2000, and provide not more than 50% glass area in the storm door. Storm doors should be wood rather than metal unless a thermal break is provided.

☐ 6. Where exterior doors are not used during the heating season and are not required as fire exits, caulk the doors, seal them with removable strips, or insulate the areas between the storm door and the exising door.

☐ 7. In new construction, where energy for cooling exceeds the amount of energy required for heating, minimize the amount of glazing in the west first, then east, south, and north exposures in that order.

☐ 8. In new construction, where the outdoor-air conditions are close to desired indoor conditions for a major portion of the year, do not caulk or install double glazing, but provide operable windows to permit natural ventilation and cooling.

☐ 9. Install insulating draperies or thermal barrier shutters to reduce heat loss through windows.

☐ 10. Treat all skylights like windows. The order of priority for treating skylights should follow the order of priority for treating windows: in cold climates first on the north wall, and in warm climates first on the west, then east, then south walls.

To determine the energy saved by double glazing:

1. Note latitude from Table 4.1 and select the appropriate graph, Fig. 6.8 or 6.9.

2. Note from Table 4.1 the heating degree-days and the mean daily solar radiation in langleys.

3. Select the orientation and determine the area of the windows to be double glazed.

4. Enter Fig. 6.8 or 6.9 at the appropriate degree-days. Follow the direction of the example line and intersect with the appropriate langleys and orientation. Read the yearly energy requirements for 1 ft² of single-glazed window. Repeat the procedure for double glazing and subtract this value from the energy required for single glazing for the load reduction in Btu per year per square foot of double-glazed window. Multiply by the area of window to determine the total yearly load reduction for the orientation considered.

5. Repeat the procedure for other orientations. (Interpolate for orientations other than the four cardinal points shown.) Note that the heat losses derived from the graph assume that the windows are subjected to direct sunshine and are not shaded by adjacent buildings, structures, or trees. If

shading occurs, then regardless of the actual orientation use the figures for north-facing windows, as these do not include a direct-sunshine component.

The total annual load reduction is the sum of all load reductions of various orientations and represents the total annual reduction in building heat load due to window treatment.

Figures 6.8 and 6.9 are based on the Sunset computer program, developed by Dubin-Bloome Associates for the Federal Energy Agency, which was used to calculate solar effect on windows for 12 selected locations. The program calculates hourly solar angles and intensities for the twenty-first day of each month. Radiation intensity values were modified by the average percentage of cloud cover taken from weather records on an hourly basis. Heat losses are based on a 68°F indoor temperature.

Additional assumptions were: (1) total internal heat gain of 12 Btu/ft²; (2) average outdoor-air ventilation rate of 10%; (3) infiltration rate of one-half air change per hour. Daily totals were then summed for the number of days in each month to arrive at monthly heat losses. The length of the heating season for each location considered was determined from weather data and characteristic operating periods. Yearly heat losses were derived by summing monthly totals for the length of the heating season. These are summarized in Table 6.3 for the 12 locations. The data were then plotted and extrapolated to include the entire range of degree-days. Figure 6.8 was derived from locations with latitudes between 25 and 35°N; Fig. 6.9 was derived from locations with latitudes between 35 and 45°N.

6.14 REDUCE HEAT LOSSES THROUGH WINDOWS WITH THERMAL BARRIERS

Most commercial, office, religious, and industrial buildings are unoccupied for more hours per week than they are occupied, and a major portion of the unoccupied hours occur at night when outdoor temperatures are at their lowest and the potential for heat loss is highest. During unoccupied hours, daylight is either not available or not required, and windows can be covered with thermally insulated barriers or shutters to decrease heat transmission losses. Thermally insulated barriers may be applied to the inside of the building and can be arranged to slide in a track or fold back into the window reveal or onto the face of the wall. When closed, the barriers should provide a reasonable seal over the window but need not be airtight. Barriers may be purchased ready-made or may be custom-made to fit the windows. The major proportion of the total cost is in labor for construction and fitting rather than the cost of materials. Thermal barriers should, therefore, be selected or designed to provide the best insulating value attainable for a given thickness.

Before installing the barrier, check the local fire code to ensure that the insulating materials chosen are not considered a fire or smoke hazard. Check also that barriers do not impede access required by the fire department.

The total insulating effect of the thermal barriers will vary with the type of window and the type of insulation used in construction. Table 6.4 gives composite U values obtained by the barriers and window assembly for various thicknesses of insulation. It is assumed that the insulation has a K of 0.25, which is typical for glass wool, beaded polystyrene, and polyurethane foam.

TABLE 6.4   THERMAL BARRIER *U* VALUES

| Insulation | Composite U Value | |
Thickness, in	Single Glazing	Double Glazing
½	0.28	0.23
1	0.18	0.16
1½	0.13	0.12
2	0.11	0.10

The costs of purchasing and installing thermal barriers will vary according to the size and type of window, the materials used in construction, and the ease of access for fitting. For order of magnitude assume \$5.00/ft² of window for folding barriers.

ACTION GUIDELINES

☐ 1. Install thermal barriers when unoccupied hours exceed 70 per week and heating degree-days exceed 3000 if existing windows are single glazed, and 4500 if double glazed.

☐ 2. Operate thermal barriers in the same manner as conventional drapes. Close them even in occupied hours when daylight cannot be used and when vision through the window is not required.

☐ 3. Do not close thermal shutters at night in the cooling season, except on the west side when there is considerable sunshine after the work day ends.

To calculate the energy saved by thermal barriers, note from Table 4.1 the heating degree-days and determine the composite U value for the barrier and window assembly. Determine the average unoccupied hours per week during the winter, enter Fig. 6.10 at the appropriate degree-days and, following the direction of the example line, intersect with the appropriate composite U value and unoccupied hours, and then read out the yearly heating load. Repeat the above procedure for a window without thermal barriers and obtain the yearly heating energy required. Subtract from this the energy required when thermal barriers are fitted to obtain the yearly heating load reduction due to 1 ft². Multiply this by the total area of treated windows to obtain the net yearly reduction in heat loss.

The development of Fig. 6.10 was based on the assumptions that:

1. Thermal barriers are closed only when the building is unoccupied.

2. The average degree-day distribution is 25% during the daytime and 75% during nighttime.

The number of degree-days occurring when the thermal barriers are closed [adjusted degree-days (DD_A)] was determined from the characteristic occupancy periods shown in the figure. This can be expressed as a fraction of the total degree-days (DD_T) by the relationship:

$$DD_A = 0.25DD_T \left(\frac{\text{unoccupied daytime hours/week}}{\text{total daytime hours/week}} \right)$$
$$+ 0.75DD_T \left(\frac{\text{unoccupied nighttime hours/week}}{\text{total nighttime hours/week}} \right)$$

Yearly heat losses can then be determined by:

$$Q \text{ (heat loss/year)} = DD_A \times U \text{ value} \times 24$$

6.15 REDUCE HEAT LOSSES THROUGH WALLS

Heat loss through walls can be significantly reduced by adding insulation. Such heat loss is a function of the wall's resistance to heat flow, modified by the effects of solar radiation which reduces heat loss and wind which increases it. The effect of solar radiation will vary with the absorption coefficient of the outside surfaces; dark colors of high absorption coefficient will reduce the heat loss more than light colors of low absorption coefficient. The wind affects the overall resistance of the wall by reducing the insulating value of the air film on the exterior wall surface.

The mass of the wall and its attendant thermal inertia have an overall modifying effect on heat loss by delaying the impact of outdoor temperature changes on the heated space. This time delay allows the wall to act dynamically as a thermal storage system, smoothing out peaks in heat flow and reducing yearly heat loss. Low-mass walls of 10 to 20 lb/ft² will have up to 5% more yearly heat loss than high-mass walls of 80 to 90 lb/ft², assuming that both walls have the same overall U value and absorption coefficient.

Table 6.7 is the result of a computer program designed to study the reaction of walls to varying climate conditions, with runs made for 12 cities chosen to provide a typical cross-section of climates. It tabulates the heat loss for 1 ft² of wall in Btu per year for two different U values and two different absorption coefficients for each of the 12 selected locations. Note that heat loss through walls is greatest in locations having high degree-days, but that for any one location heat loss will vary with the wall orientation; heat loss is highest through north walls and least through south walls, owing to the beneficial effect of the sun. The difference is more marked in walls having a high absorption coefficient. The graphs of Figs. 6.13 and 6.14 extrapolate the data of Table 6.7 and allow prediction of yearly heat loss through 1 ft² of wall based on inputs of degree-days, solar radiation, U values, orientation, and absorption coefficient.

The overall U value of walls may be decreased by the addition of insulating material to the inside surface, to the outside surface, or into cavities within the wall structure.

Addition of insulation to the outside of walls can be accomplished with relative ease in one- or two-story buildings but becomes increasingly difficult for higher stories because of the requirement for access staging. Insula-

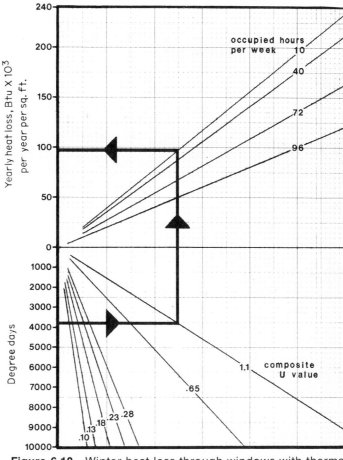

Figure 6.10 Winter heat loss through windows with thermal barriers.

tion added to the outside of walls must be weatherproof and vapor-sealed to prevent deterioration of its insulating properties due to the ingress of moisture. External insulation can be added in the form of prefabricated insulating panels or as rigid insulation over which plaster is applied, but in either case difficult design problems must be addressed. The addition of exterior insulation might seriously alter the architectural character of the building, and this may be unacceptable.

Addition of insulation to the inside surface of walls can generally be accomplished with more ease but will entail moving furniture and fittings away from the wall and could interrupt normal operation of the building. Internal insulation must be protected from moisture with a vapor barrier, and from wear and tear with an integral finished surface or with a protective finish such as wood paneling or gypsum plasterboard. Design problems likely to occur when insulation is added to the internal surfaces of walls are

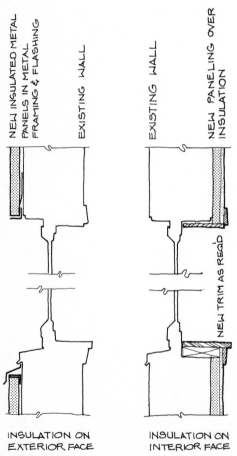

INSULATION ON EXTERIOR FACE

INSULATION ON INTERIOR FACE

Figure 6.11 Insulation techniques for existing walls.

**TABLE 6.5   VALUES FOR THE DECISION
TO INSULATE EXISTING WALLS**

Heating Degree-days	Minimum Acceptable U Value
1000	0.40
2000	0.30
3000	0.30
4000	0.20
5000	0.20
6000	0.15
7000	0.15
8000	0.10
9000	0.10

treatment of window and door openings, architraves and reveals, the junction between the insulation and the floor and ceiling, and the repositioning of equipment such as heaters and electric receptacles that are recessed in the existing wall. See Fig. 6.11.

If the wall has an internal void, the space can easily be filled with granular insulation, cellulose, or plastic foam (polyurethane). This process will not alter the appearance of the building and can be accomplished with minimum inconvenience to the occupants. It requires cutting or drilling holes into the wall to give access to the internal void. Granular insulation should be blown in under pressure to ensure a compressed fill, as it tends to pack down with time. Many firms specialize in complete insulation services. The compatibility of the insulation and the wall should be investigated to ensure that an interaction will not occur which could damage either the wall structure or the insulation. Some wall cavities in older buildings were designed to prevent driving rain from penetrating to the inside skin. In such cases, an insulating material should be used that will not interfere with this function, or the outside wall surface should be coated with epoxy resin or similar waterproofing.

ACTION GUIDELINES

☐ 1. Determine the U value of existing walls by inspection, measurement, and calculation. Table 6.5 suggests the minimum acceptable U value for existing walls in various heating degree-day zones. Walls below these standards should be insulated.

☐ 2. The cost of materials to insulate a wall is small in comparison with labor costs. Once the decision is made to insulate, add at least enough insulation to improve the U value to 0.1 or lower.

☐ 3. Insulation is most effective on the north walls, which should be given priority, followed by the east or west walls, whichever face into the winter prevailing wind, then south walls.

☐ 4. If any exposure is shaded, its priority for insulation should be modified according to the hours of shading experienced on an average winter day.

☐ 5. Give priority to spaces that are continuously occupied for the greatest length of time. Corridors, toilets, elevator shafts, and storage rooms should be assigned a lower priority regardless of wall orientation.

☐ 6. Reflective insulation is more effective on the walls than on the roof but is not as effective as other types of insulation with higher R values.

☐ 7. Where possible, install insulation on the exterior surface of masonry walls. For other types of construction, provide insulation on the interior surface or in the cavity, if there is one.

☐ 8. Check all codes before installing insulation on the interior surface of the building. Some insulation materials are not acceptable, owing to low ignition temperatures or to the formation of toxic fumes when burned.

☐ 9. When insulation is blown into a masonry wall cavity or stud space, the cavity should be completely filled under slight pressure, since fill insulation often settles with time.

☐ 10. Provide a vapor seal on the exterior surface of insulation applied to the exterior. Provide a vapor barrier on the room side of the insulation when applied to the interior.

☐ 11. When additions are built, the mass of the new walls should not be less than 80 lb/ft² in order to take full advantage of energy conservation due to thermal mass.

☐ 12. In new construction the combined U factor for opaque wall and glazing should not exceed the U values listed in Table 6.5 by more than 0.05. This can be accomplished by a combination of insulation and double or triple glazing in correct proportions.

☐ 13. When expansion occurs in retail stores and display windows are required, the space between the show window and the sales area should be blocked off by an insulated partition with a U value not less than 0.20.

To determine whether additional wall insulation is worthwhile, it is first necessary to know the U value of the untreated existing wall. By examination, determine the type and thickness of the individual components of the wall. Using this information and the thermal characteristics listed in the

ASHRAE *Handbook of Fundamentals,* compute the overall *U* value. This calculated value should be checked by the following procedure:

1. Measure the indoor and outdoor air temperatures with an accurately calibrated thermometer on a sunless day or at night.

2. Measure at the same time the inside surface temperature of the wall with an accurately calibrated contact thermometer.

3. Calculate the rate of heat flow through the wall with the inside air temperature, inside surface temperature, and the inside film coefficient (assume *R* of 0.6), using the formula:

$$\text{Heat flow} = \frac{\text{inside air temperature} - \text{inside surface temperature}}{\text{film coefficient}}$$

4. Using the calculated rate of heat flow, now calculate the overall *U* value using the formula:

$$U = \frac{\text{heat flow}}{\text{inside air temperature} - \text{outside air temperature}}$$

For example, if the indoor temperature is 68°F, the outdoor temperature is 20°F and the inside wall surface temperature is 59.4°F, then:

$$\text{Rate of heat flow} = \frac{68 - 59.4}{0.6} = 14.3 \text{ Btu/h}$$

$$\text{Overall } U \text{ value} = \frac{14.3}{68 - 20} = 0.298 \text{ or about } 0.3$$

This method allows for normal variation of the external film coefficient, but the best results are obtained if measurements are made on a day of average wind speed. The calculated and measured *U* values will not necessarily be in exact agreement, and an average of the two results should be used.

Before savings can be calculated, it is necessary to know the *U* values of the walls both before and after adding insulation; after the *U* value of the existing walls has been established, a decision must be made as to how much insulation should be added and where it should be applied. In existing buildings, this decision is governed by the practicality of coordinating the insulation with the existing structure and requires an architectural solution based on the particular circumstances of the building. Costs for several types of wall insulation are given in Table 6.6.

The new *U* value can now be derived. Refer to Fig. 6.12 and, entering the graph at the present *U* value, intersect with the appropriate insulation thickness line and then read out the new *U* value. For example: A wall

TABLE 6.6 APPROXIMATE INSULATION COSTS FOR WALLS

Rigid insulation on interior surface	$1.06/ft²
Rigid insulation on exterior surface	1.80/ft²
Rigid insulation on exterior surface with stucco finish	3.20/ft²
Cavity-fill mineral fiber	1.70/ft²
Cavity-fill plastic foam	0.80/ft²
External epoxy coating	0.42/ft²

which had an initial U value of 0.3 and 2 in of added insulation would have an improved U value of 0.08. This graph can also be used as an aid to find the required thickness of insulation needed to improve the initial U value to the desired figure. For example: If an improvement in a wall's U value from 0.7 to less than 0.08 is desired, then 3 in of insulation would be required. This graph is based on insulation having a K of 0.25, which is typical for glass wool, beaded polystyrene, foamed polyurethane, and loosely packed mineral wool.

To determine the heating energy saved by insulating walls, refer to Figs. 6.13 and 6.14, then carry out the following procedure:

1. Note latitude (Table 4.1) and select the appropriate graph.

2. Note heating degree-days and mean daily solar radiation in langleys (Table 4.1) and determine the orientation and area of the wall or walls to be insulated.

3. Select the absorption coefficient for the outside surface of the wall.

4. Using the initial U value of the wall, enter the graph at the appropriate degree-days and, following the direction of the example line, intersect the appropriate points for langleys, orientation, and U value. Read out the yearly heat loss per square foot of wall on the appropriate absorption coefficient (abbreviated a) line. Interpolate the results for absorption coefficients other than 0.3 and 0.8. Repeat this procedure using the improved U value and subtract the energy used before insulation to derive the energy saving per square foot of wall. Multiply this saving by the area of the wall insulated.

5. Repeat this procedure for each different orientation of wall to be insulated. The graph shows orientation for the four cardinal points: north, east, south, and west. Interpolate for other orientations. Note that the heat losses derived from the graph assume that the wall is not shaded from direct sunshine. If shading occurs, then, regardless of the actual orientation of the wall, treat it as a north wall, as doing this does not include a component for direct sunshine. Total the yearly savings of all walls insulated to obtain total saving of heat energy.

In addition to savings achieved by reducing winter heat losses, wall insulation will also reduce summer heat gains. This extra saving is small but nevertheless should be recognized when the economic feasibility is assessed.

Figure 6.12 is based on the addition of insulation with an R value of 4.0. The calculations to determine the U value after the addition of insulation assumed a combined inside/outside film coefficient of $R = 1$. The standard formula $U = 1/R$ was used.

Figures 6.13 and 6.14 are based on the Sunset computer program which was used to calculate solar effect on walls for 12 selected locations. The program calculates hourly solar angles and intensities for the twenty-first day of each month with radiation intensity values modified hourly by average percentage of cloud cover taken from weather records. Heat losses are based on a 68°F indoor temperature.

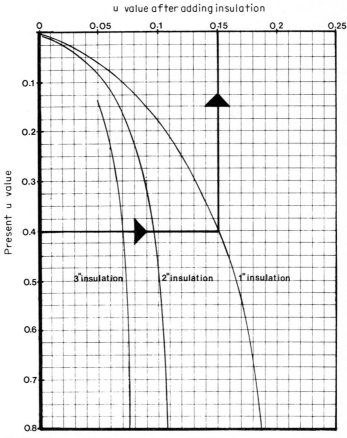

Figure 6.12 Effect of insulation on U value.

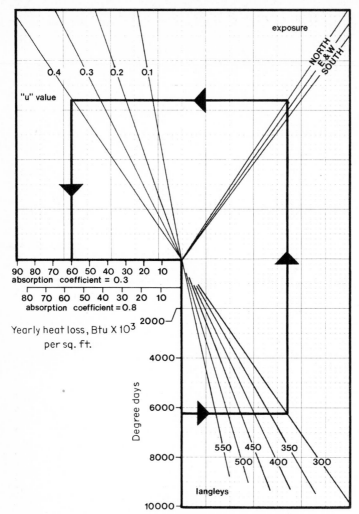

Figure 6.13 Winter heat loss through walls in latitudes 25 to 35°N.

The solar effect on a wall was calculated by the computer using sol-air temperature, and the heat entering or leaving a space was calculated with the equivalent temperature difference. Wall mass ranged from 50 to 60 lb/ft², and thermal lag averaged 4½ h. Additional assumptions were: (1) total internal heat gain of 12 Btu/ft², (2) average outdoor-air ventilation rate of 10%, and (3) infiltration rate of one-half air change per hour. Daily totals were then summed for the number of days in each month to arrive at monthly heat losses. The length of the heating season for each location considered was determined from weather data and characteristic operating periods. Yearly heat losses were derived by summing monthly totals for the length of the heating season.

Absorption coefficients and *U* values were applied and summarized for the 12 locations as shown in Table 6.7. The data were then plotted and extrapolated to include the entire range of degree days. Figure 6.13 was derived from locations with latitudes between 25 and 35°N, and Fig. 6.14 was derived from locations with latitudes between 35 and 45°N. The heat losses assume that the walls are subjected to direct sunshine. If shaded, losses should be read from the north exposure line.

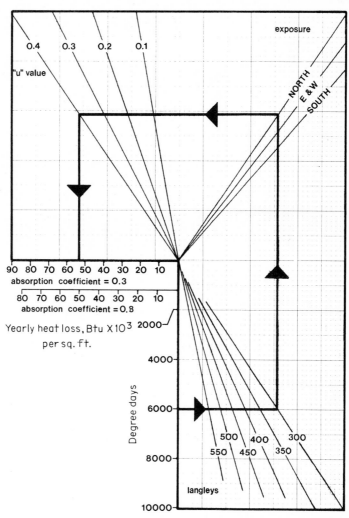

Figure 6.14 Winter heat loss through walls in latitudes 35 to 45°N.

TABLE 6.7   HEAT LOSS THROUGH WALLS

				Heat Loss, Btu/(ft²)(yr)												
				North				East and West				South			
				U = 0.39		U = 0.1		U = 0.39		U = 0.1		U = 0.39		U = 0.1	
City	Latitude	Solar Radiation, langleys	Degree-Days	a = 0.3	a = 0.8	a = 0.3	a = 0.8	a = 0.3	a = 0.8	a = 0.3	a = 0.8	a = 0.3	a = 0.8	a = 0.3	a = 0.8
Minneapolis	45°N	325	8,382	74,423	70,651	20,452	19,335	70,560	62,229	19,378	16,787	66,066	51,298	18,109	13,530
Concord, N.H.	43°N	300	7,000	68,759	64,826	18,895	17,714	64,674	55,363	17,743	14,972	59,759	43,667	16,370	11,344
Denver	40°N	425	6,283	57,337	53,943	15,755	14,824	53,726	44,937	14,763	12,198	48,780	34,095	13,405	8,720
Chicago	42°N	350	6,155	58,516	55,356	16,081	15,210	55,219	47,678	15,169	12,865	50,684	37,339	13,847	9,743
St. Louis	39°N	375	4,900	45,046	42,149	12,379	11,565	41,981	35,192	11,533	9,476	38,038	26,344	10,425	6,660
New York	41°N	350	4,871	45,906	42,950	12,615	11,804	42,843	35,368	11,774	9,594	38,385	25,231	10,548	6,406
San Francisco	38°N	410	3,015	23,258	21,120	6,392	5,803	20,916	15,631	5,748	4,118	16,948	9,812	4,645	1,743
Atlanta	34°N	390	2,983	26,922	24,803	7,398	6,771	24,475	19,206	6,716	5,103	20,639	12,399	5,562	2,587
Los Angeles	34°N	470	2,061	9,900	8,549	2,720	2,349	8,392	5,758	2,306	1,316	6,139	3,040	1,520	155
Phoenix	33°N	520	1,765	11,861	10,533	3,259	2,878	10,283	7,316	2,826	1,811	8,077	4,619	2,062	555
Houston	30°N	430	1,600	14,592	12,956	4,011	3,557	12,888	9,379	3,542	2,351	10,878	6,760	2,909	1,142
Miami	26°N	451	141	210	106	7	0	92	0	0	0	6	0	0	0

6.16 REDUCE HEAT LOSSES THROUGH ROOFS

Heat loss through roofs can be significantly reduced by adding insulation. As with walls, heat loss through a roof is a function of its resistance to heat flow, modified by the effect of solar radiation which reduces heat loss and wind which increases heat loss. The effect of solar radiation will vary with the absorption coefficient of the outside surfaces; dark colors of high absorption coefficient will reduce the heat loss more than light colors of low absorption coefficient. The wind affects the overall resistance of the roof by reducing the insulating value of the exterior roof surface air film.

The mass of the roof and its attendant thermal inertia have an overall modifying effect on heat by delaying the impact of outdoor temperature changes on the heated space. This time delay allows the roof to act dynamically as a thermal storage system, smoothing out peaks in heat flow and reducing yearly heat loss. Low-mass roofs of 10 to 20 lb/ft² will have approximately 1% greater yearly heat loss than high-mass roofs of 40 or so lb/ft², assuming both roofs have the same overall U value and absorption coefficient.

Table 6.8 shows the results of a computer study of heat gain and loss through roofs as a function of climatic variation for 12 cities chosen to give a typical cross-section of climates. It tabulates the heat loss for 1 ft² of roof in Btu per year for two different U values and two different absorption coefficients at the 12 selected locations. This table shows the marked correlation of high absorption coefficients with the beneficial effect of sunshine.

Figure 6.21 extrapolates the data from Table 6.10 and allows prediction of yearly heat loss through 1 ft² of roof based on inputs of heating degree-days, solar radiation, U values, and absorption coefficients.

TABLE 6.8   HEAT LOSS THROUGH ROOFS

City	Latitude	Solar Radiation, langleys	Degree-days	Heat Loss, Btu/(ft²)(yr) $U = 0.19$ $a = 0.3$	$a = 0.8$	$U = 0.12$ $a = 0.3$	$a = 0.8$
Minneapolis	45°N	325	8,382	35,250	30,967	21,330	18,642
Concord, N.H.	43°N	300	7,000	32,462	27,678	19,649	16,625
Denver	40°N	425	6,283	26,794	22,483	16,226	13,496
Chicago	42°N	350	6,155	27,489	23,590	16,633	14,190
St. Louis	39°N	375	4,900	20,975	17,438	12,692	10,457
New York	41°N	350	4,871	21,325	17,325	12,911	10,416
San Francisco	38°N	410	3,015	10,551	8,091	6,381	4,784
Atlanta	34°N	390	2,983	12,601	9,841	7,619	5,832
Los Angeles	34°N	470	2,061	4,632	3,696	2,790	2,142
Phoenix	33°N	520	1,765	5,791	4,723	3,487	2,756
Houston	30°N	430	1,600	6,045	4,796	3,616	2,778
Miami	26°N	451	141	259	130	139	55

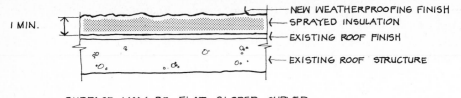

SURFACE MAY BE FLAT, SLOPED, CURVED OR IRREGULAR

Figure 6.15 Sprayed insulation applied to existing roofs.

Because of sun angles and latitudes, flat roofs receive less solar energy in the winter than east, west, and south exterior walls. On a square foot basis, insulation for roofs should be considered immediately after that for the north-facing walls, and, since the greatest heat gain in the summertime per square foot of building envelope opaque surface due to solar radiation is on the roof, the economic benefit of insulating the roof will be most favorable on a year-round basis when both heating and cooling are required.

Table 6.9 shows the minimum acceptable U value for roofs in various heating degree-day zones. Existing roofs below these standards should be insulated. The cost of insulation materials is small in comparison with the labor costs for installation; once the decision is made to insulate the roof, add at least enough insulation to improve the U value to better than 0.06. The overall U value may be decreased by the addition of insulating material to either the top of the roof or the ceiling below.

Where an existing roof has deteriorated and is scheduled for extensive repair or replacement, sprayed polyurethane foam or rigid insulation can be applied. The roof surface must be cleaned of all loose debris and, if necessary, cut back to provide a good bonding surface. It is not necessary, however, to provide a completely smooth surface for foam. The existing roof structure should be vented, and protection should be provided against overspray. Even though the insulation is not hygroscopic, it must be pro-

TABLE 6.9   VALUES FOR THE DECISION TO INSULATE EXISTING ROOFS

Heating Degree-days	Minimum Acceptable U Value
1000	0.30
2000	0.20
3000	0.20
4000	0.15
5000	0.15
6000	0.10
7000	0.10
8000 and above	0.06

tected by a weatherproof finish such as butyl. The finish selected must be applied cold, as excess heat will damage the insulation. A minimum thickness of 1 in spray foam is required to provide acceptable bond strength and integrity. See Fig. 6.15.

For rigid insulation, the roof surface must be cleaned of all loose debris and made smooth. Treated wood nailers should be used at all joints, roof penetrations, and gravel stops, and as required to fit new flashing. For sloping roofs of angles greater than 5°, treated wood nailers should be provided between insulation boards.

The rigid insulation, which may be fiberglass, fiberboard, cellular glass, or perlite boards, should be embedded in a solid mopping of hot asphalt and then protected with a built-up roof. See Fig. 6.16.

Where the existing roof is sound and the underside is directly accessible either from an attic space or ceiling void, sprayed polyurethane foam or mineral fiber or rigid insulation may be added to the inside surface. Sprayed insulation is quick and easy to apply but will entail either removal of furnishings or protection to prevent spoiling by overspray and droppings. The application of either polyurethane spray or mineral fiber spray entails health hazards, too, and the immediate building area must be evacuated while the work is done. As a side benefit, mineral fiber sprays will generally improve the fire rating of the structure and provide good attenuation of noise.

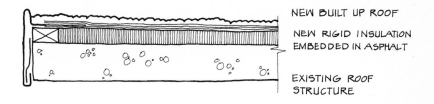

NEW BUILT UP ROOF

NEW RIGID INSULATION EMBEDDED IN ASPHALT

EXISTING ROOF STRUCTURE

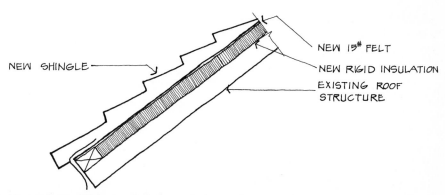

NEW 15# FELT

NEW SHINGLE

NEW RIGID INSULATION

EXISTING ROOF STRUCTURE

Figure 6.16   Rigid insulation applied to existing roofs.

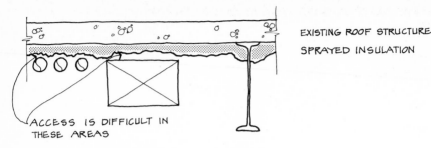

EXISTING ROOF STRUCTURE

SPRAYED INSULATION

ACCESS IS DIFFICULT IN
THESE AREAS

CEILING SPRAYED INSULATION

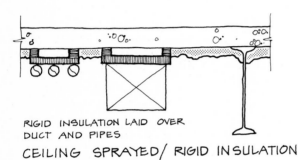

RIGID INSULATION LAID OVER
DUCT AND PIPES

CEILING SPRAYED/ RIGID INSULATION

Figure 6.17 Sprayed insulation applied to structural ceilings.

For application of sprayed foam insulation, access must be provided over the whole area. The underside must be brushed clean of all loose debris, and any ducts or pipes must be masked from the spray. A minimum thickness of 1 in sprayed insulation is required. Difficulty may be experienced in reaching roof areas where ducts and pipes are installed close to the roof and where structural members protrude. See Fig. 6.17.

To apply rigid insulation to the underside of the roof, direct access is required over the whole area. The surface should be brushed free of all loose debris and be relatively smooth. The insulation may be glued, nailed, or fixed with special clips and must be trimmed to fit around obstructions. Installation of rigid insulation is slower than spray but can be restricted to small areas at any one time for minimum inconvenience. Rigid insulation is particularly suited for large uncluttered areas without hung ceilings and will attenuate noise in addition to reducing heat loss. See Fig. 6.18.

Rigid insulation, batt insulation, or granular insulation may be laid on top of the ceiling in an attic space. This process requires access into the attic at only one point and can be easily and quickly accomplished. This addition will, however, reduce the temperature in the attic, and if pipes and ducts are run through this space, their heat loss will increase, and frost damage to

TABLE 6.10 APPROXIMATE INSULATION COSTS FOR ROOFS

Spray insulation exterior	$2.10/ft^2
Spray insulation interior	3.00/ft^2
Rigid insulation exterior flat roof	2.50/ft^2
Rigid insulation exterior sloping shingle	1.80/ft^2
Rigid insulation interior	3.00/ft^2
Attic insulation	1.60/ft^2

water pipes could result in cold climates. See Fig. 6.19. Approximate costs of various types of roof insulation are given in Table 6.10.

Although most insulating materials are of light weight and are unlikely to exceed the permissible loading of the existing structure, calculations should be made to check that the local building code is not infringed upon. The local fire department should also be queried on the acceptability of insulating materials. In some areas plastic insulation is considered a fire and smoke hazard. The local office of the Food and Drug Administration should be contacted to determine whether the insulation considered is listed as hazardous. This is particularly important in retail stores that sell food.

Soffits of roof overhangs and roof facias are frequently left uninsulated and cause high heat loss. They should be insulated at the same time as the main roof with either sprayed or rigid insulation. If the soffit is not accessible, a hole can be cut in the ceiling or in the soffit itself, and mineral fiber can be blown in. See Fig. 6.20.

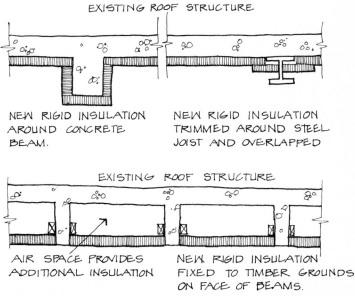

EXISTING ROOF STRUCTURE

NEW RIGID INSULATION AROUND CONCRETE BEAM.

NEW RIGID INSULATION TRIMMED AROUND STEEL JOIST AND OVERLAPPED

EXISTING ROOF STRUCTURE

AIR SPACE PROVIDES ADDITIONAL INSULATION

NEW RIGID INSULATION FIXED TO TIMBER GROUNDS ON FACE OF BEAMS.

Figure 6.18 Rigid insulation applied to structural ceilings.

ACTION GUIDELINES

☐ 1. If resources dictate only partial roof insulation, then insulate a 15-ft-wide band around the perimeter, rather than the entire roof. The roof area over the perimeter zones experiences the greatest heat loss.

☐ 2. When adding insulation to the roof outside surfaces, select a finish with a high absorption coefficient in areas where the yearly heating load exceeds the yearly cooling load. Select a low absorption coeffi-

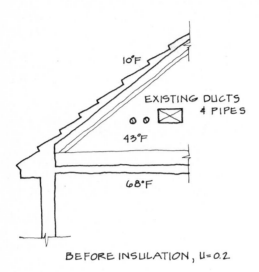

BEFORE INSULATION, U=0.2

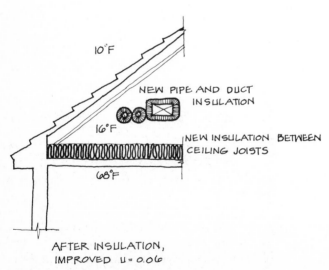

AFTER INSULATION,
IMPROVED U = 0.06

Figure 6.19 Typical *U* value and temperature changes with the addition of attic insulation.

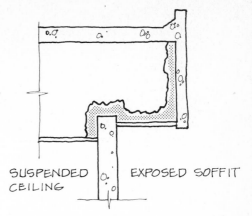

Figure 6.20 Insulation applied above ex-posed soffits.

cient in areas where the yearly cooling load exceeds the yearly heating load.

☐ 3. If the ceiling void is used as a return air plenum, insulation applied to the underside of the roof should be bonded or sealed to prevent migration of fibers and should not extend over the lighting fixtures.

☐ 4. If possible, include an air space between insulation boards and the underside of the roof to gain added insulation effect.

☐ 5. Insulate all pipes and hot ducts in an attic space if insulation is added on top of the attic floor, as the attic temperature will approach outdoor temperatures.

☐ 6. When adding insulation, provide vapor barriers on the interior surface of the roof of sufficient impermeability to prevent condensation.

☐ 7. Seal all corners to prevent infiltration.

☐ 8. Do not treat the roof with a black surface to absorb more radiation in the wintertime. It is relatively ineffective to do so, especially with good insulation.

☐ 9. Where equipment room or pipe space occurs directly under the roof, do not consider additional insulation.

☐ 10. If solar collectors are being considered for present or future installation on the roof, the amount and type of insulation must be taken into account. Where the existing roof is in good condition, consideration should be given to insulating the underside rather than the exterior of the roof or to insulating the void between the ceiling and the roof.

☐ 11. If reflective insulation is used, it should be selected with an equivalent

R value of not less than 4 and should be installed so that the shiny surface faces an air space of at least ¾ in thickness. Where possible, use board, batt, or fill-type insulation instead of reflective insulation to reduce heat loss.

To evaluate the benefits of adding insulation to a roof, it is first necessary to know the *U* value of the existing construction. A procedure for calculating and measuring existing *U* values for walls is outlined in Sec. 6.15 and should be used for roofs.

After the *U* value of the existing roof has been determined, a decision must be made as to whether additional insulation is required, how much should be added, what type of insulation should be used, and where it should be applied.

To calculate savings due to roof insulation, it is necessary to know the *U* values both before and after insulation is added. To determine the new *U* value, refer to Fig. 6.12. Enter the graph at the present *U* value, intersect with the appropriate line for new insulation thickness, and then read out the new *U* value.

To determine the heating energy saved by insulating roofs, refer to Fig. 6.21, then carry out the following procedure:

1. Note heating degree-days and mean daily solar radiation in langleys (Table 4.1).

2. Select the absorption coefficient for the outside surface of the roof.

3. Using the initial *U* value of the roof, enter the graph at the appropriate degree-days and, following the direction of the example line, intersect the appropriate points for langleys and *U* value. Read out the yearly heat loss per square foot of roof on the appropriate absorption coefficient line. Interpolate to obtain answers for absorption coefficients other than the 0.3 and 0.8 shown.

4. Repeat this complete procedure using the improved *U* value and subtract the answer from the energy used before insulating to derive the energy saving per square foot of insulated roof. Multiply this saving by the area of the insulated roof to determine the total yearly load reduction.

In addition to savings achieved by reducing winter heat losses, roof insulation will also reduce summer heat gains; this must be considered in a study of economic feasibility. In addition, when adding insulation to the outside surface of a roof, it is possible to change the absorption coefficient at the same time. A judgment must be made whether it is more advantageous to have a dark surface to enhance the heating effect of the sun in winter or to have a light surface and sacrifice the winter benefits in favor of reflecting unwanted solar gains in summer.

Figure 6.21 is based on the Sunset computer program which was used to calculate solar effect on roofs for 12 selected locations. The program calculates hourly solar angles and intensities for the twenty-first day of each month with radiation intensity values modified hourly by the average percentage of cloud cover taken from records. Heat losses are based on a 68°F indoor temperature.

The solar effect on a roof was calculated using sol-air temperature, and the heat entering or leaving a space was calculated using the equivalent temperature difference. Roof mass ranged from 25 to 35 lb/ft², and thermal lag averaged 3½ h. Additional assumptions were: (1) total internal heat gain of 12 Btu/ft², (2) outdoor-air ventilation rate of 10%, and (3) infiltration rate of 1.2

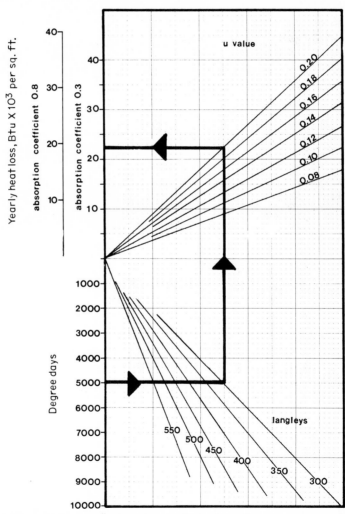

Figure 6.21   Winter heat losses through roofs.

air changes per hour. Daily totals were then summed for the number of days in each month to arrive at monthly heat losses. The length of the heating season for each location considered was determined from weather data and characteristic operating periods. Yearly heat losses were derived by summing monthly totals for the length of the cooling season.

Absorption coefficients and U values were varied and summarized for the 12 locations as shown in Table 6.10. The data were then plotted and extrapolated to include the entire range of degree-days.

6.17 REDUCE HEAT LOSSES THROUGH FLOORS

Heat loss from slab-on-grade floors occurs mainly around the perimeter of the slab; very little heat is lost from the center of the floor, owing to the insulating effect of the earth, an effect greatest in the center and least in the perimeter. Heat loss from suspended floors over an unheated space, however, occurs evenly over the whole floor area and is proportional to the difference between inside and outside temperatures.

Slab-on-grade floors should be insulated around the perimeter. This insulation should be placed vertically on the outside edge of the floor and should extend at least 2 ft below the floor surface. See Fig. 6.22.

Insulation materials may be rigid board or sprayed foam. Insulation below ground should be applied in a bedding of hot asphalt. Existing flashing at ground floor level may require extending to cover the insulation top edge.

The savings due to edge insulation cannot be predicted with any accuracy. If, however, condensation occurs on the floor perimeter in cold weather or if the floor surface temperature close to the outside wall is more than 10°F

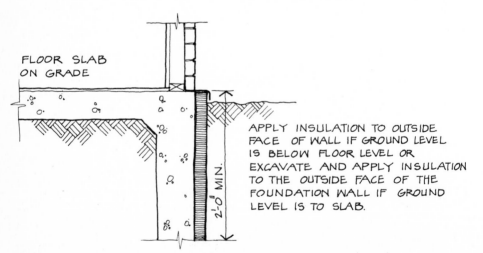

Figure 6.22 Perimeter insulation applied to exposed grade beams and slabs.

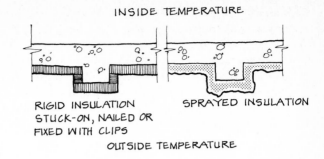

INSIDE TEMPERATURE

RIGID INSULATION
STUCK-ON, NAILED OR
FIXED WITH CLIPS

SPRAYED INSULATION

OUTSIDE TEMPERATURE

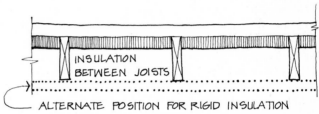

INSULATION
BETWEEN JOISTS

ALTERNATE POSITION FOR RIGID INSULATION

Figure 6.23 Insulation applied under exposed floors.

lower than the indoor temperature, then insulation will be beneficial and
should be added.

Suspended floors over an unheated space (garage, crawlway, etc.) may be
insulated on the underside by applying spray foam or rigid insulation as
described for roofs. See Fig. 6.23. Where suspended floors are over a closed-
in but unused space, apply insulation to the underside of the floor. If,
however, there may be some future use for the space, or if it contains pipes
and ducts, then apply insulation to the inside walls of the space, leaving the
floor uninsulated.

Table 6.11 shows the minimum acceptable U value for suspended floors
with exposed undersides in various degree-day zones. Existing floors below
these standards should be insulated. Approximate costs for various types of
floor insulation are given in Table 6.12.

ACTION GUIDELINES

☐ 1. If insulation is added, improve the U value to 0.08 or better.

☐ 2. Insulate slab-on-grade floors that are subject to condensation or that
have low surface temperatures.

☐ 3. Insulate slab-on-grade floors with edge insulation and protect insulation
with a waterproof seal.

TABLE 6.11 MINIMUM *U* VALUES FOR FLOORS

Heating Degree-days	Minimum Acceptable U Value
1000	0.40
2000	0.35
3000	0.30
4000	0.22
5000	0.22
6000	0.18
7000	0.18
8000 and above	0.12

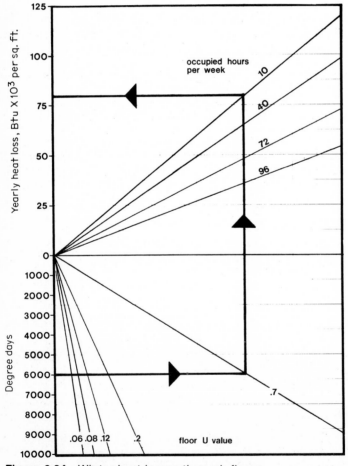

Figure 6.24 Winter heat losses through floors.

TABLE 6.12 APPROXIMATE COSTS OF FLOOR INSULATION

Edge insulation above grade	$1.70/ft²
Edge insulation below grade	2.10/ft²
Spray insulation below suspended floor	2.60/ft²
Rigid insulation below suspended floors	2.70/ft²

To evaluate the benefits of adding insulation to suspended floors, it is first necessary to know the present U value of the existing construction. A procedure for calculating and measuring these is outlined for walls in Sec. 6.15 and may be used for floors.

After the U value for the existing floor has been determined, a decision must be made as to whether additional insulation is required and if so, how much should be added and what type of insulation should be used.

1. To determine the new U value, refer to Fig. 6.12. Enter the graph at the present U value, intersect with the appropriate line for new insulation thickness, and then read out the new U value.

2. To determine the heating energy saved by insulating suspended slab floors, refer to Fig. 6.24 and carry out steps 3 to 5.

3. Note heating degree-days from Table 4.1 and determine the number of occupied hours per week.

4. Using the initial U value of the floor, enter the graph at the appropriate degree-days and, following the direction of the example line, intersect at the appropriate point for U value and occupied hours per week. Read out the yearly heat loss per square foot of floor area.

5. Repeat this complete procedure using the improved U value and subtract the answer from the energy used before insulating to derive the energy saving per square foot of insulated floor. Multiply this saving by the area of the insulated floor to find total yearly heat load reduction.

Heat losses determined from Fig. 6.24 are based on the assumption that the floor is over an unheated space which is at outdoor ambient air temperature. Since the load on the heating system during occupied hours is generally a small percentage of the total annual heating load, the figure gives heat loss during unoccupied times only, with 10 h occupied time per week being the maximum (158 h unoccupied). It was also assumed that the temperature during unoccupied time is set back to 55°F, and since degree-days are based on 65°F, all losses were multiplied by a factor of 55/65, to give this formula on which the figure is based:

$$Q \text{ (heat loss)} = \text{degree-days} \times 24 \times U \text{ value} \times 55/65 \times \frac{\text{unoccupied hours}}{168}$$

6.18 REDUCE INFILTRATION WITH BUILDING ALTERATIONS

Infiltration of cold outdoor air into a building through cracks, openings, and gaps around doors and windows increases the building heat load and is often responsible for as much as 25% of the yearly heating energy consumption.

The rate of infiltration is increased by the blast of wind on the surface of the building and in tall buildings by stack effect due to the difference between indoor and outdoor temperatures. Stack effect is enhanced by open vertical spaces such as staircases, elevator shafts, and service shafts, and in tall buildings the potential for stack effect is always present. It can be minimized only by thoroughly sealing openings through which air might enter and leave the building.

To maintain equilibrium, the quantity of air entering a building either by uncontrolled infiltration or by ventilation must equal the quantity of air leaving the building either by exfiltration or ventilation exhaust. Buildings frequently have unbalanced ventilation systems in which the rate of exhaust exceeds the rate at which air is introduced into the ventilation system. In these cases, infiltration increases to make up the balance beyond that level which would otherwise occur.

The fresh air ventilation and exhaust system should be examined, and a balance sheet drawn up. If the exhaust exceeds the outdoor air requirements, then the rate of exhaust should be reduced, which in turn will reduce infiltration. Ideally, the quantity of outdoor air for ventilation should exceed the total exhaust from the building by 10% to promote exfiltration rather than infiltration.

In buildings that experience heavy and continuous traffic through external doors, infiltration may be reduced by building vestibules for each external door to form an airlock. The vestibule should be sufficiently long so that the external door closes before the internal door is opened. Depending on the

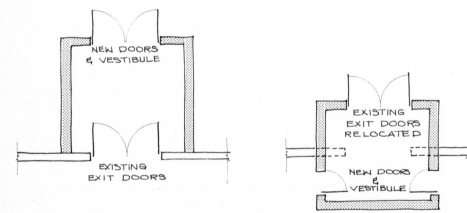

Figure 6.25 Possible vestibule alterations.

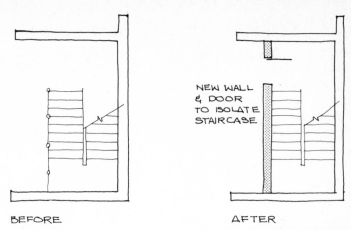

NEW WALL
& DOOR
TO ISOLATE
STAIRCASE

BEFORE AFTER

Figure 6.26 Stairwell enclosure for control of stack effect.

particular characteristic of the building, the vestibule may be constructed inside or outside the building. See Fig. 6.25.

Vestibule doors may be either manually operated and self-closing or, if traffic is particularly dense, automatically opened and closed by pressure pad or photoelectric cell.

Revolving doors may be fitted in place of existing doors, and although maximum reduction in infiltration will be achieved if they are used in conjunction with vestibules, they may also be used alone. When revolving doors are installed, it is always necessary to provide an additional hinged door for use by the handicapped and to allow passage of bulky objects. It is also necessary to ensure that installation of a vestibule or revolving door does not infringe on fire exit requirements.

For order-of-magnitude costs, a vestibule with double doors and automatic operators could be installed for $3000. To remove an existing door and to replace it with a revolving door and one single-swing door could cost $10,000 without a vestibule and $13,000 with the vestibule.

If sufficient money is not available for investment in vestibules or revolving doors, consider the installation of a wind screen to protect the external door from the direct blast of prevailing winds. Screens can be opaque, constructed cheaply from concrete block, or can be transparent, constructed of metal framing with armored glass or Lexan fill. Careful positioning can reduce infiltration through external doors by about 50% of that obtained by fitting a vestibule. The costs of wind shields vary but should be in the order of $600 for concrete block and $1000 for the glazed type.

To reduce the effect of potential stack action in vertical shafts and stairs, these areas should be sealed from the rest of the building. Open stairwells that connect with circulation spaces at each floor level should be provided with walls and self-closing doors to isolate them. See Fig. 6.26.

Access holes into vertical service shafts should be provided with gasketed covers. Holes at the shaft wall to allow the passage of pipes and ducts at each floor level should be sealed, and sleeves around the pipes and ducts should be packed with asbestos rope or other suitable material.

Elevator shafts are usually provided with a vent at the top into the equipment room. This vent is necessary as a smoke release but can be responsible for large air movements through the elevator doors up the shaft into the equipment room and then outside through badly fitted doors and windows in the equipment room. Although the equipment room is designated as unoccupied space, the windows and doors should be weatherstripped to reduce the effect of stack action in the elevator shafts. See Fig. 6.27.

Before isolating stairs from the remainder of the building, ensure that fire exit requirements are met. In addition, the walls of vertical shafts are usually required to have a high fire rating which applies equally to any access door. If access doors are replaced, make sure that these meet the fire rating requirements.

ACTION GUIDELINES

☐ 1. Weatherstrip doors and windows in elevator equipment rooms located at the top of the elevator shaft.

☐ 2. Reduce air volume to toilet rooms to 1 cfm/ft², to reduce air makeup requirements. In making the calculation, include only those areas in the toilet room proper. Do not include janitor storage, vestibules, and lounges.

☐ 3. In locations where strong winds occur for long durations, use external wind baffles at the external surface of the building to shield windows and doors.

☐ 4. Where operable windows are not used during the winter, the cracks can be sealed with removable tapes.

☐ 5. In areas where operable windows are not required, close them permanently and caulk the cracks. If windows are not required, block them off entirely with masonry or insulated wall panels. Where insulated wall panels are used, fit them tightly to effectively block infiltration.

☐ 6. Install vestibules in stores, office buildings, and religious buildings where doors are opened frequently throughout the day.

☐ 7. Install revolving doors with or inside of vestibules where exceptionally heavy traffic occurs and infiltrated air circulates freely to occupied spaces. When installing revolving doors, maintain a sufficient number of operable single or double doors to meet code requirements for safety exit.

☐ 8. Caulk cracks around window air-conditioning units or through-the-wall units which remain permanently installed year-round, and cover window air-conditioning units with plastic covers outside. Covers can be bought for about $5.00 each.

☐ 9. When installing new screens on windows and doors, select units to limit air leakage to 0.5 cfm/ft of crack when subjected to a wind pressure of 25 mi/h.

☐ 10. Seal off large openings in stair towers with masonry partitions or with a tight-fitting door.

☐ 11. Pressurize building with outdoor air, properly controlled and treated, to reduce uncontrolled infiltration.

☐ 12. In tall buildings, provide a sealed platform every five floors in vertical service shafts.

☐ 13. Weatherstrip doors and windows in elevator equipment rooms located at the top of the elevator shaft.

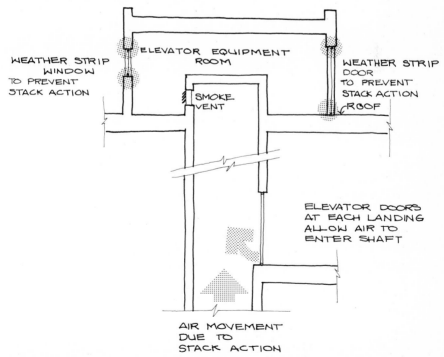

Figure 6.27 Elevator shaft isolation for control of stack effect.

To determine the energy saved by fitting vestibules or revolving doors, first determine the infiltration reduction achieved. To obtain this amount, refer to Fig. 6.28 and enter the graph at the estimated passages per hour, intersect the line for single-swing doors, and read out the rate of infiltration in 1000 cfm. Intersect also the appropriate line for a vestibule or revolving door and read out the new infiltration rate. Subtract this figure from the

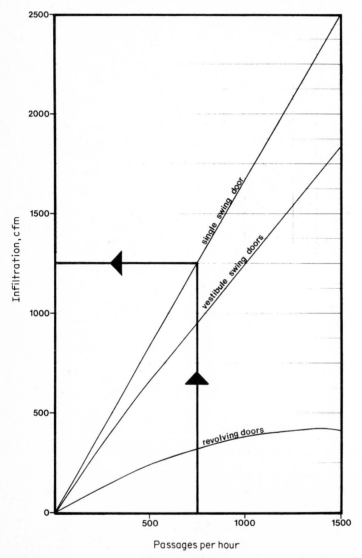

Figure 6.28 Entrance infiltration. (*Strock and Koral, Handbook of Air Conditioning, Heating, and Ventilation, 2nd ed., Table #1, 2-105.*)

infiltration rate for a single door to obtain the infiltration reduction in 1000 cfm. Refer to Fig. 6.3 and carry out the following procedure:

1. Note heating degree-days from Table 4.1, calculate the average number of hours per week in which the entrances are used, and figure the indoor temperature during occupied periods from Table 4.2.

2. Enter the graph at the appropriate degree-days and, following the direction of example line, intersect at the appropriate points with the indoor temperature and hours of occupancy.

3. Read out yearly heating energy per 1000 cfm.

4. Multiply this figure by the 1000-cfm infiltration reduction to obtain the net yearly heating load reduction.

In addition to the savings achieved by reducing winter heat loss, reduction of infiltration will also reduce sensible and latent heat gains in the summer. Depending on the building's location and the severity of the cooling season, this extra saving can be considerable and should be recognized when making an economic feasibility study.

CHAPTER SEVEN

COOLING

7.1 HEAT GAIN AND HEAT LOSS: INTRODUCTION

In the winter the sun is an asset in heating, but in the summer in almost all climates comfort dictates that it be kept out. With the changing path of the sun through the seasons, this is not particularly difficult on south facades; the sun is low and drives on through in the winter and is higher and is easily shaded in the summer. On east and west facades selected shading is far more complex. Ideally, controls should be exterior to the glass to prevent trapping with the greenhouse effect (glass is transparent to the high-frequency radiation of the sun but is almost opaque to the lower frequencies re-radiated from screen or shade), but no one set of fixed controls will adapt equally to particular needs of the day and seasonal average requirements. Indeed, because of the lag of the seasons a full month and a half behind the swing of the sun from low in the sky to high and back again, it is impossible to shade with fixed controls to satisfy both spring and fall requirements. Future building design will require ingenious devices that follow the sun or can be adjusted from within the building to meet user needs.

Windows were originally conceived to provide light and view. With the advent and common expectation of artificial light sources, however, we have come to ask of a window only view because we keep the lights on all the time. Highly reflective and absorbent glasses are well suited to this situation: they inhibit the passage of the sun's energy but do not prevent viewing. But artificial sources add heat to interior space as well as light. This energy is a significant load to the cooling system. For perimeter spaces the ideal in temperate and warmer climates is clear-glass natural lighting when conditions permit, but with protection by exterior shading from the direct rays of the sun. Several of the following subheadings speak to these interests of sun acceptance in the winter, sun rejection in the summer, and natural lighting.

Artificial light, equipment, and people all add heat to interior spaces that

161

must be removed in the summer by the cooling system. Although solar heat gain helps counter heat loss in the winter, such heat losses are more efficiently (and more cheaply) offset with boiler or furnace; thus, consideration for winter heat loss should not normally deter efforts to reduce solar heat gain.

As with heating, dramatic savings in cooling energy are available immediately by simply altering the levels of air temperature, humidity, and ventilation maintained. The following paragraphs begin with these measures.

7.2 INCREASE LEVELS OF INDOOR TEMPERATURE AND RELATIVE HUMIDITY

The air-conditioning systems in many buildings were designed to maintain 72°F DB and 50% RH during peak loads in the cooling season. They are usually operated to maintain those levels at peak conditions and to achieve even lower levels during the part-load conditions which occur most of the time; these settings are lower than necessary for comfort. Suggested dry-bulb temperature and relative humidity settings for various spaces are listed in Table 7.1 to reduce the cooling load and the energy and operating costs.

Even greater savings can be obtained without serious discomfort by maintaining higher levels in areas which are occupied only for short periods of time or by shutting down individual cooling units of unoccupied spaces such as auditoriums and cafeterias.

The entire air-conditioning system should be shut down at night and during days when the building is unoccupied. If, following shutdown, the air-conditioning system is not capable of re-establishing the desired temperature and humidity in hot spells, controls can be installed to activate the equipment a few hours before occupancy instead of operating the system all night or throughout the entire weekend.

Increasing the indoor temperature and humidity levels from 74°F DB and 50% RH to 78°F DB and 55% RH will save approximately 13% of the energy required for cooling. The exact amount saved depends upon the amount of ventilation air and infiltration which enters the building, the conduction losses, solar heat gain, internal loads of the building, and the type of air-conditioning system. With one particular system, terminal reheat, the indoor-space conditions should be maintained at lower levels to reduce the amount of reheat and to save energy. This problem is unique and is discussed in Chap. 10.

ACTION GUIDELINES

☐ 1. If spaces such as auditoriums and cafeterias are unoccupied for more than 30% of the day, shut off the cooling to such spaces when not in use. Shut down cooling or condenser fans, chilled water and condenser

TABLE 7.1   SUGGESTED INDOOR TEMPERATURE AND HUMIDITY LEVELS IN THE COOLING SEASON

	Dry-Bulb Temperature *	*Minimum Relative Humidity*
COMMERCIAL BUILDINGS, OCCUPIED PERIODS		
Offices	78	55
Corridors	Uncontrolled	Uncontrolled
Cafeterias	75	55
Auditoriums	78	50
Computer rooms	75	As needed
Lobbies	82	60
Doctors' offices	78	55
Toilet rooms	80	
Storage, equipment rooms	Uncontrolled	
Garages	Do not cool or dehumidify	
RETAIL STORES, OCCUPIED PERIODS		
Department stores	80	55
Supermarkets	78	55
Drug stores	80	55
Meat markets	78	55
Apparel stores	80	55
Jewelry stores	80	55
Garages	Do not cool	

*Except where terminal reheat systems are used.

pumps, and supply and exhaust fans as well as chillers or compressors at night and on weekends. If the cooling tower fans and pumps are interlocked with the compressors, they will shut off automatically when the compressors are shut off. Shut down boilers which supply steam or hot water to absorption units; do not maintain boiler water temperature or steam pressure when absorption units are inoperative.

□2. Install a control to prevent operation of the refrigeration system for the purpose of dehumidification until relative humidity levels exceed 60%.

□3. Do not operate the heating system during the cooling season to raise the space temperatures to 78°F, or to other temperature levels listed in Table 7.1 during occupied cycles.

□4. Install a 7-day timer programed for normal daytime operation of chillers, compressors, cooling towers, and chilled water pumps, with a cam set to shut off all equipment and blowers during holidays and weekends.

Energy is expended in air conditioning for lowering the temperature of the air and for lowering the relative humidity. A first approximation of the energy saved by raising the interior temperature to those recommended in Table 7.1 is 3000 Btu/(h)(°F) per cfm of outdoor ventilation and infiltration. To determine the annual energy saved by raising the relative humidity at a constant 78°F DB, refer to Fig. 7.1. Enter Fig. 7.1 with wet-bulb degree-hours above 66°F WB (from Fig. 4.5). Intersect the existing relative humidity, then the hours of cooling system operation per week. Read Btu per year per 1000 cfm of ventilation and infiltration air.

Repeat this operation with the new relative humidity and read a new Btu requirement. The difference in the two readings is the energy saved per 1000 cfm.

Energy used is a function of the WB degree-hours above the base of 66°F, the RH maintained, and the number of hours of controlled humidity. The base RH is 50% which is approximately 78°F DB, 66°F WB. Figure 7.1 expresses the energy used per 1000 cfm of air conditioned or dehumidified. An analysis of the total heat content of air in the range under consideration indicates an average total heat variation of 0.93 Btu/lb for each degree WB

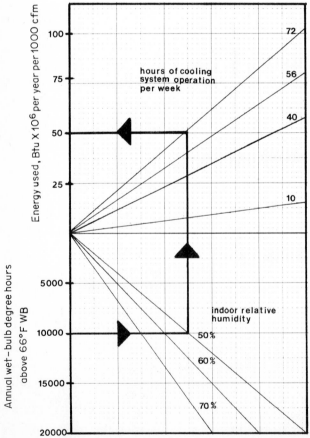

Figure 7.1 Summer energy required for dehumidification.

change at constant DB temperature and that the total heat varies nearly directly with RH. One thousand cubic feet per minute is equal to 4286 lb/h, and so 1°F WB per hour is equal to 4286×0.93, or 3986 Btu. The lower section of Fig. 7.1 shows the direct relationship from the base of 50% RH, and the upper section proportions the hours of system operation, with 56 hours per week being 100%.

7.3 REDUCE VENTILATION

Except where outdoor air can be used productively for economizer and enthalpy cooling (and the temperature and enthalpy are suitable as described in Sec. 7.9), cooling and dehumidifying outdoor air consumes large quantities of energy. Any measure that reduces the quantity of ventilation to a minimum and is compatible with physiological needs will result in substantial energy and dollar savings. Recommended ventilation standards for both summer and winter periods are listed in Table 6.2. The suggestions for ventilation discussed in Chap. 6 apply generally to the cooling season as well. It pays to shut down the ventilation system when the building is unoccupied and to reduce ventilation rates when the building is occupied.

During the cooling season, the enthalpy difference between the outdoor and indoor air determines the quantity of energy required to cool and dehumidify the ventilating air to indoor conditions. As enthalpy, or total heat, is very closely approximated by wet-bulb temperature, the difference in wet-bulb temperature between outdoor air and room conditions can be used to determine the energy required for cooling and dehumidification. Figure 7.1 expresses this relationship.

ACTION GUIDELINES

☐ 1. Refer to Secs. 6.6, 6.7, and 6.11. These recommendations for the reduction of ventilation during the heating season apply as well to the cooling season.

☐ 2. Do not reduce ventilation rates when outdoor air can be used for economizer or enthalpy cooling.

☐ 3. When the enthalpy of the outdoor air, on a seasonal basis, is lower than room conditions, outdoor-air dampers should be fully opened.

☐ 4. Check local and state codes to determine whether ventilation rates can be legally lowered when spaces are air conditioned.

☐ 5. Rebalance air supply systems to provide a maximum of 1 cfm supply air per square foot for toilet rooms and adjust the exhaust system accordingly.

☐ 6. Install charcoal filters or other air treatment devices in the exhaust air systems and circulate treated air back into the space to reduce the amount of outdoor air required for makeup for all systems handling 2000 cfm or more. Supply slightly more outdoor air than exhaust air to pressurize the conditioned space.

☐ 7. Install damper and controls to permit the ventilation system to delay the introduction of outdoor air for 1 h in the morning after opening time and to shut it off ½ h before closing time. A 7-day timer can be used in stores to reduce or shut off ventilation automatically for periods during the day with light occupancy.

☐ 8. Modify duct systems and hoods and introduce untreated outdoor air directly to exhaust hoods. Weigh this against changing hoods to new high-velocity models which require less makeup air.

Energy reduction obtained by reducing ventilation rates can be quantified by using Fig. 7.1. Enter Fig. 7.1 with annual wet-bulb hours above 66°F WB (from Fig. 4.5). Intersect indoor relative humidity, then the hours of ventilation system operation per week. Read Btu per year per 1000 cfm of reduced ventilation.

7.4 REDUCE INFILTRATION

Outdoor air introduced for ventilation imposes a cooling load on the refrigeration equipment only, but infiltration, the leakage of unwanted air directly into conditioned spaces, imposes a load on both the refrigeration cooling system and the distribution system. Air leaking into a space increases the sensible heat gain which requires an increase in chilled-water or cool-air supply with a corresponding increase in power consumption. In addition, because infiltration is a problem 24 hours a day, loads build during the night while HVAC equipment is off; the loads must be removed the next day.

In Chap. 6 recommendations are made to reduce ventilation and infiltration. Such measures taken to reduce heating loads will also reduce cooling loads. In climates with 10,000 or more wet-bulb degree-hours above 66°F WB, with coincidental DB temperatures higher than 78°F (Fig. 4.5), the measures to reduce infiltration and ventilation should be considered for cooling regardless of benefits during the heating season.

From all causes, the infiltration rate for most buildings is between ½ and 1½ air changes per hour, depending upon the condition of the doors and windows, length of cracks, height of the building, and number of door openings. Most of this infiltration is through window cracks, but the rate of infiltration in stores through door openings may be considerably higher than through windows or cracks. The rate can be quantified with Fig. 6.4, with entrants of window type and total crack on the windward sides of the building.

ACTION GUIDELINES

□ 1. Refer to Secs. 6.8 and 6.18. These recommendations for the reduction of infiltration during the heating season apply as well to the cooling season.

□ 2. Before investing any money to reduce infiltration rates, perform an analysis for both heating and cooling.

□ 3. In areas of the country which experience fewer than 10,000 wet-bulb degree-hours, energy conserved by weatherstripping windows and doors and caulking window frames does not generally justify the cost for cooling alone.

□ 4. Install revolving doors, or vestibules, where analysis shows that outdoor air passages through door openings exceed more than 10% of the cooling load.

□ 5. Seal vertical stacks in office buildings of six or more stories when there are 10,000 or more wet-bulb degree-hours during occupied periods.

Energy reduction obtained by reducing infiltration can be quantified by using Fig. 7.1. Enter Fig. 7.1 with annual wet-bulb hours above 66°F WB (from Fig. 4.5). Intersect indoor relative humidity, then the hours of cooling system operation per week. Read Btu per year per 1000 cfm of reduced infiltration.

7.5 REDUCE SOLAR HEAT GAINS THROUGH WINDOWS

Direct and diffuse solar radiation through windows is the largest single component of window heat gains. The conduction component, by comparison, is relatively small. For this reason, converting single glazing to double glazing for the sole purpose of reducing the heat gain by conduction is rarely worthwhile. Converting single glazing to double for the benefits that can be obtained in the heating season is, on the other hand, quite valuable, and if this conservation opportunity has been implemented, there will be an additional advantage of reduced conduction gains in the summer. Thus, these gains should be included in any economic analysis of double glazing made on a yearly basis.

Solar radiation places a major demand on the cooling load in all parts of the country if the ratio of window to wall area on the east, west, and south facades is 20% or more with clear unshaded glass, and 50% or more with radiation-reducing unshaded glass. In addition, solar radiation on the roof and on the east, west, and south walls increases the outside surface temperature. The increase in heat transmission through the surfaces which results adds to the conduction which would have resulted alone from temperature difference between indoors and outdoors.

Solar radiation on skylights and windows is transmitted through the glass almost instantaneously. Annually, almost 80% of all the solar radiation striking a vertical single sheet of clear glass surface is transmitted through it. This is an addition to the cooling load but a reduction to the heating and lighting loads (if lights are turned off when daylight is available). Conservation measures that reduce solar heat gain include shading the interior or exterior of the glass and treating it to increase its reflective properties. Be careful, however, of devices that prevent solar radiation from entering the building in cold weather.

Various devices are available which can be installed outside the window, inside the window, or on the window surface itself to reduce solar gain. Refer to Table 7.2 for shading coefficients, a measure of the percentage reduction of solar heat gain. External sunscreens which prevent direct sunlight from falling upon the glass surface are more effective than internal shading in controlling solar heat gain. Internal shading devices are less effective because only a small proportion of the sunlight can be reflected back through a window. External shading may be provided by eyebrows fitted over the tops of windows or fins fitted at the sides of windows; the size

TABLE 7.2   SHADING COEFFICIENTS

Glass	*Coefficient*
⅛-in Clear double strength	1.00
¼-in Clear plate	0.93–0.95
¼-in Heat-absorbing plate	0.65–0.70
¼-in Reflective plate	0.23–0.56
¼-in Laminated reflective	0.28–0.42
1-in Clear insulating plate	0.80–0.83
1-in Heat-absorbing insulating plate	0.43–0.45
1-in Reflective insulating plate	0.13–0.31

	Coefficient	
Shading Device	*With ¼-in Clear Plate Glass*	*With 1-in Clear Insulating Glass*
Venetian blinds: light colored, fully closed	0.55	0.51
Roller shade: light colored, translucent, fully drawn	0.39	0.37
Drapes: semi-open weave, average fabric transmittance and reflectance, fully closed	0.55	0.48
Reflective polyester film	0.24	0.20
Louvered sun screens		
23 louvers/in	0.15–0.35	0.10–0.29
17 louvers/in	0.18–0.51	0.12–0.45

and disposition of these must be determined for each case on the basis of the sun's altitude and azimuth for that particular location. Eyebrows are most effective over windows with a southern orientation and can be designed to provide total shading from high-altitude sun during the middle of the day. They are not as effective as side fins, however, for east and west orientations.

External louvered sunscreens may be fitted close to the outside surface of the windows and, where possible, within the window reveal itself. Operable glazing may permit installation of external sunscreens without the use of special staging. The louvered sunscreens may be fixed permanently in position, arranged to slide in channels, or made removable so that advantage can be taken of the sun's beneficial heating effect in winter.

Tinted or reflective glass may be either in addition to the existing glazing to convert it from single to double or may be used to replace the existing glazing. Reflective polyester films may be applied direct to the inside surface of the glass. These films are self-adhesive and require careful application to avoid bubbles or trapped air and formation of Newton rings. The film is easily abraded which reduces performance and appearance; film on the inside of the building subjected to normal wear and tear will have a reduced useful life. If double glazing is used, apply the film to the inside surface of the outer layer of glass.

Available in different colors, the films, depending on quality and make, allow a maximum transmission of from 55 to 90 Btu/(h)(ft^2) [The maximum heat transmitted through a clear single pane of glass is typically 215 Btu/(h)(ft^2).] The films transmit 9 to 33% of the visible light spectrum and reflect 5 to 75% of the solar radiation which strikes them. Fifty percent reflective coatings on south, east, and west exposed glass will save from 30,000 to 50,000 Btu per square foot of window per hot weather season for energy required for space cooling. (The value is higher in northern latitudes and lower in southern latitudes because of the angle at which the sun strikes vertical surfaces.)

Cooling loads can be reduced by 36,000 Btu/ft^2 of glass per year by proper use of shading devices. In southern climates, the north-facing glass can receive a surprising amount of diffuse solar radiation. If heat gain from north windows is excessive, they should be treated similarly to the other exposures.

Internal shades of various types, e.g., venetian blinds, roller shades, and drapes, may be fitted to the inside of the windows. If drapes are used, they should be fire resistant and preferably of woven fiberglass which has a high reflectance. Although internal shades are not as effective as other methods of solar control, they are relatively inexpensive and are easily adjustable. Shades need be drawn only when excessive solar gains are present; at other times they can be opened to allow full view through the window and maximum use of natural light. See Table 7.3 for costs.

Table 7.4 is the result of a computer program developed to study the reaction of single- and double-glazed windows under varying weather condi-

TABLE 7.3 SOLAR CONTROL COSTS

Solar Control Device	Average Installed Cost, dollars/ft²
External louvered screens	5.50
Tinted or reflective glass	6.00
Reflective polyester film	1.65
Venetian blinds	1.70
Vertical louver blinds	2.75
Roller shades	.85

tons for 12 cities chosen to give a typical cross-section of climates. Table 7.4 tabulates the yearly heat gain in Btu per square foot of glazing for the 12 selected locations. Heat gain through windows is maximum in areas that experience high levels of solar radiation and a large number of cooling degree-hours, but for any one geographic location heat gain is modified by orientation. Summer heat gain is greatest through windows on east and west facades, followed by south-facing windows and least in windows on the north.

Two graphs, Fig. 7.2 and 7.3, were extrapolated from the data in Table 7.4 for other climates to allow prediction of annual solar heat gain only, with entrants of cooling degree-hours above 78°F DB, solar radiation in langleys, and orientation. These two graphs were prepared for latitudes greater than 35° and latitudes less than 35°. The graphs are based on solar gains received from 9 A.M. to 3 P.M. sun time, 5 days per week, when cooling systems are in operation. When occupancy varies significantly, the solar heat gain must be adjusted proportionately. For example, if a building is occupied 6 days per week, the solar gain obtained from the graph is increased by 20% to allow for the extra day.

Total heat gain through a window in any given location is the sum of the solar and conduction loads. The graph in Fig. 7.4 indicates the conduction component of heat gain through windows and allows prediction of annual conduction gain based on cooling degree-hours above 78° DB occupied hours and single-, double-, or triple-glazed sash.

ACTION GUIDELINES

☐ 1. In hot weather, adjust existing blinds, drapes, shutters, or other shading devices on windows to prevent penetration of solar radiation into the building.

☐ 2. Install blinds, drapes, shutters, or other shading devices on the inside of all south-, east-, and west-facing windows which are subject to direct sunlight in hot weather or exposed to a large expanse of sky.

TABLE 7.4 HEAT GAIN THROUGH SINGLE- AND DOUBLE-GLAZED WINDOWS

City	Latitude	Solar Radiation, langleys	DB Degree-hours above 78°F	Heat Gain, Btu/(ft²)(yr) North Single	North Double	East and West Single	East and West Double	South Single	South Double
Minneapolis	45°N	325	2,500	36,579	33,089	98,158	88,200	82,597	70,729
Concord, N.H.	43°N	300	1,750	33,481	30,080	91,684	82,263	88,609	76,517
Denver	40°N	425	4,055	44,764	39,762	122,038	108,918	100,594	85,571
Chicago	42°N	350	3,100	35,595	31,303	93,692	83,199	87,017	74,497
St. Louis	39°N	375	6,400	55,242	45,648	130,018	112,368	103,606	85,221
New York	41°N	350	3,000	40,883	35,645	109,750	97,253	118,454	102,435
San Francisco	38°N	410	3,000	29,373	28,375	88,699	81,514	73,087	64,169
Atlanta	34°N	390	9,400	59,559	50,580	147,654	129,391	106,163	87,991
Los Angeles	34°N	470	2,000	47,912	43,264	126,055	112,869	112,234	97,284
Phoenix	33°N	520	24,448	137,771	97,565	242,586	191,040	211,603	131,558
Houston	30°N	430	11,500	88,334	72,474	213,739	184,459	188,718	156,842
Miami	26°N	451	10,771	98,496	79,392	237,763	203,356	215,382	179,376

☐ 3. Use lightweight drapes with reflective properties for effective solar radiation control.

☐ 4. Use vertical or horizontal reflective blinds. Vertical blinds are generally more effective on the west and east sides of a building, and horizontal blinds are more effective on the south side. Polyvinyl chloride (PVC) vertical louvers cost between $2.90 and $2.30/ft².

☐ 5. Add a reflective film coating to the inside surface of glazed areas on the south, west, and east windows.

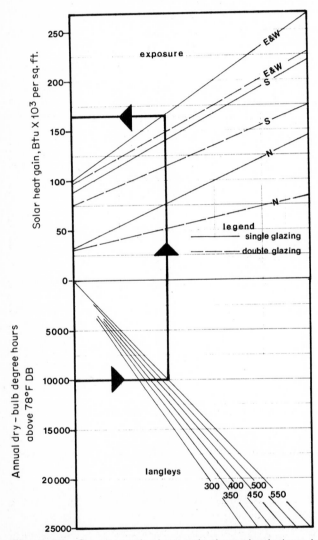

Figure 7.2 Summer solar heat gain through windows in latitudes 25 to 35°N.

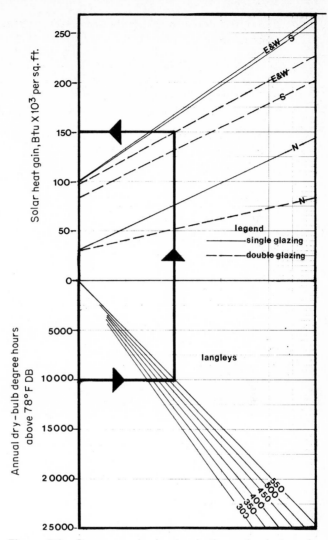

Figure 7.3 Summer solar heat gain through windows in latitudes 35 to 45°N.

☐ 6. Skylights transmit between two and four times as much solar heat in the summer as an equal area of east- or west-facing glass. Install solar controls over skylights that permit solar radiation and daylight to enter when desirable. Interior shades with Mylar polyester film reflective coatings cost about $3/ft² to install and reduce solar heat gain in the summer by up to 80%. White paint on the exterior of skylights, venetian blinds, or sunscreens can also be used to reduce solar heat gain.

☐ 7. Do not prune trees which shade the building in the summer.

☐ 8. Install horizontal fixed or movable eyebrows over south-facing glass. Install vertical fin sunscreens on the exterior or east- and west-facing windows. All exterior solar control devices should provide 100% shading from May 1 to October 31.

☐ 9. Install awnings over windows to block the summer sun. Awnings are particularly suitable to shade large display windows in stores.

☐10. Replace single clear glass on the east, west, and south facades with reflective glass when replacing broken windows if the windows are in direct sunlight 200 h or more during the cooling season. Replace all single glass on the west, east, south with reflective glass if windows are

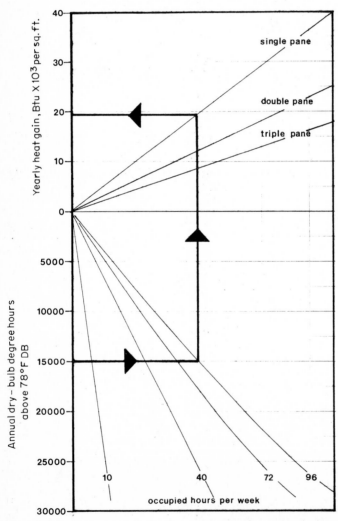

Figure 7.4 Summer conduction heat gain through windows.

in direct sunlight 500 h or more per year during the cooling season. (Reflective glass is more effective in reducing heat gain than is heat-absorbing glass.)

☐ 11. If double windows are added to reduce heat loss in the wintertime, provide the outer layers with reflective glass to reduce transmission at peak conditions to 50% of maximum.

☐ 12. When buildings are altered for expansion, consider adding a zigzag wall with the glazing in the west wall and east wall facing south in the northern climates. Such a wall can be self-shading in the summer while permitting the sun to penetrate through the glass during the heating season.

To compute the energy saved by shading devices, refer to Figs. 7.2 to 7.4 and proceed as follows:

1. Determine from Fig. 4.4 the annual degree-hours when the DB temperature is greater than 78°F.

2. Enter Fig. 7.4 at degree-hours and follow the direction of the example line to intersect with occupancy and glazing type to read yearly heat gain due to conduction.

3. Determine from Fig. 4.2 the mean daily solar radiation in langleys and select the orientation of the window.

4. Select from Fig. 7.2 or 7.3 the apropriate graph for latitude and enter the graph at degree-hours. Follow the direction of the example line to intersect with langleys and orientation.

5. Read out the yearly cooling energy requirements due to solar gain only, for 1 ft² of window, for occupancy of 5 days per week, 12 h per day. Modify this yearly energy proportionately if occupied time is different.

6. Multiply the solar gain by the shading coefficient of the selected window treatment to obtain the annual reduced solar gain and subtract this value from the original solar gain value to obtain yearly cooling load reduction for each square foot of treated window. Add the conduction heat gain.

7. Repeat this procedure for other orientations, interpolating for orientations different from the four cardinal points.

Figures 7.2 and 7.3 are based on the Sunset computer program which was used to calculate solar effect on windows for 12 locations. The program calculates hourly solar angles and intensities for the twenty-first day of each month. Radiation intensity values were modified by the average percentage of cloud cover taken from weather records on an hourly basis. Heat gains are based on a 78°F indoor temperature. During the cooling season, internal gains, ventilation, infiltration, and conduction into the building create cooling loads. The additional load caused by heat gain through the windows was calculated for each day. Daily totals were then summed for the number of days in each month to arrive at monthly heat gains. The length of the cooling season for each location considered was determined from weather data and characteristic operating periods. Yearly heat gains were derived by summing monthly totals for the length of the cooling season. These are summarized in Table 7.4 for the 12 locations. The conduction heat gain components through windows read from Fig. 7.4 were deducted from the total heat gains to derive the solar component. The solar component was then plotted and extrapolated to include the entire range of degree-hours. Figure 7.2 was derived from locations with latitudes between 25 and 35°N; Fig. 7.3, with latitudes between 35 and 45°N. The heat gains assume that the windows are subjected to direct sunshine. If the windows are shaded, gains should be read from the north exposure line. The accuracy of the graph diminishes for locations with less than 5000 degree-hours.

Figure 7.4 is based on degree-hours read from Fig. 4.4 which has a base of 56 hours per week. The following formula is used:

$$Q \text{ (heat gain)/year} = \text{degree-hours/year} \times U \text{ value}$$

U values assumed were 1.1 for single pane, 0.65 for double pane, and 0.47 for triple pane. The major portion of degree-hours occurs between 10 A.M. and 3 P.M.; hence, for occupancies between 10 and 56 hours per week, the degree-hour distribution can be assumed to be linear. However, for occupancies greater than 56 hours per week, the degree-hour distribution becomes non-linear, particularly in locations with greater than 15,000 degree-hours. This is reflected by the curves for 72 and 96 hours per week occupancies.

7.6 REDUCE HEAT GAINS THROUGH WALLS

The transmission of heat through an opaque wall is dependent on the difference between indoor and outdoor temperatures and the degree of heating of the outside surface by the sun (a function of how much sun, the orientation of the wall, and the absorption coefficient of the wall surface).

Heat flow is retarded by a well-insulated wall, measured as low in U value. Heat flow is also retarded by reflective wall surfaces of low absorption coefficient. If the building has walls of high U value and high absorption coefficient, it is worthwhile to consider reducing the absorption coefficient by refinishing the outside surface of the wall with a lighter color. Unlike the addition of insulation, refinishing with a lighter color will have an adverse effect on winter heat loss; the beneficial effect of the sunshine in winter will be reduced. Dark-colored walls with a high absorption coefficient (0.7 or greater) may be resurfaced or coated to reduce the absorption coefficient. Silver is best, but white may be more acceptable if reflections on the surrounding properties are objectionable. Theoretically, it is possible to reduce the external absorption coefficient to 0.1, but in practice dirt accumulates on building surfaces, and an absorption coefficient of 0.3 for white and 0.5 for light colors (pastel shades of yellow, green, etc.) is more realistic.

Table 7.5 is the result of a computer program developed to study the effect of varying solar conditions on heat gain through walls for 12 cities chosen to give a typical cross-section of climates. It tabulates the heat gain in Btu per year due to solar gain for each square foot of wall for the 12 selected locations with U values of 0.38 and 0.1, and outside surface absorption coefficients of 0.3 and 0.8.

Figures 7.5 and 7.6 were interpolated from data in Table 7.5 to show solar heat gain for other climates to allow predictions of annual heat gain with

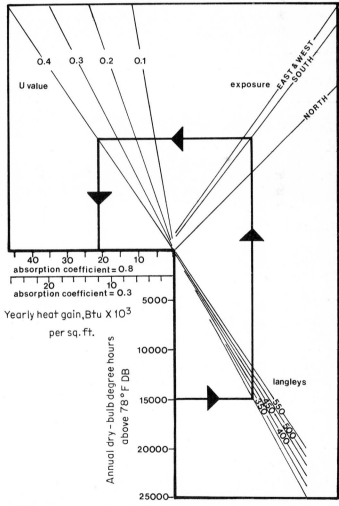

Figure 7.5   Summer solar heat gain through walls in latitudes 25 to 35°N.

TABLE 7.5 HEAT GAIN THROUGH WALLS

City	Latitude	Solar Radiation, langleys	DB Degree-hours above 78°F	Heat Gain, Btu/(ft²) (yr)												
				North				East and West				South			
				U = 0.39		U = 0.1		U = 0.39		U = 0.1		U = 0.39		U = 0.1	
				a = 0.3	a = 0.8	a = 0.3	a = 0.8	a = 0.3	a = 0.8	a = 0.3	a = 0.8	a = 0.3	a = 0.8	a = 0.3	a = 0.8
Minneapolis	45°N	325	2,500	364	2,442	19	390	1,346	7,665	164	1,747	1,601	7,439	164	1,574
Concord, N.H.	43°N	300	1,750	141	1,950	0	180	787	6,476	41	1,264	1,222	7,093	59	1,179
Denver	40°N	425	4,055	321	2,476	0	291	1,361	8,450	66	1,597	1,513	8,138	78	1,301
Chicago	42°N	350	3,100	503	2,500	46	429	1,492	7,889	233	1,835	1,698	8,088	225	1,793
St. Louis	39°N	375	6,400	2,246	5,966	419	1,386	4,165	14,116	950	3,571	3,994	12,476	779	3,074
New York	41°N	350	3,000	906	3,751	103	820	2,394	10,278	477	2,651	2,626	11,185	420	2,707
San Francisco	38°N	410	3,000	0	0	0	0	0	3,268	0	262	43	3,459	0	297
Atlantia	34°N	390	9,400	1,901	5,806	309	1,301	3,882	14,658	812	3,609	3,422	12,085	634	2,897
Los Angeles	34°N	470	2,000	0	774	0	10	180	6,575	0	889	527	7,182	0	980
Phoenix	33°N	520	24,448	17,448	24,423	4,749	6,526	21,461	36,937	5,784	9,868	20,880	34,728	5,502	9,322
Houston	30°N	430	11,500	5,002	10,687	1,178	2,643	7,895	22,431	1,981	5,521	6,985	20,893	1,605	4,713
Miami	26°N	451	10,771	7,507	15,717	1,912	4,052	12,358	31,745	3,164	8,416	11,778	29,906	2,814	8,057

entrants of cooling degree-days, solar radiation, orientation, U values, and absorption coefficients.

Figure 7.7 gives the conduction component of wall heat transmission due to the difference of indoor and outdoor temperature without solar effects.

Insulate exterior east-, west,- and south-facing walls if they are in direct sunlight during the entire year and the U value is greater than 0.20, if the wall mass is less than 10 lb/ft², or if the peak heat gain is greater than 30 Btu/(h)(ft²). Where the materials and aesthetics are not jeopardized, paint exterior sun walls with white reflective paint to reduce the absorption coefficient and hence the radiation heat gain.

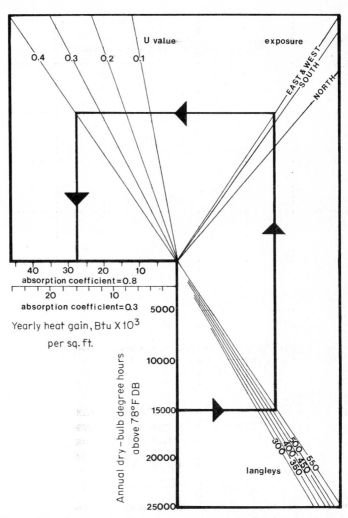

Figure 7.6   Summer solar heat gain through walls in latitudes 35 to 45°N.

To determine savings due to insulation, refer to Figs. 7.5 to 7.7 and proceed as follows:

1. Determine from Fig. 4.4 the annual cooling degree-hours above 78° DB.

2. Determine the initial U value of the wall and the improved U value after insulation with procedures from Sec. 6.15.

3. Refer to Fig. 7.7 with both U values, enter the graph at degree-hours and, following the direction of the example line, intersect occupancy and U values to read annual heat gain due to indoor-outdoor temperature differential above.

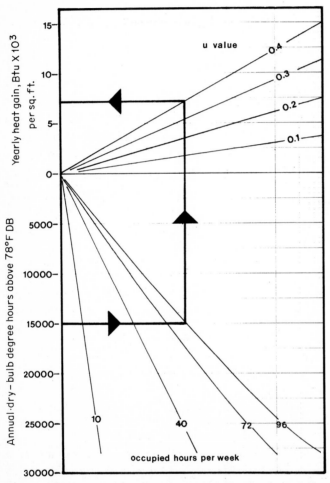

Figure 7.7 Summer conduction heat gain through walls, roofs, and floors.

4. Refer to Fig. 7.5 or 7.6, depending upon latitude, and estimate the absorption coefficient of the outside surface of the wall.

5. Enter the graph at the selected degree-hours and, following the direction of the example line, intersect langleys, orientation, and U value.

6. Read out the yearly heat gain due only to solar radiation per square foot from the appropriate absorption coefficient line. Interpolate the results for absorption coefficients other than the 0.3 and 0.8. Add the heat gain due to conduction to the heat gain due to solar radiation to obtain the total heat gain for the insulated wall.

7. Subtract the improved heat gain from the initial heat gain to obtain the annual load reduction due to insulation.

8. Repeat the complete procedure for each orientation of wall where insulation is being considered, interpolating for orientations other than the four cardinal points.

Figures 7.5 and 7.6 are based on the Sunset computer program which was used to calculate solar effect on walls for 12 selected locations. The program calculates hourly solar angles and intensities for the twenty-first day of each month. Radiation intensity values were modified by the average percentage of cloud cover taken from weather records on an hourly basis. Heat gains are based on a 78°F indoor temperature.

The solar effect on a wall was calculated using sol-air temperature, and the heat entering or leaving a space was calculated with the equivalent temperature difference. Wall mass ranged from 50 to 60 lb/ft², and thermal lag averaged 4½ h. During the cooling season internal gains, ventilation, infiltration, and conduction through the building skin create a cooling load. The thermal load caused by conduction through the walls was calculated for each day. Daily totals were then summed for the number of days in each month to arrive at monthly heat gains. The length of the cooling season for each location considered was determined from weather data and characteristic operating periods. Yearly heat gains were derived by summing monthly totals for the length of the cooling season.

Absorption coefficients and U values were varied and summarized for the 12 locations as shown in Table 7.5. Gains in Table 7.5 include both the solar and conduction components of heat gain. Values of the conduction heat gain component through walls were deducted from the total heat gain to derive the solar component, which was then plotted and extrapolated to include the entire range of degree-hours. Figure 7.5 was derived from locations with latitudes between 25 and 35°N, and Fig. 7.6 was derived for latitudes between 35 and 45°N. The heat gains assume that the walls are subjected to direct sunshine. If shaded, gain should be read intersecting with the north exposure line.

Figure 7.7 is based on degree-hours read from Fig. 4.4, which has a base of 56 h/week. The following formula is used:

$$Q \text{ (heat gain)/year} = \text{degree-hours/year} \times U \text{ value.}$$

The major portion of degree-hours occurs between 10 A.M. and 3 P.M.; hence, for occupancies between 10 and 56 hours per week, the degree-hour distribution can be assumed to be linear. However, for occupancies greater than 56 hours per week the degree-hour distribution becomes nonlinear, particularly in locations with greater than 15,000 degree-hours. This is reflected by the curves for 72- and 96-hour per week occupancies.

7.7 REDUCE HEAT GAINS THROUGH FLOORS AND ROOFS

The difference between outdoor and indoor dry-bulb temperatures is smaller during the cooling season (especially if indoor conditions are maintained at 78°F or higher), and adding insulation to existing floors is not usually worthwhile. Insulation added to reduce heat loss in the winter through floors exposed to outdoor conditions, in buildings on stilts, for instance, however, will have a small added benefit in the summer.

Roofs offer more potential. Conduction and solar gains through roofs can form a considerable part of the cooling load. For buildings located in high degree-hour zones which have large roof areas in proportion to floor areas, such as supermarkets and other single-story buildings, consideration should be given to insulating the roof and to reducing its absorption coefficient.

Adding light-colored coatings or cooling the roof with a roof spary are two options for reducing absorption coefficients. However, measures to reduce absorption of solar energy for walls which have a mass of 50 to 100 lb/ft² and are insulated to a U factor of 0.1 or lower are ineffective. Light colors on roofs are also quite difficult to maintain in urban areas.

Paints or reflective finishes must be compatible with the existing roof and must be capable of withstanding abrasion. The absorption coefficient of roofs may also be reduced by adding a surface layer of white pebbles or gravel. The weight of the additional layer must not exceed the structural bearing capacity, and gravel stops should be fitted around rainwater drains and the perimeter of the roof.

Figure 7.7 gives yearly heat gain through roofs for conduction only (no solar effect) with entrants of annual degree-days, occupied hours per week, and U value. Table 7.6 is the result of a computer program developed to study heat gain through roofs for both conduction and solar heat gains. Figure 7.8 was extrapolated from the data in Table 7.6 and gives only yearly solar heat gain through roofs, with entrants of annual dry-bulb degree-hours above 78°F, solar radiation in langleys, U value, and absorption coefficient. A cursory study of Fig. 7.7 and 7.8 indicates the predominance of solar heat

TABLE 7.6   HEAT GAIN THROUGH ROOFS

City	Latitude	Solar Radiation, langleys	DB Degree-hours above 78°F	Heat Gain, Btu/(ft²)(yr) U = 0.19 a = 0.3	U = 0.19 a = 0.8	U = 0.12 a = 0.3	U = 0.12 a = 0.8
Mineapolis	45°N	325	2,500	2,008	8,139	1,119	4,728
Concord, N.H.	43°N	300	1,750	1,891	7,379	1,043	4,257
Denver	40°N	425	4,055	2,458	9,859	1,348	5,680
Chicago	42°N	350	3,100	2,104	7,918	1,185	4,620
St. Louis	39°N	375	6,400	4,059	12,075	2,326	7,131
New York	41°N	350	3,000	2,696	9,274	1,543	5,465
San Fancisco	38°N	410	3,000	566	5,914	265	3,354
Atlanta	34°N	390	9,400	4,354	14,060	2,482	8,276
Los Angeles	34°N	470	2,000	1,733	10,025	921	5,759
Phoenix	33°N	520	24,448	12,149	24,385	7,258	14,649
Houston	30°N	430	11,500	7,255	20,931	4,176	12,369
Miami	26°N	451	10,771	9,009	24,594	5,315	14,716

gain over conduction heat gains and the importance of roofing materials that reflect the sun.

ACTION GUIDELINES

☐ 1. Do not consider insulating floors to reduce heat gain alone.

☐ 2. Insulate roofs to U values listed in ASHRAE Standard 90-75, *Design Criteria for Energy Conservation in New Buildings* in all locations where there are more than 500 hours per year above 85°F, if the roof or combined roof-ceiling has a U value in excess of 0.20.

☐ 3. Provide insulation on top of the roof or below the roof and above the ceiling whenever the interior surface temperature of the ceiling (roof if there is no ceiling) is above 82°F on clear days with 50% or more sunshine.

☐ 4. Cover the exterior surface of the roof with white pebbles, light-colored tiles, or other durable materials to reduce the absorption coefficient in direct sunlight to 0.3 or less.

☐ 5. Install a roof spray system to reduce the temperature of the exterior surface of the roof if calculations indicate that sol-air temperatures exceed 100°F for 250 h or more, the roof has a U value larger than 0.2, and the roof has a mass less than 50 lb/ft².

☐ 6. Install a roof pond in place of roof sprays if the roof construction permits and analysis shows that the cost benefits are greater than for insulation or roof sprays.

To determine the cooling energy saved by insulating floors, refer to Fig. 7.7 and carry out the following procedure:

1. Determine from Fig. 4.4 the annual degree-hours above 78°F DB for the location.
2. Determine the initial U value and the improved U value after insulation with procedures from Sec. 6.15.

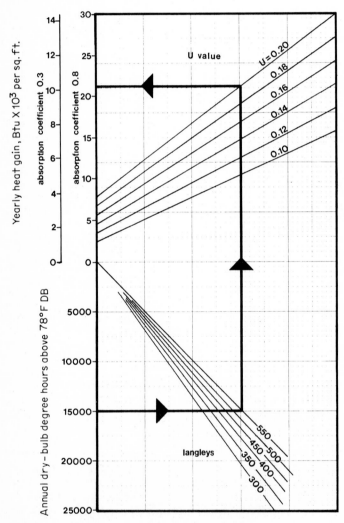

Figure 7.8 Summer solar heat gain through roofs.

3. Using the initial U value of the floor, enter Fig. 7.7 at the selected degree-hours and following the direction of the example line, intersect at occupied hours and U values.

4. Read out the yearly heat gain per square foot of floor area.

5. Repeat this complete procedure using the improved U value and by subtraction, obtain the reduction in heat gain per square foot of insulated floor.

To determine the cooling energy saved by insulating roofs or by changing absorption coefficients, refer to Figs. 7.7 and 7.8 and carry out the following procedure:

1. Determine from Fig. 4.4 the cooling degree-hours above 78°F DB for the location.

2. Determine from Fig. 4.2 the mean solar radiation in langleys for location.

3. Determine the initial U value and improved U value after insulation with procedures from Sec. 6.15.

4. Determine the initial absorption coefficient and the improved absorption coefficient.

5. Using the initial U value of the roof, enter the graph at the selected degree-hours and follow the direction of the example line, intersecting at occupancy and U value. Read out the yearly heat gain due only to conduction per square foot of roof.

6. Now, refer to Fig. 7.8 and, using the initial U value of the roof and the initial absorption coefficient of the roof, enter the graph at the selected degree-hours and, following the direction of the example line, intersect at the appropriate points for solar radiation in langleys and U value. Read out the yearly heat gain per square foot of roof on the appropriate absorption coefficient line. Interpolate to obtain answers for absorption coefficients other than the 0.3 and 0.8 shown.

7. Repeat the procedure, using the improved U value or the improved absorption coefficient and, by subtraction, obtain the reduction in cooling load due to insulation or modified absorption coefficient per square foot of roof.

Figure 7.8 is based on the Sunset computer program which was used to calculate solar effect on roofs for 12 selected locations. The program calculates hourly solar angles and intensities for the twenty-first day of each month. Radiation intensity values were modified by the average percentage of cloud cover taken from weather records on an hourly basis. Heat gains are based on a 78°F indoor temperature.

The solar effect on a roof was calculated using sol-air temperature, and the heat entering or leaving a space was calculated with the equivalent temperature difference. Roof mass ranged from 25 to 35 lb/ft², and thermal lag averaged 3½ h. During the cooling season, internal gains, ventilation, infiltration, and conduction through the building skin create a cooling load. The additional load caused by heat gain through the roof was calculated for each day. Daily totals were then summed for the number of days in each month to arrive at monthly heat gains. The length of the cooling season for each location considered was determined from weather data and characteristic operating periods. Yearly heat gains were derived by summing monthly totals for the length of the cooling season.

Absorption coefficients and U values were varied and summarized for the 12 locations as shown in Table 7.6. Gains in Table 7.6 include both the solar and conduction components of heat gain. Values of the conduction heat gain component through roofs were deducted from the total heat gains to derive the solar component. The solar component was then plotted and extrapolated to include the entire range of degree-hours.

7.8 REDUCE INTERNAL HEAT GAINS FROM LIGHTS AND EQUIPMENT

Heat gain from lights is often the major part of the cooling load in an office building or store. Because the level of illumination is often far in excess of that required for visual acuity, many steps may be taken to reduce the lighting level and increase the efficacy of the lighting system ranging from simply switching off unnecessary lights to changing lighting fixtures for more efficient types. By rearranging tasks and removing fluorescent tubes from noncritical areas, it is usually possible to maintain the same footcandles of illumination on critical tasks.

Reducing power for lighting by 1 kWh will reduce the cooling load by approximately 3400 Btu. The resulting reduction in quantity of cooling supply air will reduce the energy required for fans as well as refrigeration. In addition to savings related to the cooling system, energy for lighting is saved as well throughout the year.

Electrically operated business machines such as typewriters and photocopiers generate heat in direct proportion to the quantity of electricity they consume. The heat output of office equipment cannot be significantly reduced while it is being used, but hours of operation can be reduced by making sure that each piece of equipment is turned off when it is not required for immediate use.

Equipment such as ovens, ranges, fryers, and other kitchen equipment emit much heat. Hoods should be provided, and a separate air supply should be installed to reduce the makeup air load to the hoods. Provide shields, baffles, and insulation for other equipment which has surface temperatures exceeding 100°F for 2 or more hours per day.

ACTION GUIDELINES

□ 1. Turn off unnecessary lights and heat-producing equipment.

□ 2. Reduce lighting levels by removing lamps.

□ 3. Exhaust the heat from ovens, ranges, and motors directly outdoors when the enthalpy of the outdoor makeup air is lower than the enthalpy of the space.

□ 4. Insulate hot surfaces of tanks, piping, and ducts which are in air-conditioned spaces.

□ 5. Where possible, relocate near north and east windows all equipment and processes that emit heat to facilitate transfer of heat from indoors to outdoors.

□ 6. Install automatic controls to deenergize dry-type transformers which are located in air-conditioned spaces when there is no operating load.

□ 7. Locate vending machines, office duplicating equipment, and other equipment and appliances out of air-conditioned areas. (It may be possible to relocate them in areas requiring heat in the wintertime.)

□ 8. Install hoods, baffles, and insulated panels to minimize heat gain to the space from equipment which emits a large amount of heat.

Electrical equipment such as electric lighting and coffee makers, rated in watts, contributes to the cooling load directly (1 kW equals 3413 Btu/h). Motors require about 0.8 kW, or about 2500 Btu per horsepower hour.

7.9 USE OUTDOOR AIR FOR COOLING

Economizer cooling brings outdoor air into conditioned areas to remove internal heat gain without operating the refrigeration compressor (or, in marginal conditions, to reduce the load on the refrigeration compressor).

When nighttime outdoor temperatures are below indoor temperatures by 5°F or more, shutting off the refrigeration system and using outdoor air for night cooling will save energy in most areas of the country. There will be no advantage to bringing in outdoor air above that temperature because heat from the fan motors will raise the air temperature 2 to 3°F, and this plus the power required to drive the fans will outweigh the savings. The cutoff temperature at which night air cooling is advantageous may be even lower in large buildings which have high-velocity systems, as the power requirements for fan motors in these systems are higher.

To fully utilize outdoor air for cooling, it may be necessary to install return-air and outdoor-air dampers and provide a means of relieving air pressure. Some possible options include partially opening windows, operating an exhaust system, or installing propeller-type exhaust fans in the wall.

During occupied periods, the opportunities to use outdoor air for cooling depend on the outdoor WB temperature as well as DB temperature. If outdoor air is brought into the building above 60°F WB (the equivalent to the wet-bulb temperature when the spaces are maintained at 78°F DB and 55% RH), the cooling load is increased. The wet-bulb temperature is a measure of the total heat content of the air, and if the wet-bulb temperature is lower than 66°F, outdoor air can be introduced through the cooling coil to reduce the cooling load.

For many existing systems, however, the cooling coils are not designed to handle outdoor air at temperatures above 85°F DB and still maintain room conditions at 78°F DB despite the fact that outdoor air at less than 66° WB actually has a lower total heat content than room air. Figure 4.6 indicates the number of hours when the dry-bulb temperature is greater than 85°F in the summer in the United States. Denver, Colorado, on a seasonal basis, has low wet-bulb temperatures, and it would appear to be more economical to utilize 100% outdoor air through the air-conditioning coils in the daytime than to recirculate air. Summer outdoor dry-bulb temperatures in Denver, however, often rise above 90°F. If the existing coils have not been selected to reduce the temperature of outdoor air at those times they will be incapable of handling the conduction, solar, and internal heat gains which occur. To conserve energy in this type of climate, the outdoor-air intake should be opened fully except when indoor dry-bulb temperatures cannot be maintained.

Operating systems manually to reflect temporary conditions and conserve energy is sometimes difficult, but the effort yields significant savings. Automatic controls are available to optimize the operation of most systems and to meet varying and selective conditions. There are two suitable economizer systems: The first type monitors and responds to dry-bulb temperature only. It is suitable in areas where wet-bulb degree-hours greater than 66°F are fewer than 8000 per year. The second type monitors and responds to both WB and DB temperatures (enthalpy) and is suitable in locations which annually experience more than 8000 WB degree-hours.

The first economizer system, which responds to dry-bulb only, operates as follows:

1. When the outdoor-air DB temperature is lower than the supply-air DB temperature required to meet the cooling load, the compressor and chilled-water pumps are turned off, and the outdoor-air, return-air, and

exhaust-air dampers are positioned to attain the required supply-air temperature.

2. When the outdoor air DB temperature is higher than the supply-air temperature required to meet the loads, but is lower than the return-air temperature, the compressors and chilled water pumps are energized and the dampers are positioned for 100% outdoor air.

3. Minimum outdoor air is brought in when the outdoor DB temperature exceeds the return air DB temperature.

4. When the relative humidity in the space drops below desired levels and more energy is consumed to raise the relative humidity than is saved by the economizer system, the economizer cooling is shut down.

The second economizer system, which responds to both DB and WB, operates in a similar sequence with the exception that enthalpy is the measure rather than DB conditions. It must be noted that economizer systems are less effective if used in conjunction with heat recovery systems. Trade-offs must be analyzed. See Fig. 7.9.

ACTION GUIDELINES

□ 1. Use outdoor air for economizer cooling when the enthalpy is lower than room conditions during occupied periods and when the dry-bulb temperature is 5°F lower than indoor design conditions during unoccupied periods.

□ 2. Use operable windows, without fan operation, for outdoor-air cooling. Outdoor temperature during unoccupied periods should be below room DB conditions and during occupied periods below room DB and WB conditions. (Unfavorable acoustics and air quality may preclude implementation of this option.)

□ 3. If cool outdoor air is available, consider cooling the building well below normal during the night and early morning hours preceding any day that is expected to be extremely hot.

□ 4. When the volume of outdoor air is increased for economizer cooling, and if there is no exhaust system which can handle an equal quantity of air, provide pressure relief.

□ 5. Install dampers in the fresh-air duct and the return-air duct at air-handling units, and interlock the dampers so that one opens when the other closes.

□ 6. Install an exhaust fan with relief louvers (or for large multifloor buildings a return-air and exhaust-air fan) with dampers in the exhaust duct interlocked with the fresh-air and return-air dampers.

□ 7. Provide controls to open outdoor-air-dampers when outdoor temperatures are 5° or more below indoor conditions, or install enthalpy control to open outside-air dampers when the outdoor wet-bulb temperature is below 65°F and the dry-bulb temperature is below 85°F.

It is not possible in this manual to provide blanket rules for determining energy saved or costs for economizer and enthalpy control systems since too many variables affect an individual application. By use of methods described here, however, the amount of cooling which can be achieved with outdoor

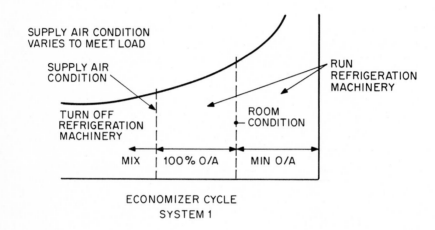

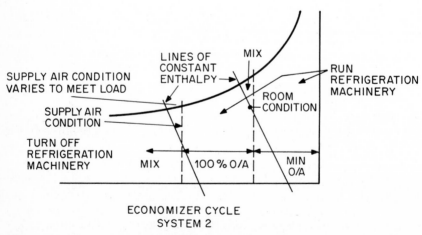

Figure 7.9 Dry-bulb and enthalpy economizer cycles.

air cooling can be assessed. The costs of installation and efficiency of the systems must be determined on an individual basis.

7.10 USE EVAPORATIVE COOLING OR DESICCANT DEHUMIDIFICATION

In geographic areas which experience high DB temperatures but low coincidental WB temperatures (65°F WB or less), evaporative cooling may be used to reduce the temperature of the outdoor air. For instance, outdoor air at 95°F DB, 65°F WB can be adiabatically cooled by water sprays to 77°F DB, 65°F WB, giving a relative humidity of 50%. This effectively eliminates the ventilation cooling load and under certain circumstances will allow the use of economizer cooling, even though the outdoor DB temperature exceeds the design DB temperature within the building. To determine the benefit of evaporative cooling, an analysis must be made of weather data for each case to assess the number of hours of possible use and the quantity of energy saved.

Evaporative cooling can be achieved by installing sprays within the main HVAC units or in duct extensions together with drain pans and eliminators. It is often effective to spray the water directly onto the upstream face of the coil and use the existing downstream eliminators, if the design of the cooling coil is suitable.

Spray water can be continuously recirculated from the drain pan through the spray headers and back to the drain pan by a small pump, but continuous makeup and blowdown should be provided to limit precipitation of solids. The quantity of blowdown and makeup will depend on local water analysis. If low-temperature well water is available, this can first be used in the cooling coil and then used as feed to the evaporative cooler sprays.

Even in relatively humid climates, there are a significant number of hours per year in which adiabatic cooling can be used and operation of refrigeration equipment can be minimized.

Desiccants such as silica gel can be used as adsorbents to reduce the moisture content of air. Dehumidification is achieved adiabatically, and the air DB temperature increases proportionally with the quantity of moisture removed; e.g., air at 85°F DB, 75°F WB, would change adiabatically to 122°F DB, 75°F WB, with a corresponding reduction in moisture content of 56 gr/lb. Air can then be reduced to 60°F, 50% RH, by sensible heat removal only with a relatively high-temperature cooling medium other than chilled water in the cooling coils. Cooling-tower water, well water, or waste cool water can be utilized in the coils.

Desiccant dehumidification is particularly applicable in geographic locations experiencing long periods of high WB temperatures. The desiccant eventually becomes fully charged with moisture and has to be regenerated by heat to drive the moisture off. When the effectiveness of desiccants is calculated, this heating energy requirement must be taken into account. If possible, recaptured waste heat or solar energy should be used for regeneration. See Fig. 7.10.

Desiccant dehumification can be usefully applied to buildings or areas of buildings subject to high internal latent heat gains or areas with requirements for a high percentage of outdoor air. Basically, desiccant dehumidification converts latent cooling loads to high-temperature sensible cooling loads, a proportion of which can easily be handled without resorting to mechanical refrigeration.

For example, refer to the psychrometric diagrams in Fig. 7.10. A space requires supply air at 58°F DB/55°F WB to maintain 78°F/68°F. The refrigeration cooling load to reduce 5000 cfm outdoor air from 85°F/72°F to 58°F/

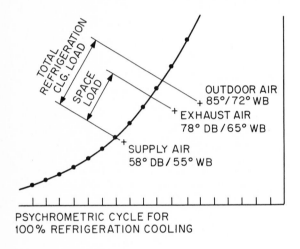

PSYCHROMETRIC CYCLE FOR
100% REFRIGERATION COOLING

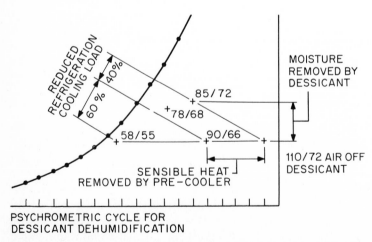

PSYCHROMETRIC CYCLE FOR
DESSICANT DEHUMIDIFICATION

Figure 7.10 Refrigeration cooling and dessicant dehumidification.

55°F is 22 tons. Using desiccant dehumidification, the outdoor air condition changes from 85°F/72°F to 110°F/72°F. High-temperature dry air is now sensibly cooled to 90°F/66°F by a precooling coil (the cooling medium can be well water, condenser water, waste water, etc.) The cooling load to reduce the 5000 cfm air from 90°F/66°F is 12.75 tons, resulting in a refrigeration load reduction of 9.25 tons, or 40%.

ACTION GUIDELINES

☐ 1. Reduce the temperature of incoming outdoor air by using evaporative coolers in geographical areas with high DB temperature coincident with low WB temperatures (65°F or less).

☐ 2. Use desiccant dehumidification to reduce the moisture content of incoming outdoor air in areas with long periods of high WB temperatures.

CHAPTER EIGHT

DOMESTIC WATER HEATING

8.1 THE NATURE OF THE LOAD

Domestic water heating does not fit neatly into any one of the categories of building load, distribution load, or conversion load: it spans all three. Energy is expended for hot water as an end use building load—it is used, and perhaps partially reclaimed, then dumped. But energy for hot water is also expended in distribution leaks and heat losses and in heater fuel conversion losses. Thus, in the following sections conservation follows the order of reduction of hot water quantity and temperature requirements, then reduction of piping losses, then reduction of conversion losses for hot-water generation.

8.2 REDUCE DOMESTIC HOT-WATER TEMPERATURES

Lowering the temperature of the hot water reduces the energy required for the domestic hot-water system. The energy required for the system is computed by the following formula:

$$\text{Yearly Btu} = Q \times Td$$

where Q = quantity of domestic hot water used per year in pounds and Td = magnitude of the difference, in degrees Fahrenheit, between the temperature of cold water entering the heater and the temperature of the hot water leaving it.

Figure 8.1 indicates the energy added to domestic hot water in the heater at various generation temperatures and usage rates (a temperature of 50°F for the incoming water and an occupancy of 251 days per year are assumed). The total amount of fuel energy required to supply the domestic hot-water system depends upon the seasonal efficiency of the heater, E, which varies with the

195

type of heater and the fuel used. On a seasonal basis, the following are average efficiencies:

System	Efficiency
Oil-fired heating boilers used year-round but with domestic hot water as the only summer load	0.45
Oil-fired heating boilers used year-round with absorption cooling in the summer	0.7
Gas-fired heating boilers used year-round but with domestic hot water as the only summer load	0.50
Gas-fired heating boilers used year-round with absorption cooling in summer	0.75
Separate oil-fired hot-water heaters	0.70
Separate gas-fired hot-water heaters	0.75
Separate electric water heaters	0.95

To determine actual energy consumption, divide the value obtained from Fig. 8.1 by the appropriate efficiency. If incoming temperature differs from 50°F, adjust the value from Fig. 8.1 before dividing. If incoming temperature is 60°F, for instance, at a generation temperature of 150°F, multiply the value by $\dfrac{150 - 60}{150 - 50}$. Actual energy consumption is also given by this formula:

$$\text{Yearly Btu} = \frac{Q \times Td}{E}$$

ACTION GUIDELINES

☐ 1. Where possible, use cold water alone for hand washing in lavatories when the cold water temperature is 75°F or above. This is most readily accepted in retail stores, religious buildings, owner-occupied small office buildings, and in washrooms used primarily by the public on an infrequent basis.

☐ 2. Where tenants insist upon hot water for hand washing, heat tap water to 90°F.

☐ 3. Do not maintain an entire hot-water system at the same temperature required for the most critical use. Do not heat water for hand washing, rinsing, or cleaning to the same temperature required for dishwashing sterilization.

☐ 4. If the space-heating boiler is also used to supply domestic hot water, lower the aquastat setting in the summertime to 100°F. The same setting should be used for storage tank temperature control, summer and winter.

☐ 5. Where higher temperatures are required, at a dishwasher, for example, use a small gas or electric booster activated only as needed.

☐ 6. Use cold water detergents for laundries and laundromats and set water temperature at 70°F.

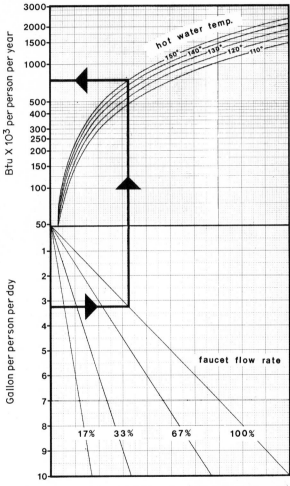

Figure 8.1 Energy variables in domestic hot-water heating usage, faucet flow rate, and water temperature. (This chart is constructed with the assumption that users will open faucets for a set amount of time, for instance, that hand-washing is a function of time, not water quantity.)

☐ 7. Install a 7-day timer clock to operate a water heater to further reduce temperatures during unoccupied periods.

8.3 REDUCE DOMESTIC HOT-WATER QUANTITY

The energy required to provide domestic hot water is a direct function of the way the water must be heated and how much is heated. Reducing the quantity of hot water consumed saves energy. A secondary benefit is the reduction in raw source energy required to treat the water supply, whether on-site or off-site. In areas where there is a charge based on total water consumption flowing into the sewer, the reduction in consumption of water will result in direct savings as well. Water consumption can be lowered to 1¼ or 1½ gpd per person in office buildings without inconvenience to the occupants. Table 8.1 gives typical domestic hot-water usage by building type, and, as noted above, Fig. 8.1 gives annual energy savings for reduction in faucet flow rate.

ACTION GUIDELINES

☐ 1. Take measures to reduce consumption when usage rates exceed 1½ gpd per person in offices and 1 gpd per person in stores or religious buildings.

☐ 2. Install flow restrictors in the supply branch to groups of taps when existing faucets have flow rates greater than 1½ gpm.

☐ 3. Insert orifices in the hot-water pipes to reduce flows.

☐ 4. Install spray-type faucets that use only ¼ gpm instead of 2 or 3 gpm (at an installation cost of about $50 a unit).

☐ 5. Install self-closing faucets on hot-water taps.

☐ 6. In buildings with cooking facilities that are used only periodically, such as meeting rooms in religious buildings, shut off the hot-water heating system, including gas pilots where installed, when the facilities are not in use.

☐ 7. Reexamine the need to heat an entire tank of water when only a small quantity of hot water is needed.

☐ 8. Simplify menus to reduce the need for large pots and pans that require large amounts of hot water for cleaning. Where practical, use short dishwashing cycles and fill machines fully before use.

☐ 9. Reduce the number of meals served or serve more cold meals to reduce the hot-water requirements for dishwashing.

☐ 10. In areas where water pressure is higher than a normal 40 to 50 lb/in², restrict the amount of water that flows from the tap by installing pressure-reducing valves on the main service. Do not reduce pressure below that required for fire protection or for maintaining adequate pressure on the top floor for flushing.

☐ 11. For new additions, install foot-operated paddle valves.

☐ 12. Install a flowmeter on the cold-water line supplying the water heater when hot-water consumption is more than 3000 gpd.

☐ 13. Replace obsolete kitchen equipment such as dishwashers with new ones that have minimal water requirements.

TABLE 8.1 DOMESTIC HOT WATER USAGE (AVERAGE)

OFFICE BUILDINGS

Without kitchen and cafeteria services	2 to 3 gpd per capita for hand washing and minor cleaning (based on an average permanent occupancy which includes daily visitors)

DEPARTMENT STORES

Without kitchen and cafeteria services	1 gpd per customer

KITCHENS AND CAFETERIAS

Dishwashing, rinsing, and hand washing	3.0 gpd per meal plus 3 gpd per employee

SCHOOLS

Boarding	25 gpd per capita
Day	3 gpd per capita (does not include cafeteria or athletic facilities)

APARTMENTS

High rental	30 gpd per capita
Low rental	20 gpd per capita

HOSPITALS

Medical	30 gpd per capita
Surgical	50 gpd per capita
Maternity	50 gpd per capita
Mental	25 gpd per capita

HOTELS

All types	30 gpd per capita

8.4 REDUCE DOMESTIC HOT-WATER SYSTEM LOSSES

Heat losses from uninsulated hot-water system distribution piping can be substantial. The magnitude of these losses depends on the temperature differential between pipe and ambient air, on pipe size, and on length of piping in the system.

Exposed piping in basements and equipment rooms is relatively simple to insulate. Piping in ceiling spaces may also be readily accessible by removing ceiling panels. Preferably, the entire piping system should be insulated, but inaccessible portions may be left bare providing they are a small percentage of the total, as this will have little effect on the total savings possible.

The costs for sectional insulation installed with jacket, listed in Table 8.2, are offered as an order-of-magnitude guide.

For costing purposes, add to the total lineal feet of piping 3 lin ft for each fitting or pair of flanges to be insulated. Savings achieved by insulating hot-water piping can be determined from Fig. 11.11 for domestic hot-water temperatures ranging from 100 to 180°F.

The loss of heat from the domestic storage tank must continuously be offset by the addition of heat to maintain a ready supply of hot water. This heat loss occurs 24 hours per day whether the building is occupied or not. Storage tanks should be covered with a minimum of 3 in of insulation ($K =$

TABLE 8.2 PIPING INSULATION COSTS

| Pipe Size, in | Price Installed, $/lin. ft Insulation Thickness | |
	1 in	1½ in
½	1.35	2.05
¾	1.40	2.10
1	1.45	2.15
1¼	1.50	2.20
1½	1.55	2.25
2	1.60	2.35
2½	1.65	2.45
3	1.70	2.50
4	2.00	3.05
5	2.25	3.15
6	2.55	3.30
8	3.15	4.20
10	3.85	5.00
12	4.50	5.60
14	5.20	6.45
16	6.00	7.20
18	6.70	7.60
20	8.25	8.50
24	9.00	9.70

TABLE 8.3 TANK INSULATION VALUES

Insulation Thickness, in	Tank Size, gal	Btu (Millions per Year) Lost at Water Temperatures of		
		100°F	120°F	160°F
1	50	1.9	3.0	5.2
	100	3.0	4.7	8.2
2	250	3.1	4.9	8.4
3	500	3.1	4.9	8.4
	1000	5.2	8.2	14.1

0.3). Calculate energy savings by determining the heat lost from the tank before and after insulation. Assume that water temperature and ambient air temperature are constant. Install a tank jacket for additional insulation (for a 40-gal tank the cost will be about $20). Table 8.3 gives values (in Btu) of tank insulation.

Costs for insulating hot- or cold-water tanks with 3-in density fiberglass (foil scrim craft facing, finished with presized glass cloth jacket) are given in Table 8.4.

ACTION GUIDELINES

☐ 1. Repair insulation of hot-water piping and tanks, or install where missing (unless piping and tanks are located in areas which require space heating).

☐ 2. Where forced circulation of hot water is used, shut off the pump when the building is unoccupied. When hot-water usage is light, consider using gravity circulation without the pump.

☐ 3. Flush water heater during seasonal maintenance of heating systems.

☐ 4. Repair leaky faucets.

☐ 5. Repack pump packing glands of recirculation hot-water heaters to reduce leaking of hot water.

TABLE 8.4 TANK INSULATION COSTS

Material Thickness, in	Cost, $/ft² Surface Area
1	2.60
1½	2.70
2	2.95
3	3.60

☐ 6. For boilers with immersion tankless domestic hot-water coils, make sure boiler water covers coils.

☐ 7. Insulate hot-water storage tanks when insulation is less than the equivalent of 3-in fiberglass or when insulation is in need of repair.

☐ 8. Insulate the exterior jacket of tankless or tank heaters which are not immersed in the hot-water boiler or hot-water tank.

☐ 9. When installing new storage tanks or making major modifications to the building, relocate the hot-water tank as close to the load as possible.

☐ 10. Insulate hot-water piping whenever there is less than the equivalent of 1-in fiberglass.

8.5 IMPROVE DOMESTIC HOT-WATER SYSTEM PERFORMANCE WITH EQUIPMENT MODIFICATIONS

Commercial hot-water systems frequently require hot water for short periods of heavy use at various locations within the building. It is often more efficient to provide water heaters close to the usage points than to maintain central generation and long runs of hot-water piping.

Patterns of hot-water use within the building should be analyzed to determine whether installation of local units is advantageous. The energy saved is the sum of reduced distribution losses and the increase in the average generation efficiency of local units over central units.

When multiple temperature requirements are met by a central domestic hot-water system, the minimum generation temperature is determined by the maximum usage temperature; lower temperatures are attained by mixing with cold water at the tap. Where the majority of hot-water usage is at the lower temperatures and higher temperatures are required at a few specific locations only, booster heaters or separate heaters for high temperatures can be installed.

In many buildings, the heating-system boilers provide primary heat for the domestic hot-water system. While this is satisfactory during the heating season when boilers are firing at high efficiency, demand for boiler heat in summer will probably be limited to hot-water generation only. Operating large heating boilers at light loads to provide domestic hot water results in low boiler efficiency. To reduce energy losses due to low boiler efficiency in summer, a separate hot-water heater should be installed. Then the heating boiler can be shut down in the summer, and domestic hot water can be generated at improved efficiency.

ACTION GUIDELINES

☐ 1. Repipe hot-water storage tanks if the cold-water makeup supply is connected to the upper half of the tank or if the hot-water outlet from the tank is in the lower portion of the tank.

☐ 2. When hot-water demands are increased owing to expansion or change in occupancy, provide either oil- or gas-fired water heaters or heat pumps rather than electric resistance heating. If the heating boiler has sufficient capacity and is in operation year-round for air conditioning as well, install a tank or tankless heater in place of a separate hot-water generator. If the additional requirements for hot water are to serve facilities remote from the boiler and usage is small, install a separate heater in the cold-water line directly at the fixtures rather than serving them from a central system.

☐ 3. Install local hot-water heating units when domestic hot-water usage points are concentrated in areas distant from the central generation and storage point.

☐ 4. Use a hot-gas heat exchanger in the hot-gas line of refrigeration units or heat pumps to heat domestic hot water.

☐ 5. Where diesel or gas engines are in use, install a heat exchanger and utilize waste heat from the engine.

☐ 6. Where incinerators are handling more than 1 ton of solid waste per day, utilize the waste heat from the incinerator to generate hot water.

☐ 7. Install a heat exchanger in hot-water drains from kitchens and laundries where the flow exceeds 2000 gpd.

☐ 8. Use hot condenser water from refrigeration systems to preheat domestic hot water.

☐ 9. Install a heat exchanger in condensate lines from steam equipment to preheat domestic hot water.

☐ 10. Install a heat pipe or heat exchanger to extract heat from boiler breechings to preheat hot water.

☐ 11. Install a solar water heater to replace or supplement the existing hot-water generator (see Chap. 15).

☐ 12. Replace gas pilots with electrical ignition.

☐ 13. Improve the efficiency of boilers with the measures explained in Chap. 12 and apply these measures to hot-water heaters as well.

☐ 14. Install a storage water heater for summer use when the existing space heating boiler is used for hot-water generation, when there is little or no demand for steam or hot water during the summer, and when hot water is 20% or more of the load.

☐ 15. Replace electric hot-water heaters with heat pumps to improve the coefficient of performance from 1 to approximately 3. Use hot drain water as a heat source for the heat pump, or use an air-to-water heat pump.

CHAPTER NINE

LIGHTING

9.1 THE DEMAND FOR LIGHT: INTRODUCTION

In the past we have taken full advantage of cheap energy by using it lavishly to produce light. Now, with energy costly and at times unavailable, we question the quantities of illumination that we have come to accept and expect, especially in office and store.

Some 20% of all electricity generated in the United States is used for lighting, and this figure is conservative considering that additional electricity is used in cooling systems which must remove the heat of the light from occupied spaces. For each kilowatt of lighting, an additional 0.4 kW is required for cooling if lighting and air-conditioning load occur simultaneously. And although any reduction in lighting will reduce the amount of useful heat available in the heating season, heat can be supplied more efficiently by the heating system than by the lighting system.

The sections below give recommendations for reducing existing and new lighting levels, suggest methods of avoiding simple waste, and offer direction for improving the efficacy of lamps and fixtures and installation. Lighting is fertile ground for conservation measures.

9.2 REDUCE ILLUMINATION LEVELS

The first step in reducing energy requirements for the lighting system is the reduction of lighting levels that are too high. Table 9.1 suggests adequate lighting levels for various office building areas.

One major cause of excessive lighting levels is uniform lighting, especially in offices and stores. Uniform lighting maintained at a level necessary for the most critical task wastes energy when other, less critical, tasks do not require the same amount of illumination. Uniform lighting should be converted to selective lighting. Each distinct functional area within the building and the discrete tasks which occur within the same room should be lighted only to the level and quality required for each task, and only for the time span during which the task occurs. A uniform modular lighting pattern of general illumination, throwing light equally on all areas, regardless of task, may waste up to 50% of the energy actually required for lighting. Orient lighting to suit the tasks to be performed.

Tasks requiring quite different lighting levels are not infrequently mixed more than need be. Lighting efficiency is enhanced if like visual tasks can be grouped together. If this is not possible, the simple expedient of a portable

TABLE 9.1 SUGGESTED LIGHTING LEVELS

Circulation areas between work stations	20 fc.
Background beyond tasks at circulation areas	10 fc.
Waiting rooms and lounge areas	10–15 fc.
Conference tables	30 ESI fc, with 10 fc for background lighting.
Secretarial desks	50 ESI fc with auxiliary localized (lamp) task lighting directed at paper holder (for typing) as needed. 60 ESI fc in secretarial pools.
Area over open drawers of filing cabinets	30 fc.
Courtrooms and auditoriums	30 fc.
Kitchens	Nonuniform lighting with an average of 50 fc.
Cafeterias	20 fc.
Snack bars	20 fc.
Testing laboratories	As required by the task, but background not to exceed 3:1 ratio footcandles.
Computer rooms	As required by the task. Consider two levels, one-half and full. In computer areas, reduce general overall lighting levels to 30 fc and increase task lighting for areas critical for input. Too-high a level of general lighting makes reading self-illuminated indicators difficult.
Drafting rooms	Full-time, 80 ESI fc at work stations. Part-time, 60 ESI fc at work stations.
Accounting offices	80 ESI fc at work stations.

light at the critical task, a drafting lamp, for instance, may allow a reduction in existing general space illumination. This is a cheap answer—portable lights should average less than $25 per lamp (and should be fluorescent rather than incandescent). Mating the light to the task rather than to the room is a valuable direction for good lighting design. Furniture-mounted task lighting is a logical resultant; it provides illumination of specific tasks to the extent necessary, controlled by the user of the furniture, while circulation areas between tasks can be maintained at much lower lighting levels. Power for furniture-mounted fixtures can be supplied from underfloor electrical ducts, from existing lighting circuits in the ceiling through power poles, or from additional surface-mounted raceways. Existing ceiling fixtures can be dimmed or reduced in number for low-level general illumination. And one final advantage—furniture-mounted task lighting moves with the furniture.

Light levels can be determined with a portable illumination meter such as a photovoltaic cell connected to a meter calibrated in footcandles. The light meter should be accurate to about ±15 percent over a range of 30 to 500 fc and ±20 percent from 15 to 30 fc. The meter should be color corrected and cosine corrected. The levels cited in Table 9.1 refer to average maintained horizontal footcandles at the task or in a horizontal plane 30 in above the floor. Task lighting surveys should be taken without daylight for a true evaluation of the lighting system. The lighting levels suggested for offices agree closely with new standards recommended by the General Services Administration for public office buildings. With proper attention to quality, these suggested levels should generally be adequate for tasks of good contrast.

A reasonable lighting energy budget for stores and offices is 2 W/ft^2 of gross floor area. The most obvious approach in reducing lighting levels in existing buildings to meet such budgets is to remove excess lamps where an analysis of daily work tasks indicates they are not needed. Another approach, where functions requiring different light levels take place in the same space at various times, is to install multilevel ballasts or dimmers. These several approaches are treated below.

Many general lighting systems in use today are made up of two-lamp fluorescent fixtures. To reduce the lighting level within these fixtures, both lamps must be deactivated. With four-lamp fluorescent fixtures, two lamps can be removed, and with three-lamp fixtures, one lamp can be removed. For two- and four-lamp fixtures the black and white power leads to the associated ballast should be removed as well, because even when the lights are not in place, the ballasts will continue to consume energy. The additional savings through removal of ballasts noted in Table 9.2 can be added to the wattage of the lamps removed to measure the electricity saved.

Two lamps in a 2-ft-wide fluorescent fixture will actually give more light per lamp than will four lamps in the same fixture. If both inside 430-mA lamps are removed from a typical four-lamp fixture, the resultant lighting levels will be about 10% greater than half the four-lamp level.

**TABLE 9.2 LAMP TYPE AND BALLAST
WATTAGE**

Lamp Type	Ballast Savings, W
F40	7
Slimline	11–13
High output (800 mA)	12
High output (1500 mA)	13.5

In high-intensity discharge fixtures, removing a lamp from a two-lamp mercury or metal halide ballast will generally cause no adverse effect. However, there will still be a current flow consuming as much as 20 W in a 400-W lamp ballast or 50 W in a 1000-W lamp ballast. However, removing a lamp from a two-lamp high-pressure sodium ballast for more than a short period of time will damage the starting circuit in the ballast because the circuit will operate continuously with the lamp removed. A failed lamp can cause the same type of damage.

If more than one lighting level is required for activities that occur in the same space at different times, multilevel ballasts with two- or three-level lamp controls are available. These controls, designed for 430-mA fixtures, allow reduction in illumination levels without sacrificing the symmetry of the lighting fixture pattern, and the percentage reduction in power can actually exceed the percentage reduction in lumen output at the lower level.

Two-level ballasts can replace single-level ballasts in existing fluorescent installations without a change in the fixture and with only slight modifications, if any, to mounting and wiring. A three-level ballast can also be used with existing fixtures but requires mounting and wiring changes. At present, multilevel switching is available only at the ballast, and wall switches with relays must be used if remote switching is desired. Other equipment for remote control of multilevel ballasts without relays is currently in development.

The lumen output of 400-W mercury vapor lamps can be reduced by 50% through modifications to the ballast system which permit switching from a 33-μF capacitor to a 22-μF capacitor. The lumens are reduced by 50% and the wattage by about 40%; the reduced power factor (PF) prevents a one-to-one reduction. If the PF of the entire building is high, the lower PF of the lamps will not be a serious problem. Multilevel fluorescent ballasts cost about twice as much as standard ballasts, but if the low level is used most of the time, the costs are repaid within a few years.

When new electrical fixtures are in order, a three-level lighting fixture can be used rather than a three-level ballast. These are offered as three-lamp 3 ft × 4 ft fixtures with standard 40-W lamps, with the center lamp wired in tandem. A separate switching circuit is required for each light level. One manufacturer offers the following data for typical office installations:

3 lamps: 124 fc at 3.7 W/ft²

2 lamps: 83 fc at 2.5 W/ft²

1 lamp: 37 fc at 1.2 W/ft²

When frequent light level changes are required in areas of transient occupancy, dimmers are a better alternative than multilevel switching, multilevel ballasts, or lamp replacement. Dimmers for incandescent lights are not expensive—materials cost about $30.00 for a 600-W dimmer and about $125.00 for a 2000-W dimmer. Fluorescent lamp dimming is more expensive because ballasts must be replaced and a wall control unit must be installed; new ballasts cost about $15.00 each, and the dimmer control module which will handle up to twenty 40-W lamps costs about $70.00. Dimmers for 400- and 1000-W mercury lamps, which also require ballast replacement and special control modules, are more costly than fluorescent dimmers. Costs for mercury lamps are highly variable but may approximate $75.00 for an additional ballast and from $400.00 to $2500.00 for the control unit, depending on the number of lamps dimmed.

In the process of converting from uniform lighting to selective lighting, it may be necessary, after removing some lamps, to use increased-output lamps in place of some of the remaining lamps. Where change is in order, it may be possible to relamp the fixtures with more efficient lamps to provide more lumens per watt—increased illumination levels without increased wattage. Refer to Fig. 9.1 to determine appropriate footcandle output with some common lamp replacements and Fig. 9.2 to determine the reduction in power with dimming ballasts.

Lamp efficacies vary with color, shape, gaseous fill, cathode construction, and internal coating. For example, a fluorescent 40-W, T-12 lamp, rated 430 mA, can have an output of around 53 lm to around 84 lm/W. As a further example, a natural white fluorescent lamp at 2100 lm (53 lm/W) provides one-third less foot candles than a cool white fluorescent lamp at 3200 lm (80 lm/W). In some cases, it may be desirable to relamp with lower wattage lamps rather than actually removing lamps.

ACTION GUIDELINES

☐ 1. Turn off incandescent lighting over top display of meat cases in a supermarket.

☐ 2. Turn off flood lighting which is strictly decorative.

☐ 3. Direct security lighting where it is needed, such as at windows and entrances, and reduce it where security problems are minimal.

☐ 4. For display or merchandising lighting establish grouped, highlighted

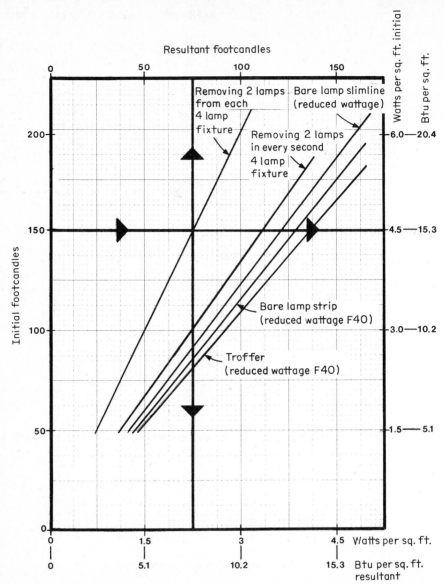

Figure 9.1 Footcandle general illumination as a function of wattage per square foot for various lighting-energy conservation tactics. (Derived from calculations based on manufacturers' data, and should only be used to obtain order-of-magnitude savings.)

display islands where many products can be lit with the same lighting sources, and thus reduce the total number of display islands.

☐ 5. Substitute small table or floor-mounted lamps in lounge areas or waiting rooms and turn off modular ceiling fixtures.

☐ 6. Remove unnecessary lamps when those remaining can provide the

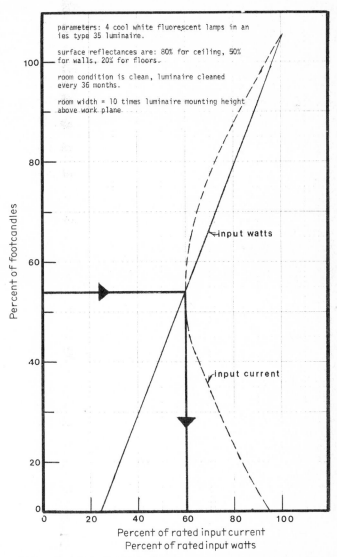

Figure 9.2 Reduction in fluorescent power requirements with the use of dimming ballasts. (Derived from manufacturers' data typical of fluorescent dimming ballasts.)

desired level of illumination. When removing fluorescent or high-intensity discharge lamps, remove the ballast also (or disconnect it in place).

☐ 7. If lighting requirements are not constant, install multilevel ballasts or multilevel fluorescent fixtures.

☐ 8. If frequent changes in the level of lighting are necessary, install dimmers.

☐ 9. When relamping, use more efficient lamps.

9.3 IMPROVE LIGHTING SYSTEM OPERATIONS

Contrary to common belief, leaving on a fluorescent lamp rather than turning it off never saves electrical energy. When electric lighting is not required, switch it off. Figure 9.3 indicates the potential energy savings for an office lighting system of 1000 fluorescent luminaires, when lights are on only when needed.

But to switch off the lights requires a switch, and many buildings, especially office buildings built in the 1950s and 1960s, switch whole floors or whole groups of rooms with a single switch. Fine tuning of the lighting system requires a capacity to switch off lights in unoccupied spaces and areas, and if such switches were not a part of the original design of a building, they should be added.

Many types of surface-mounted flat-ribbon conductors are available for installation in existing spaces; these can be installed with a minimum dislocation of existing wiring or damage to interior decorations. New switches should be located near doors, if possible, or where they will be most convenient for occupant use. Switches in inconvenient locations will not be used. If switches are group-mounted, each switch should be labeled to indicate the area that it controls.

Time switches should be provided for areas which are commonly used for short periods and in which lighting is inadvertently but frequently left on, such as reference rooms and stock rooms. At a predetermined time after the switch has been turned on, it will automatically shut off; if the area is to be used for a long period of time, the switch can be manually overridden.

For large areas, remote-control lighting contactors or lighting relays (to operate multiple circuits) can be added above the ceiling or at the lighting distribution panel. Remote switches can be either line or low voltage.

Site lighting and parking lot lighting should be controlled with time switches or with photoelectric cells. The latter are also effective in controlling light fixtures at the interior perimeters of the building where daylight is sufficient for illumination.

A frequency-controlled relay, now available, can be added at individual fixtures or at the circuit breakers which control a number of fixtures. The relay is controlled by an activator which superimposes a special command frequency over the existing wiring system. At present, two-command frequencies are available. Each activator will handle three hundred 40-W rapid-start ballasts located within 500 ft of the activator, which is mounted at the panel. The activator can be controlled by remote-control or local switch, time clock, or photoelectric switch. One activator can control lights which are on any number of separate circuits.

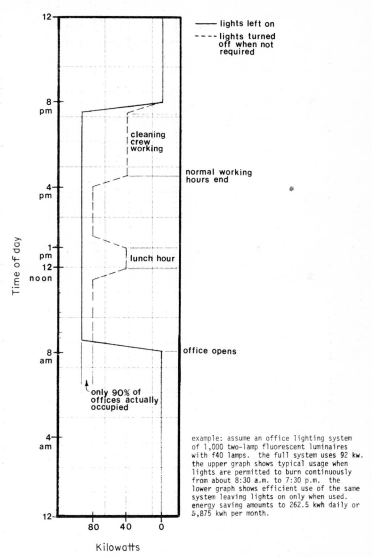

Figure 9.3 Possible power reductions for office lighting by turning out unnecessary lights. (Developed from theoretical calculations based on typical office building practices.)

Pilot lights should be installed outside all rooms that are infrequently used and where there is no other external indication that lights have been left on. Pilot lights should also be added to indicate when loads are energized at remote locations such as site lighting, sign lighting, snow melting, ovens, blueprint and reproduction equipment, mechanical spaces, and penthouses. Pilot lights can be surface mounted or recessed and can usually be added in parallel with the circuit or the load to be monitored. To conserve energy, neon-type pilot lights are better than incandescent lights.

The cost of adding switching depends upon the type of switching system selected and the existing fixture circuiting. Time switches cost from $30.00 to $150.00. Photoelectric cells cost about $75.00. Frequency-controlled relays cost about $20.00 per relay, and the associated controller costs approximately $300.00 per channel. A wall switch costs about $25.00, a low-voltage relay costs as little as $6.00, and a low-voltage transformer costs approximately $25.00. A surface-mounted neon pilot light can be installed over a door (with wiring in surface raceway to a ceiling space and BX cable to the nearest fixture in a room) for as little as $25.00.

Switching is important to controlling the energy used for the lighting system, but switching is worthless if it is not used. Perhaps more than with any other aspect of energy conservation, with switching, the attitude of the people using the building makes the difference. Switching off lights when leaving a room or space has to be a habit, like brushing teeth after meals.

The value of cleanliness is obvious from a cursory glance at Fig. 9.4. Fixtures should be cleaned, at the least, at relamping, for if relamping and cleaning are combined, the additional labor required is very small. In extremely dirty atmospheres, however, the fixtures should be cleaned between lamp replacements as well. The approximate cost of cleaning ceiling-mounted 48-in four-tube fluorescent fixtures (or incandescent fixtures of similar size) varies as follows:

Number of Fixtures	Price, $
10	14.00
50	58.00
100	115.00
200	224.00

The value of relamping before burnout is not so obvious as the value of cleanliness but is well worth consideration. There is added expense for the lamps themselves, but with new lamps there will be an improved value in lumens per watt of lighting. In addition, with group relamping there will be some reduction in labor cost per light in comparison with on-call-at-burnout replacement. This is a nice problem in cost effectiveness, but generally a

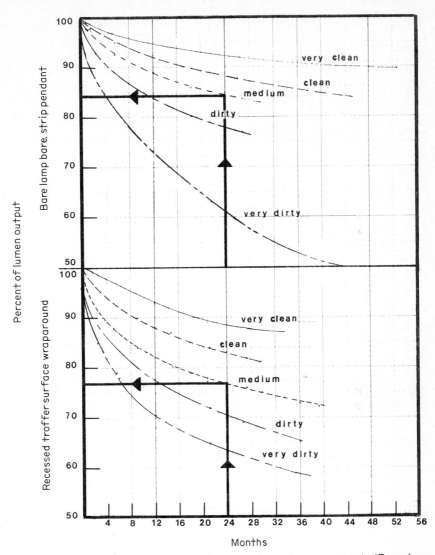

Figure 9.4 The effect of maintenance on lamp-lumen output. (Based on actual measurements of lumen output of bare lamps in various environments and for various lengths of time.) (*IES Lighting Handbook, 5th ed., 1972, pp. 9-17, Fig. 9-7.*)

group replacement plan at 80% of the rated hours of fluorescent lamps is advisable. Such a plan is indicated graphically in Fig. 9.5.

Switching is important for effective operation of the lighting system, and so is maintenance. If lamps are kept in use until burnout and fixtures are dirty, a revised relamping schedule and cleaning may actually allow a reduction in lamp wattage in each fixture without any reduction in lighting level. Two prime variables are the measure of maintenance: *lamp lumen depreciation* (LLD) and *light fixture dirt depreciation* (LDD). LLD is a drop in lamp lumen output over the life of the lamp and is a characteristic of the lamp type. LDD is a drop in fixture lumen output from design standards due to the buildup of dirt on the fixture's reflective surfaces and lens.

A third designation, *light loss factor* (LLF), combines LLD and LDD plus any other elements such as voltage variations, temperatures, and atmo-

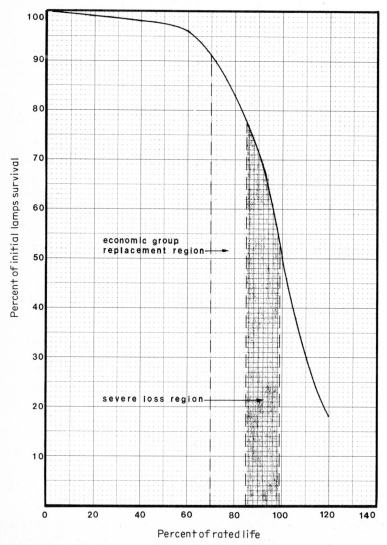

Figure 9.5 Fluorescent group relamping. (Based on manufacturers' information of statistics of fluorescent lamp life.)

spheric quality that finally alter the delivered lumen output of a light fixture. The term *maintenance factor* (MF) is the reciprocal of LLF and is thus the actual delivered percentage of initial lumen output for a given fixture and set of conditions. Figure 9.4 gives the MF for two types of fluorescent fixtures under various conditions of lamp life and cleanliness.

ACTION GUIDELINES

☐ 1. For cleaning which must be done at night, turn on lights only in that portion of the building which is being cleaned at the moment.

☐ 2. Switch off lights in each area when moving to the next.

☐ 3. In kitchens, avoid leaving infrared food warming lamps on when no food is being kept warm.

☐ 4. Turn off lights in areas of religious buildings which are not used during the week.

☐ 5. Turn off all lights other than those needed for security when the building is unoccupied.

☐ 6. Provide manual switches or photocells to shut off lights when available daylight is adequate to replace the required lighting.

☐ 7. Install additional switches to permit shutting off lights in unoccupied areas of the building.

☐ 8. Install time switches which will automatically turn off lights after a preset period in areas where occupancy is slight or occasional, and which will require manual switching to energize the lights again after that time.

☐ 9. Provide time switches to turn off internal fixture lighting in food cases or over merchandise displays in supermarkets when the premises are unoccupied.

☐10. Install photocells or electric timers to shut off outdoor parking lot lighting when the lot is not in use. Schedule the operation of photocells and electric timers with the working shifts of employees.

☐11. Install neon pilot lights to alert personnel that lights are on in adjacent areas.

☐12. Consider installing a master switching system, using low-voltage switching, permitting an operator at one or more stations to turn off all lighting at the end of occupied periods. Seven-day timers can also be used to automatically program the lights.

☐13. Mark all ganged switches to identify the lights controlled. Color code the switches and institute a program of use (for example, blue, 7 A.M. to 6 P.M.; red, 9 A.M. to noon; and 1 P.M. to 4:30 P.M.).

☐ 14. Instruct occupants and maintenance personnel to switch off all lights which are not required, even for small portions of the day.

9.4 USE DAYLIGHT

In the halcyon days of cheap energy, we were persuaded that windows were for a view in or a view out, but that lighting was more effectively and comfortably obtained with artificial means. We can no longer afford such a dismissal of the value of daylighting. Too many contemporary office buildings are curtain walled with surfaces of glass, yet burn their electric lights in perimeter spaces, rain or shine, night or day without discrimination.

Part of the pressure toward such indiscriminate use of perimeter electric lighting is the extreme variation in the availability of natural lighting. Daylight varies with the time of day, time of year, location, and weather. The amount of daylight that actually reaches a task surface is a further function of window size and location, exterior reflections and blockages, interior reflectances, and window cleanliness. Design with daylight is difficult; design with electric lighting is much easier. With daylight there is on occasion too much light and on occasion not enough. Frequently there is the added problem of glare, in part because the window light source is close to eye level.

To obtain the best of natural light, the designer or user needs common sense and a few controls. Blinds and drapes which are already installed should be adjusted through the day to make the most effective use of daylight, and in this instance the user of the space is the expert. In winter, direct sunlight can be welcome additional heat if it is not borne by any one person, that is, if it can be allowed to fall on circulation space or unoccupied space. Otherwise the direct sun must be screened and only the indirect light from the sky or reflected sun should be allowed into the space.

Electric lighting should be layered away from sources of daylight such that each layer, or bank, of lighting can be switched independently. As the sun sets, or the fog rolls in, lights can be turned on in banks toward the outside windows; as the sun rises, or the sky clears, banks of lights can be switched off away from the natural light. In existing buildings with good natural light, such layered switching should be installed if at all possible.

If daylight is available, less electric light will be necessary for general illumination or specific task lighting. But people, however well intentioned with respect to energy conservation, are not wholly reliable in turning out unnecessary light when the sun shines. One particularly useful investment is a photosensitive switch which can turn off outer banks of light when daylight is available.

In summer and in hot climates, the sun can be a problem, and use of daylighting can become very complicated. If direct sunlight is not properly controlled, the heat gain imposed on the cooling system can easily outweigh savings from turning off electric lights. On east and west exposures it is difficult to screen the direct sun without also sharply diminishing the amount of useful daylight, but this can be done. These issues are more thoroughly explored in Sec. 7.5, but generally the goal must be to keep out the sun, without overdoing it. Exterior controls are more effective but are generally fixed; interior controls are less effective because of the greenhouse effect but are easily adjusted. Reflective coatings or tinted screens on unprotected glass are useful in reducing heat gain through the glass from the sun but cut down in proportional measure at all times the amount of daylight for lighting—this can be counterproductive.

The ideal for any climate is adjustable control for windows that in winter will allow the maximum benefit for both heating and lighting and in summer will keep the direct sun out but let skylight in.

The penetration of daylight into any space is in part dependent on the color and texture of exterior materials near the window. The actual illumination at a task within a space can be materially improved by increasing the reflectances of nearby exterior surfaces. In light wells and courts, walls should be painted light colors. White stone or concrete at the base of a building will bounce light into ground floor ceilings. In like manner outside horizontal reflectors at the windowsill can increase room daylight by 25%. Vertical reflectors can be equally useful but can also be difficult-to-control sources of glare.

Skylights are windows in the roof. They have the potential for admitting much light (as much as 40 W equivalent fluorescent lighting per square foot of skylight), but much difficult-to-control sun as well. In many buildings the skylights have been closed off because of this latter problem of the sun. These skylights can be cleared to admit valuable daylight if controls are added to keep out the sun (unless it is wanted in the winter). Such control might be an operable louver installed above the skylight. If, when the direct sun is admitted, glare is a problem, the skylight can be painted a diffusing white. A better solution is to add a prismatic lens at the base of the skylight. See Fig. 9.6.

ACTION GUIDELINES

☐ 1. Clean windows and skylights.

☐ 2. Where practical, schedule periods of occupancy, cleaning, and meetings to maximum use of daylight.

☐ 3. Locate tasks that need the best illumination closest to the windows, with the task-viewing angle parallel to the windows.

□ 4. Switch off electric lights in areas when natural light is available.

□ 5. To reduce glare, rearrange work stations so that side wall daylight crosses perpendicular to the lines of vision.

□ 6. Refinish interior room surfaces with lighter colors that are more reflective to increase the efficacy of all light sources.

□ 7. Modify existing skylights to use available daylight. Modifications should include removing paint or coverings which may have been installed to reduce glare or heat, and substituting proper solar control devices to keep out direct sun but allow diffuse light to enter.

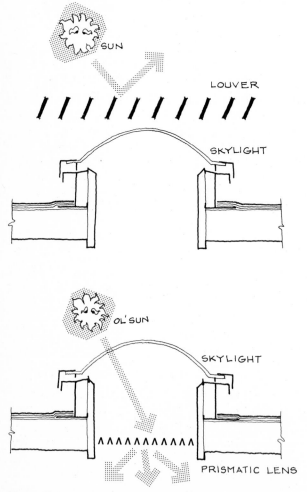

Figure 9.6 Solar controls for skylights.

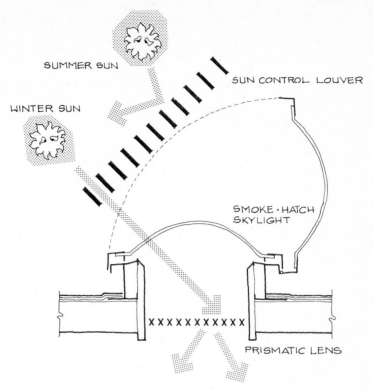

SUMMER SUN

SUN CONTROL LOUVER

WINTER SUN

SMOKE·HATCH
SKYLIGHT

X X X X X X X X X X X

PRISMATIC LENS

Figure 9.6 (*Continued*)

☐ 8. When remodeling or extending the building, install skylights and/or windows for the maximum practical use of natural lighting; be guided in doing this by an analysis indicating the reduction in electrical energy for lighting which will not result in an increase of the heating or cooling loads by the same or larger amount of energy.

☐ 9. Install reflectors on the exterior of the building or treat the horizontal surface below windows with light-reflective materials to increase the intensity of daylight illumination at the window surface.

☐ 10. Install an operable shutter above the skylight to shade the sun but let in the light in the summer and let in both in the winter.

☐ 11. Paint the interior of the light well with a light color to increase illumination. Where skylights are used as an automatic smoke hatch, modifications must not change this function.

☐ 12. Paint skylights with white or light-colored paint to reduce glare.

☐ 13. Add prismatic lighting lenses below skylights to reduce glare.

9.5 IMPROVE SPACE CONDITIONS FOR LIGHTING

Much of the light from a fixture is reflected from room surfaces and furnishings before it reaches a visual task. Reflected light will have less value, or strength, than direct light, by the amount absorbed by the walls and other reflecting surfaces.

The larger a room, and the lighter the room finishes, the lower will be the amount of reflected light absorbed and the greater will be actual task illumination. As an example of the importance of room size, a study for the General Services Administration building in Manchester, New Hampshire, by Dubin-Bloom Associates indicated that a 50 ft × 80 ft office space designed for open landscape planning would use 25% less energy for illumination than an equal amount of area divided into 30 smaller-sized offices.

As an indication of the importance of reflectance, if wall, ceiling, and floor reflectances of 50-30-20 are increased to 80-60-40, the lighting level will go up by about 15%; if reflectances of 50-10-10 are increased to 80-60-40, the lighting level will go up by about 35%.

The following examples show the increase in illumination with increased room size and reflectance. Note in both examples that the increase in efficacy is specifically expressed as an increase in the term *coefficient of utilization* (CU).

Example 1

Effect of Room Size on Wattage per Square Foot To Produce 60 fc, Maintained

$$\text{Number of fixtures} = \frac{\text{area} \times \text{footcandles}}{\text{no. of lamps} \times \text{lumens} \times \text{CU} \times \text{MF}}$$

$$= \frac{\text{area} \times 60}{4(3150) \times \text{CU} \times 0.75}$$

$$= \frac{\text{area} \times 0.0064}{\text{CU}}$$

For a 12 ft × 12 ft room:

$$\text{Number of fixtures} = \frac{144 \times 0.0064}{0.43} = 2.19 \text{ fixtures} = 438 \text{ W}$$

$$\frac{\text{Watts}}{\text{Square foot}} = \frac{438}{144} = 3.04 \text{ W/ft}^2$$

For a 16 ft × 24 ft room:

$$\text{Number of fixtures} = \frac{384 \times 0.0064}{0.56} = 4.39 \text{ fixtures} = 878 \text{ W}$$

$$\frac{\text{Watts}}{\text{Square foot}} = \frac{878}{384} = 2.28 \text{ W/ft}^2$$

For a 40 ft × 60 ft room:

$$\text{Number of fixtures} = \frac{2400 \times 0.0064}{0.70} = 21.9 \text{ fixtures} = 4388 \text{ W}$$

$$\frac{\text{Watts}}{\text{Square foot}} = \frac{4388}{2400} = 1.83 \text{ W/ft}^2$$

The electrical energy load, in watts per square foot, for lighting in rooms with one-half or three-quarter height partitions (with or without glass from top of partition to the ceiling) falls between that for individual office layout and that for open space.

Example 2

Effect of Room Finish on Power Wattage Requirements

Basic formula: Assume a 16 ft × 24 ft room with MF = 0.75.

$$\text{Number of fixtures} = \frac{\text{area} \times \text{footcandles}}{\text{lamps} \times \text{lumens} \times \text{CU} \times \text{MF}}$$

$$= \frac{384 \times 60}{4(3150) \text{ CU} \times 0.75} = \frac{2.44}{\text{CU}} = \text{no. of fixtures}$$

With 80-60-40 interreflectances, CU = 0.63; therefore $\frac{2.44}{0.63} = 3.85$ fixtures, for a total of 770 W.

$$\frac{\text{Watts}}{\text{Square foot}} = \frac{800}{384} = 2.08 \text{ W/ft}^2$$

With 50-30-20 interreflectances, CU = 0.53; therefore $\frac{2.44}{0.53} = 4.58$ fixtures, for a total of 916 W.

$$\frac{\text{Watts}}{\text{Square foot}} = \frac{916}{384} = 2.38 \text{ W/ft}^2$$

With 50-10-10 interreflectances, CU = 0.47; therefore $\frac{2.44}{0.47} = 5.17$ fixtures, for a total of 1034 W.

$$\frac{\text{Watts}}{\text{Square foot}} = \frac{1034}{384} = 2.69 \text{ W/ft}^2$$

One additional direction can be explored in improving task illumination, and that is actually lowering the entire ceiling including the lights, or lowering the mounting heights of the fixtures alone. If a room were infinite in breadth and width, doing this would add nothing, but in average rooms the walls absorb much of the light before it reaches the task, even if the walls are white. When the mounting height of the light is lowered, less light is reflected by the walls and more is placed on the task.

By dropping the lighting mounting height from 14 to 9 ft in a 30 ft × 40 ft space, the same illumination level can be maintained with 10% less energy. In supermarkets, warehouses, and high-ceilinged lobbies fixtures are often mounted on the ceiling, 14 to 18 ft above the floor. Conversion to pendant mounting can be very productive. But in most cases, this will be an extreme move, and is mentioned simply to cover all possibilities. If ceilings are already 10 ft or less, there is little to be gained in lowering them further. With pendant lighting in high spaces the merit in lowering the light level is greater. Increased glare from lowered lights is a factor, though, and so is the possible damage done to a carefully designed space. With older buildings, furniture-mounted lighting is frequently a fine solution to improving task lighting without aesthetic dismemberment. More than one fine old library has returned to table lighting after misadventures with low arrays of over-head-suspended luminaires for task lighting.

ACTION GUIDELINES

☐ 1. Clean and wash walls, ceilings, and floors.

☐ 2. When recarpeting or retiling, use lighter-colored carpets or tiles.

☐ 3. Paint light-colored reflective finishes on interior room surfaces to increase interreflectances and to improve the efficacy of both natural and artificial illumination: walls first in priority, then ceilings and floors.

☐ 4. When undertaking alterations for other purposes, use open landscape planning to reduce the amount of lighting energy which is absorbed by walls.

☐ 5. Where partitions cannot be removed entirely, install low partitions to reduce light absorption and to permit the use of "borrowed" light from adjacent spaces.

☐ 6. Select lighter-colored furnishings that do not have a glossy surface or give specular reflections.

☐ 7. When renovating, consider lowering the fixture mounting height.

☐ 8. Relocate existing lighting fixtures to minimize veiling reflectances. Where possible, light tasks from the side rather than from the front.

□ 9. Where appropriate use furniture-mounted lighting, then reduce background lighting levels.

9.6 IMPROVE LAMP AND FIXTURE EFFICACY

The efficacy of lamps is measured in lumens per watt. Selecting lamps with higher lumen-per-watt output permits the removal of some lamps or the raising of light levels. More efficient lamps will impose smaller heat loads on the air-conditioning system; in supermarkets and produce markets reduced lighting will reduce loads on commercial refrigeration systems as well. In winter, any heat lost by a reduction in lighting wattage can generally be supplied more efficiently by the heating system.

The extreme variation in efficacy of lamps is clearly shown in the listing below, taken from Table 9.4.

Sodium vapor	83–140 lm/W
Metal halide	80–115 lm/W
Mercury vapor	20–63 lm/W
Fluorescent	31–84 lm/W
Incandescent	4–25 lm/W

A first reading would indicate that all fixtures should be relamped for sodium vapor, but good lighting design and the use of higher-efficiency lamps are not always compatible for all functions. A high-pressure sodium vapor lamp is a very bright, concentrated light source and is difficult to use in, say, a low-ceiling office without excessive glare, aside from questions of color rendition.

Fluorescent lamps with outputs up to about 84 lm/W are easier to handle. In another case, a 20-W incandescent lamp in a fixture directly attached to a paper holder at a typist's desk could provide 60 fc on a manuscript; 100 W of fluorescent lighting might be required from a ceiling fixture to give the same illumination, even though the fluorescent lamp produces more lumens per watt. Aesthetics, size, efficacy, color, initial cost, and cost of operation and maintenance all help to determine choice of lamps. Energy conservation in lighting is not synonymous with a sacrifice in quality; good lighting is obtained by an intelligent application of many interrelated factors. Correctly applied, energy conservation can increase the quality of lighting while reducing operating costs. Table 9.3 suggests some energy-conserving lamp applications for various building functions, Table 9.4 lists lamp characteristics, and Table 9.5 suggests substitutions in incandescent lamps to reduce wattage requirements without modifications to luminaires.

The ballast for electrical discharge lights requires wattage in addition to that needed for the lamps. Ballasts, like lamps, vary in efficacy. When the standard ballasts in existing systems fail or must be replaced, it is possible to save 2 to 4 W per ballast by using a line of premium-priced ballasts. For instance, in an installation of 1000 two-lamp 40-W fluorescent luminaires, each ordinary ballast consumes 12 to 14 W, amounting to an annual energy consumption of 24,000 to 28,000 kWh in buildings operating 2000 hours per year. A more efficient ballast which consumes 10 W will use only 20,000 kWh annually, for a savings of 4000 to 8000 kWh a year. Figure 9.7 shows lamp efficacies. It is based on the following types of lamps:

- High-pressure sodium: 70, 100, 150, 250, 400 and 1000 W

- Metal halide: 175, 250, 400, 100 and 1500 W

- Fluorescent
    430 mA: 4-84 W, cool white
    800 mA: 20-110 W, cool white
  1500 mA: 110-220 W, cool white

- Mercury vapor: 40-3000 W, deluxe white

- Tungsten halogen: 43-1500 W

- Incandescent: 3-1500 W, inside frosted

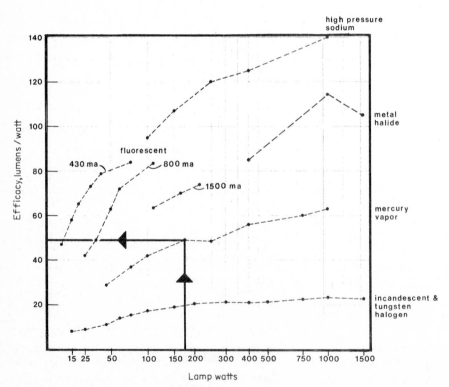

Figure 9.7 Lamp efficacies by lamp type and wattage. (Developed from manufacturers' information and is typical for generic types of lamps.)

TABLE 9.3   EXAMPLES OF ENERGY-CONSERVING LAMP APPLICATIONS

Lamp Type	Applications
Incandescent	1. Decorative display lighting
	2. Religious worship halls
	3. Work closets or other very confined spaces
	4. Stage spotlighting
	5. Tasks which require a small light source
Fluorescent	1. Offices and other relatively low-ceiling applications
	2. Flashing advertising signs
	3. Islands at service stations
	4. Display cases in stores
	5. Desk lamps
	6. Classrooms or training centers
	7. Cafeterias
High-intensity discharge	1. Stores and some office areas
	2. Auditoriums
	3. Outdoor area lighting
	4. Outdoor floodlighting
	5. Outdoor building security lighting
	6. Marking of obstructions

Most fluorescent installations currently in use operate at the power supply frequency of 60 Hz. Higher efficacy with fluorescent lighting is possible with higher frequencies. When remodeling or expanding all or a portion of an office building or store with fluorescent lighting (or when changing from incandescent lighting to fluorescent), consider high-frequency lighting. Existing fixtures can be modified to be used with high-frequency systems, and new fixtures are available with high-frequency ballasts. Though conversion costs are quite high, high-frequency system lamps produce about 10% more lumens per watt; thus fewer lighting fixtures are required, and ballasts can be located out of the conditioned area, reducing the load on the air-conditioning system.

The higher frequencies most commonly considered are 400, 800, and 3000 Hz. Future installations may include 30,000 Hz; the higher the frequency,

TABLE 9.4 LAMP CHARACTERISTICS

Characteristics	Incandescent (including tungsten halogen)	Fluorescent	High-Intensity Discharge (HID)		
			Mercury Vapor	Metal Halide	High-Pressure Sodium
Wattages normally available	3 to 1500	4 to 219	40 to 3000	175, 400, 250, 1000, 1500	70, 100, 150, 250, 400, 1000
Efficacy, lm/W, lamp only	4 to 25	31 to 84	20 to 63	80 to 115	83 to 140
Life, h	750 to 12,000	9000 to 30,000	16,000 to 24,000	1500 to 15,000	10,000 to 20,000
Light control	Very good to excellent	Fair	Good to very good	Good to very good	Very good
Relight time	Immediate	Immediate	3 to 5 min	10 to 20 min	Less than 1 min
Color rendition	Very good to excellent	Good to excellent	Poor to very good	Good to very good	Fair
Initial installation cost	Low because of simple fixtures	Moderate	Higher than incandescent and fluorescent	Generally higher than mercury vapor	Highest
Comparative operating cost	High because of relatively short life and low efficacy	Lower than incandescent; replacement costs higher than HID because of greater number of lamps needed: energy costs generally lower than mercury vapor	Lower than incandescent; replacement costs relatively low because of relatively few fixtures and long lamp life	Generally lower than mercury vapor; fewer fixtures required, but lmap life is shorter and lumen maintainence not quite as good	Generally lowest; fewest fixtures required

TABLE 9.5 IMPROVED EFFICACY WITH RELAMPING: INCANDESCENT LAMPS

Present Light Source*						Replacement Light Source*				
Watts	Description	Volts	Lumens	Watts	Hours Life	Watts	Description	Lumens	Hours Life	Reduction, W
40	40 A/99	130	323	35	7,000	25	25 A	235	2,500	10
60	60 A/99	130	597	53	7,000	40	40 A	455	1,500	13
						40	40 A/99	420	2,500	13
75	75 A/99	130	770	66	7,000	54	54 A	775	3,500	12
						55	55A	670	2,500	11
						60	60 A	870	1,000	6
						60	60 A/99	775	2,500	6
100	100 A/99	130	1,147	88	7,000	75	75 A/99	1,190	750	13
100	100 A21/99	130	1,109	88	7,000	75	75 A/99	1,000	2,500	13
150	150 A23/99	130	1,779	132	7,000	90	90 A	1,290	3,500	42
150	150/99	130	1,771	132	7,000	92	92 A	1,490	2,500	40
						100	100 A	1,750	750	32
						100	100 A/99	2,500	1,490	32
150	150 R/FL	120	1,870	150	2,000	75	75ER30	900	2,000	75
						75	75PAR/FL	765	2,000	75
150	150PAR/FL	120	1,740	150	2,000	250	Q 250 PAR38	3,220	6,000	50
200	200 A/99	130	2,626	176	7,000	150	150 A	2,880	750	26
200	200/99 IF	130	2,510	176	7,000	150	150 A23/99	2,310	2,500	26

TABLE 9.5 IMPROVED EFFICACY WITH RELAMPING: INCANDESCENT LAMPS (*Continued*)

| Present Light Source° | | | | | Replacement Light Source° | | | | |
Watts	Description	Volts	Lumens	Hours Life	Watts	Description	Lumens	Hours Life	Reduction, W
					135	135A	2,100	3,500	41
					138	138 A	2,300	2,500	38
					160	HSB160/SS/M	2,700	20,000	16
60	60/99IF Extended Service	120	740	2,500	54	54/99IF Extended Service	645	2,500	6
60	60A19/35 Industrial Service	120	670	3,500	54	54A19/35 Industrial Service	590	3,500	6
100	100/99IF Extended Service	120	1,480	2,500	90	90/99IF Extended Service	1,230	2,500	10
100	100A21/35 Extended Service	120	1,280	3,500	90	90A21/35 Extended Service	1,090	3,500	10
150	150A/99IF Extended Service	120	2,350	2,500	135	135A/99IF Extended Service	1,990	2,500	15
150	150A25/35 Industrial Service	120	2,150	3,500	135	135A25/35 Industrial Service	1,790	3,500	15

300	300	120	4,900	300	3,000
1,000	1,000	120	18,300	1,000	3,000
1,500	1,500	120	28,400	1,500	3,000
300	300 M/99 IF	130	3,996	264	7,000
300	300/99 IF	130	3,996	264	7,000
300	300	120	5,820	300	1,000
500	500/99 IF	130	6,984	440	7,000
500	500/99 IF	120	9,070	500	2,500
750	750/99	130	10,934	660	7,000
750	750 R 52	120	13,000	750	2,000
1,000	1000/99	130	15,246	880	7,000
100	Reflector Floodlight R-40	800	5,000	75	

250	250PS-35 Self-Ballasted Mercury	4,800	11,000	50
750	750R-57 Self-Ballasted Mercury	17,650	15,000	250
1,250	1250 Bt-56 Self-Ballasted Mercury	38,000	15,000	250
200	200A	4,010	750	64
200	200A/99	3,410	2,500	64
300	HSB 300/SS/M	7,800	20,000	0
300	HSB 300/SS	7,800	20,000	0
300	HSB 300/SS	7,800	20,000	140
450	HSB 450/SS	9,500	16,000	50
500	500	10,850	1,000	160
500	500/99	9,070	2,500	160
450	HSB 450/SS	9,500	16,000	210
750	HSB 750R/120	14,000	16,000	0
750	750	17,040	1,000	130
750	750/99	14,200	2,500	130
	Projector Floodlight PAR-38	1,430	5,000	25

TABLE 9.5 IMPROVED EFFICACY WITH RELAMPING: INCANDESCENT LAMPS (*Continued*)

	Present Light Source*				Replacement Light Source*				
Watts	Description	Volts	Lumens	Hours Life	Watts	Description	Lumens	Hours Life	Reduction, W
150	Reflector Floodlight R-40	1,200	5,000		100	Projector Floodlight PAR-38	2,230	5,000	50
150	Reflector Floodlight R-40	1,200	5,000		75	Projector Floodlight PAR-38	1,430	5,000	75
150	Reflector Floodlight R-40	1,200	5,000		100	Floodlight BR-40	1,200	5,000	50
200	Reflector Floodlight R-40	1,600	5,000		150	Projector Floodlight PAh-38	3,450	5,000	50
300	Reflector Floodlight R-40	2,450	5,000		200	Projector Floodlight PAR-38	4,560	5,000	100
500	Reflector Floodlight R-40	3,600	5,000		250	Projector Floodlight PAR-38	5,850	5,000	250

*All lamps operating at 120 V.

the greater the advantage. Figure 9.8 graphs these advantages; note, however, that they apply only to fluorescent systems, not high-intensity discharge (HID) or incandescent systems. Table 9.6 suggests substitutions in fluorescent lamps to reduce wattage requirements without modifications to luminaires. Table 9.7 does the same for high-intensity discharge lamps.

Two types of converters commonly used are rotary-phase converters, which are about 70 to 85% efficient, and solid-state converters, which are approximately 90% efficient. The increase in lamp efficacy and the reduction in power for air conditioning exceed these converter losses. The converter and series capacitors can be mounted in each circuit or at a central location. With a central system, the circuits serving the fluorescent lighting fixtures

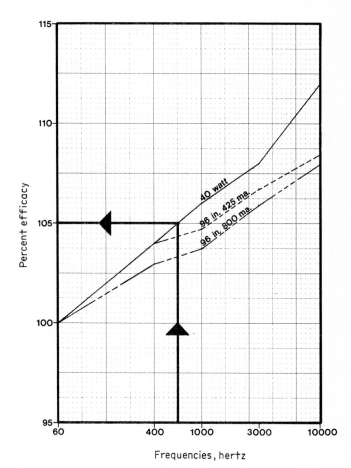

Figure 9.8 The improvement of efficacy of fluorescent lamps with increased frequency. (Based on data published by the IES obtained from measurements of lumen output over a range of frequencies.)

TABLE 9.6   IMPROVED EFFICACY WITH RELAMPING: FLUORESCENT LAMPS

	Present Light Source				Replacement Light Source			
Watts	Description	Color	Lumens	Nominal Watts	Reduced Wattage Fluorescent Description	Color	Lumens	Possible Reduction, W
40	F40T12-48″	CW	3150	35	F40/RS	CW	2800	5–7
40	F40T12-48″	WW	3200	35	F40/RS	WW	2900	5–7
75	F96T12 Slim line	CW	6300	60	F96T12	CW	5220	13–18
75	F96T12 Slim line	WW	6400	60	F96T12	WW	5340	13–18
110	F96T12/HO	CW	9200	95	F96T12	CW	8500	12–17
110	F96T12/HO	WW	9200	95	F96T12	WW	8500	12–17

must be isolated from circuits supplying other loads. As a result, major wiring revisions are often required in existing installations and must be accounted for in cost and economic feasibility analyses. Individual converters which can be mounted directly on a fluorescent light fixture (eliminating the need for separate wiring) are in development and may be available soon. If a total energy system providing electricity to the building is installed, high-frequency lighting can be generated directly, and the costs for converters can be eliminated.

Lamp and fixture efficacies together determine the quantity of light transmitted into a space for each watt of power consumed. When a more efficient lamp source suits the application, but conversion of the fixture to handle this source is not possible, consider replacing the fixture itself. Select the most efficient lamp for the application, then choose a fixture with good performance and reasonable brightness control. Fixture performance is indicated by the CU under specific room conditions and includes the effect of lenses or reflectors. The CU for various conditions is given in the manufacturer's data.

The price of fixtures unfortunately is typically inversely related to efficacy; incandescent fixtures are generally less expensive than fluorescent, which are less expensive than the higher-output HID fixtures. But the life-cycle cost is what counts, as clearly indicated in Fig. 9.9. Examples of conservation of energy by replacing fixtures and lamps are given in Table 9.8.

Many fixtures have lenses that diffuse an intense light source, or improve distribution of the light, but in so doing sharply reduce the fixture efficacy. If lenses are not really required for glare control or distribution, they can simply be removed. This is often the case in corridors, toilets, and storage

TABLE 9.7 IMPROVED EFFICACY WITH RELAMPING: HIGH-INTENSITY DISCHARGE LAMPS

	Present Light Source					Replacement Light Source				
Watts	Description	Color	Lumens	Hours Life	Watts	Description	Color	Lumens	Hours Life	Reduction, W
400	Mercury-vapor H33CD-400	Clear	21,000	24,000	300	Mercury vapor H33CD-300	Clear	14,000	16,000	100
	H33G1-400/DX	Deluxe white	23,000	24,000	300	H33GL-300/DX	Deluxe white	15,700	16,000	100
400	Mercury vapor		23,000	24,000	360	High-pressure sodium		34,200	12,000	40
175	Mercury vapor		8500	24,000	150	High-pressure sodium		12,000	12,000	25

rooms. If lenses are inefficient but nonetheless necessary, they can be replaced with more efficient types.

To evaluate the efficiency of lenses in existing fixtures, use the fixture manufacturer's CU. Compare the lens and CU to the resulting CU with contemplated replacements. The improvement in light level will be directly proportional to improvement in CU. The relative efficiency and general characteristics of the most common lenses, compared with prismatic plastic, are as follows:

Prismatic Plastic and Glass: Generally the most efficient of its type for the degree of glare control. Best suited for all finished office space applications. A nonyellowing type of plastic should be selected.

Plastic Louver (½ in × ½ in or ⅜ in) : Very inefficient. Should be used only where extreme dirt buildup is a problem. About 33% less efficient than prismatic plastic.

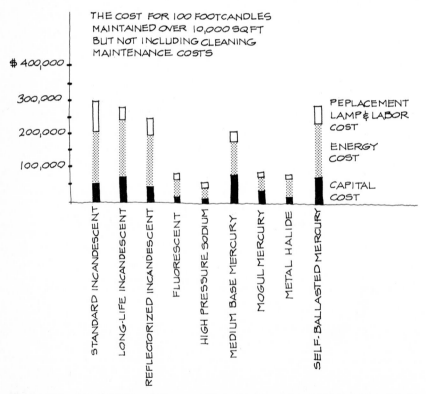

Figure 9.9 An estimate of the overall costs for various lamp types due to capital, energy, and replacement costs.

TABLE 9.8   EXAMPLES OF CONSERVATION OF ENERGY BY REPLACING FIXTURES AND LAMPS

Existing	*Replacement*
STORE	
Sixty-six 500-W incandescent down lights	Twenty 400-W metal halide down lights
Total kWh/yr = 122,100	Total kWh/yr = 34,040
Total savings = 88,060 kWh/yr = 72% savings	
Seventeen 1000-W incandescent pendant fixtures	Fourteen 400-W metal halide pendant fixtures
Total kWh/yr = 62,900	Total kWh/yr = 24,050
Total savings = 38,850 kWh/yr = 62% savings	
PARKING AREA	
Twelve 1500-W tungsten halogen lamp floodlights	Four 1000-W metal halide cluster
Total kWh/yr = 27,000	Total kWh/yr = 6480
Total savings = 20,520 kWh/yr = 76% savings	
SIGN	
Two 1500-W tungsten halogen floodlights	Two 400-W metal halide floodlights
Total kWh/yr = 12,000	Total kWh/yr = 3680
Total savings = 8320 kWh/yr = 69% savings	
SECURITY	
Six 500-W tungsten halogen floodlights	Six 150-W high-pressure sodium lights
Total kWh/yr = 12,000	Total kWh/yr = 3920
Total savings = 8080 kWh/yr = 67% savings	
SUPERMARKET	
Two hundred fifty 4-lamp 40-W strip fluorescent fixtures	Seventy 400-W high-pressure sodium down lights (where change in color rendition is acceptable)
Total kWh/yr = 146,250	Total kWh/yr = 84,000
Total savings = 62,250 kWh/yr = 43% savings	

TABLE 9.8 EXAMPLES OF CONSERVATION OF ENERGY BY REPLACING FIXTURES AND LAMPS (*Continued*)

OFFICE

Building area	1 million ft²
Office area (fluorescent)	700,000 ft²
Support areas (incandescent)	
Mechanical areas	150,000 ft²
Circulation and storage	150,000 ft²

	Existing	*Replacement*
Mechanical area Lighting energy Consumption	Primarily incandescent lamps used	90% fluorescent lamps used
	117,500 kWh/yr	70,500 kWh/yr
Circulation and storage area lighting energy consumption	85% incandescent 15% fluorescent	15% incandescent 85% fluorescent
	518,400 kWh/yr	324,00 kWh/yr
Yearly energy consumption	635,900 kWh/yr	394,500 kWh/yr
Total energy saved	635,900 kWh/yr − 394,500 kWh/yr 241,400 kWh/yr = 38% savings	

Parabolic Wedge Louver: Very inefficient in transmitting light but provides excellent visual comfort and glare control. This lens should be used for special effects for very low ceiling brightness and minimum distraction.

Fresnel-Type Lens: The most efficient of its type, best suited for all recessed fixture applications.

Opal or White Diffusers: Much less efficient in delivering light to a work surface. Generally 30% less efficient than Fresnel-type lens. Good light diffusion but may become glare source.

Labor costs for lens replacement in existing fixtures are about $10.00 per unit. The materials costs of various replacement lenses are approximately as follows:

Primsatic plastic (acrylic 2 ft × 4 ft)	
Extruded	$ 8.00
Injection molded	12.00
Prismatic glass (2 ft × 4 ft)	20.00
Louvered (2 ft × 4 ft)	
Plastic	12.00
Aluminum	10.00
Parabolic wedge louver (2 ft × 4 ft)	40.00
Dropped opal (acrylic 2 ft × 4 ft)	12.00
Fresnel (8 in × 8 in)	8.00
White diffuser (8 in × 8 in)	5.00

ACTION GUIDELINES

☐ 1. Use a single, larger incandescent lamp where possible, rather than two or more smaller lamps. Higher-wattage general-service incandescent lamps are more efficient than lower-wattage lamps.

☐ 2. Avoid multilevel lamps when light levels are not often changed. The efficacy of a single lamp is higher per watt than a multilevel lamp.

☐ 3. Avoid replacing incandescent lamps with self-ballasted mercury vapor lamps; they give less light than incandescent lamps of the same wattage.

☐ 4. Avoid using extended-service lamps except in special situations in which short lamp life is a problem.

☐ 5. When fluorescent ballasts burn out, replace them with high power factor (90% or more) low-watt ballasts. (Circuits with 50% power factor ballasts have 240% more energy losses in wiring than those with 90% power factor ballasts.)

☐ 6. When undertaking major alterations or additions to a space or when replacing groups of lighting fixtures, consider high-frequency lighting to obtain more lumens per watts.

☐ 7. When undertaking major alterations or expansions in an air-conditioned building, use air extract fixtures in a "dry heat of light" system to reduce the room cooling load and to increase the lamp efficacy.

☐ 8. When undertaking major alterations or expansions with 480/277-V service, install 277-V fluorescent lighting in preference to the 110/120-V type, and provide dry-type transformers to serve convenience outlets.

□ 9. Replace all incandescent parking lighting with high-pressure sodium or metal halide lamps.

□ 10. Remove display lighting fixtures in retail stores, bulletin boards, and other areas where inefficient incandescent lamps are used, and substitute fluorescent fixtures.

□ 11. Replace existing lamps with new ones which have a lower lamp lumen depreciation over rated life (initial light output is often increased as well). Fluorescent lamps with better LLD factors may cost somewhat more, but the investment can generally be recovered by lower expenditures for energy and maintenance.

□ 12. Modify existing fixtures to accommodate higher-efficiency lamps.

□ 13. Replace incandescent fixtures with more efficient fluorescent or mercury fixtures, depending upon the application.

□ 14. When replacing fixtures, select a type with a higher coefficient of utilization.

□ 15. Where visual comfort is critical, it may be better to use a less efficient lens with better glare control. The quality of lighting will be improved, permitting a reduction in footcandles (and wattage) for the same visual comfort.

□ 16. Remove all louvers and lenses in areas where the CU will be improved and the additional glare can be tolerated. Where lenses are necessary to control glare, use efficient types.

CHAPTER TEN

POWER

10.1 THE NATURE OF THE LOADS

The uses and losses related to electricity are not easily set out as building, distribution, or conversion loads; thus, for simplicity, all things electrical (except lighting) are included in this common chapter. Of interest are electrical equipments of an assortment that include elevators, coffeepots, soft-drink machines, and typewriters, and operations that control their consumptive use. Refrigeration equipment, which may account for 50% of the energy usage of supermarkets, is especially important to this interest. Of additional concern are measures that limit peak electrical demands for the whole building (to drastically reduce electricity bills where utility rates include demand charges), measures that improve the efficiency of motors, and measures that reduce transformer losses.

10.2 REDUCED ENERGY REQUIREMENTS FOR ELEVATORS AND ESCALATORS

Elevators and escalators account for about 1 to 4% of the electrical energy required for office buildings and large department stores. The amount of power required annually to operate an elevator is a function of the height of the building, the number of stops, passenger capacity and load factors, and the efficiency of the hoisting mechanism. A 2500-lb-capacity "local" elevator making 150 stops per car-mile consumes 5 kWh per car-mile; an "express" elevator making 75 stops per car-mile consumes 4 kWh. A 4500-lb-capacity elevator in a 12-story department store stopping at every floor will use 13 kWh per car-mile. Consumption will vary between elevators of the same capacity depending on the type of hoisting motor and control, whether the elevator is hydraulic, geared or gearless, and the kind of service and the amount of load offset by the counterweight. Speeds should be selected that are as slow as possible, while keeping maximum waiting time to no more than 2 min. The elevator manufacturer or a consulting engineer can provide help in analyzing traffic patterns and in selecting automatic programs for providing lowest speeds and heaviest loadings (reducing the number of elevator units traveled per year) to conserve energy. Where motor generator (MG) sets (which draw power whether or not the elevator is in operation) are installed in existing buildings, they should be deenergized with 7-day timers for periods when one or more elevators are not required. Consideration should be given to changing from MG sets to more efficient silicon controlled rectifier (SCR) controllers.

Escalators, unlike elevators, consume energy whether they are carrying passengers or not. Assuming 35% equivalent full-load operation, escalator energy consumption may vary from 1.3 kW/h for a 32-in-wide model operating at 90 ft/min with a 14-ft vertical rise, to 3.0 kW/h for a 48-in-wide model operating at 90 ft/min with a 25-ft vertical rise.

ACTION GUIDELINES

☐ 1. Reduce the number of elevators in service during hours when a majority of persons are not leaving or entering the building.

☐ 2. Turn off the motor-generator set located in the elevator machine room when not in use: nights, weekends, holidays, and slack periods during the day.

☐ 3. Turn off escalators to unoccupied floors of offices or retail stores during renovations.

☐ 4. Operate demand escalators only during peak periods.

☐ 5. Reduce speed of escalators and elevators.

☐ 6. Where security arrangements permit, encourage employees to walk up and down one flight of stairs rather than use vertical transportation systems.

☐ 7. Consider turning off all "down" escalators during periods of light traffic.

☐ 8. Consider turning off "up" escalators on alternate floors during periods of light traffic.

☐ 9. Deenergize MG elevator sets or use SCR controllers. Select more efficient elevators and control devices when replacing elevators or expanding elevator service.

10.3 REDUCE ENERGY CONSUMPTION FOR EQUIPMENT AND MACHINES

Most buildings contain many electrically driven machines which are left switched on and idling but are used only for short periods of time or only when the building is occupied. Equipment not in use should be turned off. Automatic timers with remote-control switching are valuable to deenergize equipment which is not required at night and during weekends.

It is natural to think first of large motors and large loads only when embarking on an energy conservation program, but the power consumption of many small, inefficient motors, in aggregate, can exceed the same amount of connected horsepower of one larger motor. Where circuiting permits, the operation of small motor-driven equipment should be controlled by a single timer. Where multiple circuits supply equipment, recircuiting should be considered.

In addition to motors, other loads, such as resistance heating equipment, electric cooking equipment, meat preparation rooms in supermarkets and restaurants, electric signs, and business machines, can be deenergized for a substantial portion of the week.

The costs to deenergize equipment are for the control system only, unless other equipment on the same circuit must be recircuited. Automatic time switches can be installed at a distribution panel for about $100; remote-control switches vary in cost depending upon the length of interconnected wiring between switch and contactor. Relays or contactors vary greatly in cost depending on electrical rating and number of poles, but a 100-A contactor and time clock combination costs about $450 installed at the load panel.

Improving the insulation on equipment such as fryers, ovens, food warmers, kilns, refrigerators, and freezers provides additional opportunities to conserve energy. For example, a U.S. Navy study indicates that a 9-kW insulated fryer has the same capacity as a 12-kW noninsulated unit. In addition to savings in power for operation, insulating equipment reduces the heat gain to spaces to a minimum, resulting in lower installation and operating costs for the air-conditioning system.

In existing facilities where newer types of equipment have not been installed, shielding with aluminum or asbestos barriers and insulation of hot surfaces are effective and should be utilized to reduce unwanted heat gains and eliminate sources of radiant heat. Where heating and cooling equipment are within inches of each other, use rigid insulation between them.

When choosing new items, select equipment which is well-insulated, which shortens preparation time (more efficient ovens, pressure cookers), and which has a surface temperature no greater than 90°F. Additional costs will be quickly recovered, and the kitchen working conditions will be improved.

As an example, in a fast-food restaurant, a deep-fat fryer was used 7 hours per day, 200 hours per month. If an insulated 9-kW fryer were used rather than a noninsulated 12-kW unit, 600 kWh could be saved each month. The yearly savings would be $216, which would amortize the cost of the additional fryer in a short period of time.

ACTION GUIDELINES

☐ 1. Turn off coffeepots and food warmers when not in use.

☐ 2. Turn off refrigerated drinking fountains at the end of normal business hours.

☐ 3. Turn off refrigeration units.

☐ 4. Turn off vending machines at the end of the week where food spoilage is not a problem. Use time clocks to turn on the vending machines in time for the soft drinks to reach 45°F by Monday morning when employees arrive.

☐ 5. Turn off portable electric heaters, portable fans, typewriters, calculators, and reproduction machines when not in use.

☐ 6. Turn off automatic window displays and revolving signs at the end of normal business hours (and consider further reductions in operating time).

☐ 7. Turn off electric heat tracing when there is no fluid flow in pipes and when the outdoor temperature is above freezing.

☐ 8. Turn off elevator fans where smoking is not permitted and where applicable codes allow.

☐ 9. Encourage employees to go to the cafeteria or canteen for coffee breaks rather than operating coffee percolators in offices.

☐ 10. Encourage chefs to preheat ovens no earlier than necessary and to forgo preheating completely except for baked goods.

☐ 11. Consider reducing the number of electrically powered business machines in use.

☐ 12. Insulate cooking equipment in kitchens, when possible.

☐ 13. Prohibit use of portable electric heaters and encourage employees to move to a different location on the floor if drafts or cold radiations from windows are causing them discomfort.

☐ 14. Where practical, substitute manual labor for electrical power, such as using manual labor to remove snow and ice rather than electric resistance snow-melting systems.

□ 15. Install manual or automatic controls to disconnect loads when not required.

□ 16. Insulate electric heating and cooling equipment. Purchase new equipment with better insulation.

10.4 REDUCE ENERGY REQUIREMENTS FOR COMMERCIAL REFRIGERATION

Internal lights in refrigerated cabinets heat the cold air and increase the load on the refrigerating machine. In many cases, the internal lights can be removed and the contents of the cabinet lit either from repositioned existing fixtures or from new fixtures positioned to shine into the cabinet. The reduction in load is in direct proportion to the wattage of the lamp removed and is affected by type (fluorescent lamps give off 65% of their rating as heat, incandescent lamps give off 90% of their rating as heat). Each kilowatt of "lighting heat" removed will reduce the refrigeration load by 0.28 ton. Some reduction in display effectiveness will result when internal lights are removed, and the trade-off between this and the energy saved should be considered in stores and supermarkets.

Some existing cold cabinets and deep-freeze chests in stores and supermarkets, because of their design, have to be open to allow visual display and public access to the produce. Heat exchange takes place at the interface between the warm room air and the cold freezer air owing to mixing and cold air spill. While this heat exchange or gain must be tolerated during occupied hours, access and display are not required in unoccupied hours, and the cabinet can be modified to close by installing custom-made thermally insulated covers (see Fig. 10.1). Night covers should be constructed in easily handled sections with sufficient thermal insulation to prevent condensation.

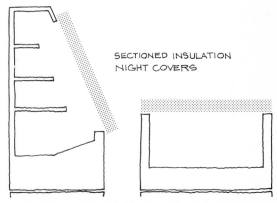

SECTIONED INSULATION
NIGHT COVERS

Figure 10.1   Cold cabinets with night covers.

The energy savings obtained by fitting night covers will vary with each type of cabinet and will be greatest on vertical display cabinets where the cold air tends to spill over the bottom lip and spread over the floor. The utility of covers can be seen by measuring near-floor temperatures adjacent to the cold cabinets, or by using a smoke tracer to show the magnitude of air spill. When assessing the energy savings, bear in mind that the heat transfer into open cabinets results in a reduction of building space temperature, thus increasing the building heating load.

Closed cabinets should be selected when open cabinets require replacement. Vertical display-type cold cabinets and freezers are available as totally closed units with self-closing glass doors to provide visibility and access. These units have considerably lower heat gains than the open types but still provide acceptable access and display.

Their initial cost does not warrant replacing new or almost-new open types, but where existing equipment is at or near the end of its useful life or where remodeling is contemplated, closed cabinets should be used in preference to open. Select new equipment on the basis of efficiency and seasonal coefficient of performance (COP).

Further energy can be saved by limiting distribution losses within the system. The resistance to refrigerant flow in long pipe runs between the compressor and condenser raises the head pressure and temperature. This results in less refrigeration output, increased power input, and longer cycles of operation to meet a given load. The efficiency of the refrigerating machines can be increased by relocating air-cooled condensers to minimize lengths of pipe runs. In winter the hot air from air-cooled condensers can be usefully employed to meet part of the heating load.

If the air-cooled condensers can be relocated, it is possible to duct hot air off the coil into the building. Depending on the configuration, the existing condenser fan may have sufficient capacity reserve to provide adequate airflow through the new duct system, or an additional fan can be added. If possible, arrange duct and dampers so that cold exhaust air from the building can be directed through the condenser coil in summer. See Fig. 10.2.

ACTION GUIDELINES*

☐ 1. Do not permit refrigerated products to stand in the aisles, on docks, or anyplace where they will warm up and create an additional refrigeration load when they are placed in fixtures. Reduce the volume of items requiring cold storage.

☐ 2. Avoid setting controls (pressure and temperature) any lower than

*Many of these suggestions are from *Retail Food Store Energy Conservation,* from the Commercial Refrigerator Manufacturers Association.

necessary. Too often, a freezer may be operating at −30°F air temperature when most often a −10°F or higher is all that is necessary.

☐ 3. Keep products below clearly marked load lines. An overloaded display case decreases product quality and increases energy use as much as 10 to 20% for each fixture.

☐ 4. Turn off refrigeration for cutting rooms, preparation rooms, and some display fixtures, such as meat cases, when not in use. Put all food products in coolers where possible.

☐ 5. Where possible, construct a tight partition or hang a heavy drape from roof to floor between sales and storage areas to prevent interchange of air and maintain storage areas at 60°F or lower in winter.

☐ 6. Keep return grilles of fixtures clear of stacked products; otherwise refrigerated air will flow into aisles.

☐ 7. As customers shop from fixtures throughout the day, repack product displays and keep them below load lines.

☐ 8. With multishelf fixtures, follow the recommendations of manufacturers in regard to shelf position and size to prevent increased refrigeration loads.

☐ 9. Consider reducing (or turning off entirely) the internal shelf lights to reduce both refrigeration requirements and the lighting load.

☐10. Automatically shut off all preparation rooms at night and on weekends. Arrange for automatic startup when required.

☐11. Consider unloading meat and produce display cabinets and shut off

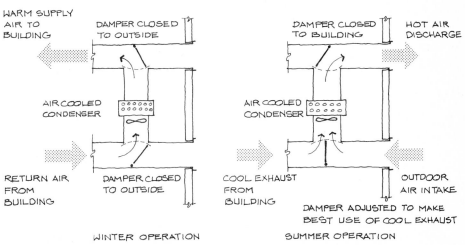

Figure 10.2 Space heating with refrigeration condenser cooling air.

refrigeration in them at night and on weekends. Set up a time clock to automatically turn them off at closing and on in the morning, early enough to allow fixture temperatures to drop to the required level.

☐ 12. If refrigeration load is decreased by reducing the lights, for cleaning, or for any other reason, recheck temperature and pressure control settings to avoid freezing of products or short cycling of compressors. Enforce rules to close the door when case is not in use, even for short periods of time.

☐ 13. In multishelf low-temperature equipment, check all fixtures for inoperative fan motors, often unnoticed, which affect efficiency.

☐ 14. Minimize head pressure to increase compressor capacities and reduce energy use by increasing air supply over condenser. Clean condenser coils regularly. Maintain lowest head pressure at which the commercial refrigeration system can operate without short-cycling or impairing expansion valve and coil efficiency.

☐ 15. Set fan cycling with small differential and low cut-in point. Where more than one fan is used, consider cycling one or more fans.

☐ 16. To prevent pressure drop and loss of compressor capacity, avoid the use of suction line controls.

☐ 17. Insulate suction and liquid line together, except where not recommended (hot gas defrost), to increase system efficiency.

☐ 18. For systems where product is required at 32°F or above, use time defrost rather than an added heat source. With existing equipment, disconnect the heater and reset the controls. Energy will be saved while the compressor is not operating and because no electricity is used for defrost heaters.

☐ 19. On forced-defrost systems (electric or hot gas) use defrost-terminating thermostat on each fixture to avoid over-defrosting individual fixtures, and bring compressor back on as soon as all fixtures are satisfied.

☐ 20. Consider demand defrost for all types of defrost systems. The number of defrosts is normally set up for the most adverse store conditions that may occur. These conditions usually exist for only short periods of the year. Demand defrost compensates for these periods by causing fixtures to defrost only when required. Consult the fixture manufacturer before specifying demand defrost.

☐ 21. Consider separate wiring circuits for anticondensate heaters. Energy use by these heaters, which operate 24 hours a day, 365 days a year, is high, particularly on freezers with glass doors. Heaters are required only when humidity is high. Be aware, however, that most existing installations have fan and anticondensate heaters on the same circuit.

Do not turn off the fan. Consider eliminating incandescent lighting over top-display meat cases. In particular, minimize spotlighting.

☐ 22. Clean display fixture and cooler coils regularly. Be sure to shut off refrigeration before using water for cleaning.

☐ 23. Use pressure spray to clean flues.

☐ 24. Remove discharge grilles to thoroughly clean inlet side (unseen from outside). A reduction of airflow can result in a rise of as much as 10°F in product temperature and cause more frequent defrost.

☐ 25. Drains should be pressure-flushed regularly to prevent buildup in fixture bottoms.

☐ 26. Clean back or inlet side of cooler units.

☐ 27. Check cooler door seals for loss of refrigerated air. Install spring-loaded door closers, reminder sign in plain view, buzzers, lights, etc., to be sure employees keep doors closed.

☐ 28. Check all electric circuits for power leak to ground. A leak to the ground may be small enough to go undetected for years with substantial accumulated loss of energy.

☐ 29. Check all systems for correct refrigerant charge to avoid excessive compressor operation. Shortage will usually show up when low ambient air conditions exist.

☐ 30. Where possible, provide staged cooling controls. Consider replacing one large compressor and coil with two or more circuits. Where possible, consider removal of hot gas bypass capacity control.

☐ 31. Redirect outlets which are discharging into refrigerated fixtures.

☐ 32. Convert air-cooled compressor and condenser installations to water-cooled systems when ambient wet-bulb temperatures on a seasonal basis are favorable.

☐ 33. Install a heat reclamation system to salvage heat rejected from condensers for use in space heating.

☐ 34. When replacing frozen food cases, refrigerators, and cold display cabinets, select equipment on the basis of cooling efficiency rather than first cost.

☐ 35. When replacing refrigeration equipment, select equipment to give the highest seasonal COP.

☐ 36. Replace internal lights in cold cabinets with external lights.

☐ 37. Provide night covers for open cold cabinets but be careful to avoid frost buildup on product.

☐38. Choose closed cabinets when replacing existing open-type cold cabinets.

☐39. Move condensers close to compressors to minimize length of pipe runs.

☐40. Relocate condensers to utilize hot air off the coil.

10.5 REDUCE PEAK LOADS

Utility rate structures are based not only on the building's total usage of electricity but also on the peak demand—which may occur for only a few hours once or twice each year but which establishes demand charges for the rest of the month. The major purpose of load shedding, or leveling, is to reduce the peak electrical loads to decrease these electrical demand charges. Since many of the largest loads are deactivated for considerable periods of time, load shedding conserves energy in addition to reducing electrical equipment power and line losses.

Where manual load shedding cannot be instituted because of cost and demand upon personnel, consider automatic load shedding. First, tabulate all electrical loads. Note the periods when they must be in operation and the duration of time that they can be disconnected without impairing the safety of the occupants or condition of the structure and equipment. Then organize the load-shedding program to shut down nonessential loads by a timing device (when the hours that operations can be suspended are predictable) in a way that large loads are not allowed to occur simultaneously. Peak loads which are not predictable in advance can be monitored, and equipment operation can be automatically programmed to respond to load conditions by deenergizing selected loads by priority and reactivating them when reduced demand permits (new loads are also precluded on a priority sequence).

Available automatic load-limiting devices range in complexity from a simple thermal sensor (which works much like a circuit breaker thermal element and switches off a low-priority load when the building load reaches a preset point) to a unit that reads current and provides for shedding and restoring loads. Automatic load shedding is also commonly accomplished with a watthour meter combined with time switches and load controllers.

The simplest way to control a 25-A circuit is to install a time clock with adjustable start/stop times. This costs about $50 installed. The larger, more complex, load limiters vary greatly in cost depending on the sophistication of the system and the size of the loads controlled. The cost of equipment for a four-priority load controller with automatic shed and restore is about $6000.

In new or fully renovated buildings a fundamental addition to the HVAC system can go a long way to the leveling of the building electrical demand through a full 24-h day. If storage tanks are installed for the air-conditioning system, the chillers can store chilled water in them during the night and off periods when other primary electrical loads are secured (and when coefficients of performance for the chillers are increased with the night cool). The chilled-water reservoir is available for cooling the next day, and daytime loads due to chiller operations can be limited. In addition to a reduction in peak loads, the chilling-condensing equipment can be reduced in size (by working through a longer period for the same load), and capital costs are reduced.

10.6 REDUCE TRANSFORMER LOSSES

Transformers reduce transmission and distribution voltage to equipment operating voltage. Heat generation and dissipation, due to electrical resistance in the transformer, result in electrical energy losses. Efficiencies of most dry-type transformers range from 93 to 98%; the losses occur from the core (magnetizing) and coils (resistance and impedance). Even when equipment served by the transformer is inoperative, some energy is lost unless primary power to the transformer is switched off. When the transformer serves loads which are not required for relatively extensive periods of time, complete disconnection from the primary power may be feasible. Take care, however, to avoid disconnecting transformers that feed clocks, heating-control circuits, fire alarms, or critical process equipment. Potential savings are 3 to 4 watts per/kilovolt-ampere (W/kVA).

Example

Switch off a 150 kVA transformer for 12 h overnight five nights a week and 48 h over the weekend (or a total of 108 h/week and 5615 h/yr). Assuming savings of 4 W/kVA, energy saved in 1 year will be:

4 W/kVA × 150 kVA × 5616 h = 3,369,600 Wh or approximately 3370 kWh

At a price of 4¢ kWh, savings will be:

$$3370 \text{ kWh} \times 4¢/\text{kWh} = \$135.80$$

ACTION GUIDELINES

☐ 1. Select the most efficient dry-type transformer when replacement due to load changes or breakdown is required.

☐ 2. Dry-type transformers are available in many temperature-rise classifications. The lower the temperature-rise rating, the lower will be the coil losses owing to the larger conductors used in the winding. The usual temperature-rise ratings are 80, 115, and 150°C; select the 80°C rise transformer for greater efficiency. Transformers with lower heat-rise ratings also have longer life-expectancy which may offset their higher cost (about 10%).

☐ 3. If larger liquid-filled transformers are required for additions, select those with the highest efficiency.

☐ 4. Deenergize transformers supplying unused offices or other areas.

☐ 5. Deenergize refrigeration chiller transformers during the heating season.

☐ 6. Deenergize heating equipment transformers during the cooling season.

☐ 7. Where there is a bank of two or more transformers, operate transformers at the most efficient loading point.

☐ 8. Reduce copper losses in the wiring (which increase with ambient temperature) by ventilating transformer vaults to reduce the ambient temperature.

☐ 9. Shade outdoor transformer banks from solar radiation.

☐10. Deenergize dry-type transformers serving convenience outlets when there is no load at night and during weekends and holidays.

☐11. When adding or replacing transformers, select efficient dry-type transformers.

10.7 IMPROVE THE EFFICIENCY OF MOTORS

Because original calculations of loads are usually conservative and after loads have often been reduced through conservation measures, most motors will be oversized for the load they are serving. If the ratio of the motor's load to its horsepower rating is small, the power factor will be low, and the motor will operate inefficiently.

Undertake a comprehensive study of all the motors in the building to determine their load factors. Gather and record the following information:

1. The equipment served by the motor.

2. The nameplate information for each motor, including: brand, type, frame, horsepower, speed or speeds, voltage phases and frequency, mounting, full load rating.

3. Measure loads with an ammeter and record the full load running current of each phase.

4. Measure with voltmeter and record the voltage at the motor when measuring running current.

5. Record pully size and type, e.g., V belt or chain.

Find motor loading by multiplying the current draw, as recorded in item 3, above, by the voltage, item 4, and then dividing by 1000 to convert to kilowatt input. Convert nameplate horsepower rating, item 2, to kilowatts by multiplying horsepower by 0.746. Then take the ratio of actual input to nameplate rating to determine the load factor on the motor.

Motors that are not loaded to at least 60% of their potential are relatively inefficient and reduce the power factor of the entire electrical system. Underloaded motors can be exchanged with others in the building to achieve as close to full loading on each motor as possible. If a motor needs replacing, a new motor more closely matched to the load should be selected. For equipment which cycles on and off at short intervals, heat buildup is a less critical problem than in other cases; the motor service factor establishes the maximum overload possible without exceeding the motor's temperature rating. Utilize this information (available through the manufacturer) when in the process of exchanging and interchanging the motors currently in use; then, for cycling equipment, select smaller motors, rated slightly below maximum load requirements, and allow some overloading to occur.

ACTION GUIDELINES

☐ 1. Small split-phase or shaded-pole motors, often used in perimeter fan coil units, have low power factors and are inefficient. These motors should be turned off during mild winter nights to permit the fan coil to operate by natural convection.

☐ 2. When evaluating the potential for larger savings, compare the cost of replacing inefficient motors with the savings in electrical operating cost which would be achieved with new motors.

☐ 3. Tighten belts and pulleys at regular intervals to reduce losses due to slip.

☐ 4. Lubricate motors and drives regularly to reduce friction.

☐ 5. Replace worn bearings.

☐ 6. Check alignment between motor and driven equipment to reduce wear and excessive torques.

☐ 7. Keep motors clean to facilitate cooling.

☐ 8. Where it is impracticable to replace motors which have low load and power factors, use capacitors at motor terminals to raise the power factor to 90%.

☐ 9. When selecting new motors, match the phase and the frequency with the characteristics of the electrical distribution system.

☐ 10. When replacing worn or defective motors, replace with motors sized as close to load as possible and use motors of the highest efficiency available.

☐ 11. When new motors cannot be exactly matched to system voltage, select for loadings greater than 75% of their capacity motors rated at just slightly more than system voltage. (But problems may arise if the nameplate rating of the motor selected is in excess of system voltage by more than 5%—a likely situation in case of brownouts and voltage cutbacks, which actually increase the difference.)

10.8 CORRECT POWER FACTOR

Low power factor (PF) increases losses in electrical distribution and utilization equipment, such as wiring, motors, and transformers, and reduces the load-handling capability and voltage regulation of the building's electrical system. At a unity power factor, losses are zero. When PF is below a designated level, utilities often charge a penalty, and although their methods of computing the charge vary, the magnitude is about the same. As an example, one utility company charges a penalty of $0.25 per kilovolt-ampere reactive (kvar) over 50% of total demand. Poor power factor also reduces the capacity of the electrical service. Even with a PF of 80%, spare capacity is

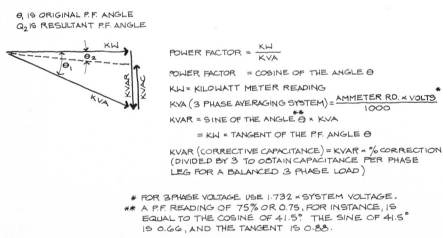

Figure 10.3 The correction for power factor.

TABLE 10.1   CAPACITOR LOCATIONS

Advantages	*Disadvantages*
INDIVIDUAL EQUIPMENT	
Increased load capabilities of distribution system	Small capacitors cost more per kvar than larger units (economic break-point for individual correction is generally at 10 hp)
Can be switched with equipment; thus no additional switching is required	
Better voltage regulation because capacitor use follows load	
Capacitor sizing is simplified	
Capacitors are coupled with equipment and can be moved with equipment if rearrangements are instituted	
GROUPED EQUIPMENT	
Increased load capabilities of the service	Switching means may be required to control amount of capacitance used
Reduced material costs relative to individual correction	
Reduced installation costs relative to individual correction	
MAIN SERVICE	
Low material installation costs	Switching means will usually be required to control the amount of capacitance used
	Does not improve the load capabilities of the distribution system

decreased by an additional 25% of the load, and transmission losses are increased by 56% when compared with unity PF.

Power factor is the ratio of actual power (kW) to apparent power (kVA); it also may be expressed as the cosine of the phase angle between the impressed voltage and the current. Figure 10.3 illustrates these relationships and indicates the effect of corrective capacitance.

A number of devices are available for the purpose of power factor correction. Three of the more common types are capacitors, synchronous motors, and synchronous condensers. Synchronous condensers are the most expensive and are not a practical solution for problems in building services.

Synchronous motors, too, are generally impractical for commercial buildings. But correction by use of capacitors is relatively inexpensive in both material and installation costs. Capacitors can be installed at any point in the electrical system and will improve the power factor between the point of application and the power source (but the power factor between the equipment and the capacitor will remain unchanged). Capacitors are usually added at each piece of offending equipment, ahead of groups of small motors (ahead of motor control centers or distribution panels), or at main services. Refer to the National Electric Code, 1975, Article 460, for installation requirements. The advantages and disadvantages of each type of installation are listed in Table 10.1.

TABLE 10.2   CAPACITOR RATING AND MOTOR HORSE POWER*

Motor Rating, hp	Nominal Motor Speed, rpm							
	3600		1800		1200		900	
	kvar	%AR†	kvar	%AR†	kvar	%AR†	kvar	%AR†
3	1.5	14	1.5	15	1.5	20	2	27
5	2	12	2	13	2	17	3	25
7.5	2.5	11	2.5	12	3	15	4	22
10	3	10	3	11	3.5	14	5	21
15	4	9	4	10	5	13	6.5	18
20	5	9	5	10	6.5	12	7.5	16
25	6	9	6	10	7.5	11	9	15
30	7	8	7	9	9	11	10	14
40	9	8	9	9	11	10	12	13
50	12	8	11	9	13	10	15	12
60	14	8	14	8	15	10	18	11
75	17	8	16	8	18	10	21	10
100	22	8	21	8	25	9	27	10
125	27	8	26	8	30	9	32.5	10
150	32.5	8	30	8	35	9	37.5	10
200	40	8	37.5	8	42.5	9	47.5	10
250	50	8	45	7	52.5	8	57.5	9
300	57.5	8	52.5	7	60	8	65	9
350	65	8	60	7	67.5	8	75	9
400	70	8	65	6	75	8	85	9
450	75	8	67.5	6	80	8	92.5	9
500	77.5	8	72.5	6	82.5	8	97.5	9

*Partial listing of suggested maximum capacitor ratings when motor and capacitor are switched as a unit. For use with three-phase, 60-Hz NEMA classification B motors to raise full-load power factor to approximately 95%.

†%AR: The percent reduction in line current due to capacitors. If a capacitor of lower kilovar is to be used, the new %AR will be approximately proportional to the actual capacitor rating divided by the kilovar value in the tables. Selection of the motor overload relay should be based on the nameplate full load current of tne motor reduced by the %AR value.

Where loads contributing to low power factor are relatively constant and system load capabilities are not a factor, correcting at the main service could provide a cost advantage. When the low power factor is derived from a few selected pieces of equipment, correction of individual pieces of equipment is cost effective. Most capacitors used for PF correction have built-in fusing, but if not, fusing must be provided.

The costs of capacitors for power factor correction range from about $200 for 1 kilovolt-ampere capacitance (kvac) to $550 for 50 kvac. If more than 60 kvac is required, use banks of capacitors. The cost of installation depends on the location and type of mounting.

In order to determine the capacitance required to correct PF, follow one of two procedures:

When billings include a PF penalty, a reading on a kvar meter can be obtained from the utility. Determine the required corrective capacitance (kvac) by multiplying the number of kvars by the percent correction desired.

As an alternative, use a PF meter to determine the PF for individual items of equipment, then use an ammeter for an ampere reading (at normal loading). Use the PF and ampere readings to calculate kvars and required kvac.

When a PF meter is not available, the power factor can be determined by dividing the wattage (found with a wattmeter) by the product of the average of the ammeter readings of the phases and the system voltage. See Table 10.2.

$$PF = \left(\frac{W \text{ (wattmeter reading)}}{\text{(phase A + phase B + phase C ammeter readings)} \times \text{system voltage}} \right) \Big/ 3$$

Correct low power factor with capacitors to reduce losses and increase the capacity of the electrical distribution system.

When undertaking additions or major remodeling involving the installation of new equipment, use 480/277-V systems in preferance to 120/208-V where the electrical service is approximately 500 kVA or more. Whenever possible use the highest practicable voltage available (for example, 480 V instead of 120 V).

CHAPTER ELEVEN

HEATING, VENTILATING, AND AIR-CONDITIONING SYSTEMS

11.1 THE DISTRIBUTION LOAD FOR HEATING AND COOLING: INTRODUCTION

It takes energy to move energy for heating or cooling. It takes less energy, however, if the following changes can be made:

- Improve the performance of the terminal devices to reduce their resistance to fluid flow and increase their heat transfer characteristics.

- Lower the resistance to flow in duct and piping systems to reduce the required horsepower for fans and pumps.

- Modify the control systems and modes of operation of air-handling and piping systems to reduce simultaneous heating and cooling.

- Decrease fluid leaks and thermal losses from piping, air-handling equipment, and other vessels holding hot or cold water, air, or steam.

- Improve the performance of fans, pumps, and motors by maintenance and operating procedures.

- Reduce the hours of fan and pump operation.

To make these changes, it is necessary to define the HVAC system, and for this purpose the following rates and temperatures should be recorded (as nearly simultaneously as possible) as benchmarks for evaluation and modification:

- Airflow rates
    Total outdoor air
    Total return air
    Total supply air
    Through trunk ducts

Through terminal units
Through air-cooled condensers

- Water flow rates
  Through boilers
  Through chillers
  Through cooling towers
  Through heat exchangers
  Through coils and terminal units

- Temperatures
  Outdoor air, DB and WB
  Return air, DB and WB
  Mixed air entering coils, DB and WB
  Supply air leaving coils DB and WB
  Hot deck
  Cold deck
  Air at terminals
  Conditioned areas, DB and WB (typical for each functional use)
  Boiler supply and return
  Chiller supply and return
  Condenser supply and return
  Heat exchanger supply and return
  Coil supply and return

- Refrigerant temperatures
  Hot gas line
  Suction line

The characteristic of any given HVAC system is predetermined by the length and size of pipes or ducts and the size and shape of fittings (bends, tees). The resistance to flow is a function of velocity, fluid density, and the

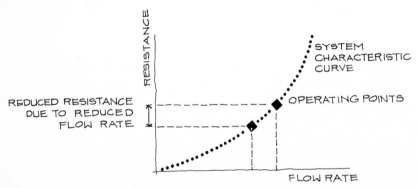

Figure 11.1 The piping characteristic curve with variables of flow rate and resistance.

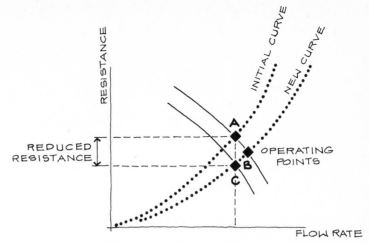

Figure 11.2   The effect of a changed piping characteristic curve.

system characteristic. With any change in flow rate, the system resistance will change (according to the laws of fluid flow), and the operating point will move accordingly on the system characteristic curve. See Fig. 11.1. For instance, a piping system with a flow rate of 2000 gpm at 150 ft equivalent resistance changed to 1800 gpm will have a new resistance of:

$$150 \times \left(\frac{1800}{2000}\right)^2 = 121.5 \text{ ft}$$

If it is not possible to reduce the flow rate, the remaining option to reduce power required for a given flow rate is to alter the system characteristic curve.

As Fig. 11.2 shows, modification of piping system characteristic will change the operating point from A to B along the pump curve and result in an increased flow rate at a lower resistance. To reduce the flow rate to its initial point requires reducing the pump speed so that the new pump curve intersects the new characteristic curve at point C, giving the desired flow rate. These principles apply equally to fan/duct systems and pump/pipe systems.

When flow rates and system resistance are changed, adjustments and modifications should be implemented in the following logical sequence:

- Measure initial flow rates and resistances and construct the system characteristic curve.

- Reduce loads and calculate the new reduced flow rate required to meet them.

- Modify the system to reduce resistance to flow and from measurements construct the new system characteristic curve.

- Determine from the new characteristic curve the resistance at the new reduced flow rate.

- Reduce the fan or pump speed or reduce the pump impeller diameter so that the new fan or pump curve crosses the new system characteristic curve at the desired operating point.

11.2 REDUCE RESISTANCE TO FLOW IN AIR DISTRIBUTION SYSTEMS

The total resistance to airflow in a duct system is the sum of the resistances of the individual parts of the index circuit—that path of flow from fan to room register which offers the greatest resistance to flow within the air distribution system. The index circuit must account for resistances from the duct work and from filters, coils, and dampers. Although straight lengths of duct generally offer little or no opportunity for resistance reductions, the potential for other items is more substantial.

The resistance to airflow through filters is a function of filter construction, type of medium, area of medium per unit volume of air passed through it (proportional to actual velocity through the medium—not to be confused with face velocity), and the dirt load at any given time. In general, resistance to airflow increases with filter efficiency, although there are some high-efficiency filters which also have low resistance (see Table 11.1). To determine the "dirty" resistance limit, the manufacturers of the particular filter should be consulted. A manometer can be installed across each filter bank to indicate when the filters should be changed. If the building contains many air systems, an alarm system can be used to report dirty filter conditions. Such an alarm can be an element of a computer control system.

Filters may impose high resistance by their inherent characteristics or because they exceed required filtration standards. Installation and operating costs of 25 different filter types in common use are shown in Table 11.1. Filtration requirements should be determined in terms of the National Bureau of Standards (NBS) Atmospheric Dust Spot Efficiency, taking into account the ambient air conditions and space air quality requirements. Except for special applications such as computer rooms and food service areas, a 50% NBS dust rating is generally adequate. Present filter installation should be checked to determine whether an alternative type of filter will meet building needs at operating costs sufficiently reduced to provide an economic payback.

Resistance to airflow through coils depends upon area and largely upon the number of rows in depth required for adequate heat transfer. Cooling coils are usually many rows deep because of the small temperature differences between coil and air; consequently, they offer a high resistance to airflow. Heating coils, which have a higher temperature difference, are

TABLE 11.1 AIR FILTER TYPE COMPARISON

	1	2	3	4	5	6
Filter Type	*Cleanable 2 in Thick (High Vel.)*	*Throw-away 2 in Thick*	*Automatic Roll Type*	*Pleated Filter 2 in Thick*	*Bag-type Cartridge*	*Bag-type Cartridge*
Size of Filter Bank						
No. high × no. wide	2 × 3	2 × 5		2 × 4	2 × 2	2 × 2
Height × width	4 ft × 6 ft	4 ft × 10 ft	5 ft 4 in × 5 ft	4 ft × 8 ft	4 ft × 4 ft	4 ft × 4 ft
Average Efficiency						
NBS atmospheric						
Dust spot	8–10%	10–15%	20–25%	36%	38–40%	45%
Other methods						
Filter Life, h						
Prefilter	600	480	3750	2000	6600	2920
Afterfilter						
Pressure Drop, in WG						
Initial	0.08	0.10	0.40	0.14	0.30	0.25
Final	0.40	0.30	0.40	0.60	0.75	1.00
Average	0.24	0.20	0.40	0.37	0.525	0.625
Initial Installation[a]						
Equipment	$328	$107	$ 780	$124	$113	$182
Installation	78	91	91	91	78	78
Electric wiring			215			
	$406	$198	$1086	$215	$191	$260
Cost of One Set-Replacement Medium						
Prefilter[b]		$ 21	$ 87	$ 49		
Afterfilter[b]						$ 46
Replacement Time Man-Hours						
Prefilter	4	2	2	3	4	2
Afterfilter						
Annual Operating Cost						
Material		$381	$203	$216	$ 83	$137
Labor @ $7/h	$533	333	43	120	25	55
Electric power[c,e]	126	104	210	194	276	336
Miscellaneous						
Total[d]	$659	$818	$456	$530	$384	$528

[a]Spare set of replacement cells was included in installed cost of cleanable filters.

[b]Allowance for replacement filter price increase averaging 15% over 20 years was included.

[c]Fan power was figured at $0.03 per kWh and 59% overall efficiency (fan efficiency typically varies from 50 to 70%).

[d]Operating costs based on 8760 hours of operation per year.

[e]Power absorbed by filter to overcome its resistance to airflow.

TABLE 11.1 AIR FILTER TYPE COMPARISON (*Continued*)

Filter Type	7 Bag-type Cartridge	8 Bag-type Cartridge with Prefilter	9 Auto-Roll Bag-type Afterfilter	10 Rigid Cartridge	11 Rigid Cartridge with Prefilter	12 Bag-type Cartridge
Size of Filter Bank						
No. high × no. wide	2 × 3	2 × 3		2 × 3	2 × 3	2 × 3
Height × width	4 ft × 6 ft	4 ft × 6 ft	4 ft 8 in × 6 ft 8 in	4 ft × 6 ft	4 ft × 6 ft	
Average Efficiency						
NBS atmospheric Dust spot	50–55%	50–55%	50–55%	55–60%	55–60%	80–85%
Other methods						
Filter Life, h						
Prefilter	5500		3,750	3500		
Afterfilter		9000	10,000		4500	4000
Pressure Drop, in WG						
Initial	0.30	0.30	0.70	0.20	0.20	0.37
Final	0.80	0.80	1.25	1.00	1.00	1.00
Average	0.55	0.55	0.975	0.60	0.60	0.685
Initial Installation[a]						
Equipment	$351	$351	$1175	$332	$332	$378
Installation	78	78	169	78	78	78
Electric wiring			215			
	$429	$429	$1559	$410	$410	$456
Cost of One Set-Replacement Medium						
Prefilter[b]	$306		$ 72	$278		
Afterfilter[b]		$306	306		$278	$337
Replacement Time, Man-Hours						
Prefilter	3		2	3		
Afterfilter		3	3		3	3
Annual Operating Cost						
Material	$486	$298	$ 434	$ 697	$546	$ 737
Labor @ $7/h	44	26	68	69	53	60
Electric power[c,e]	288	288	512	316	316	360
Miscellaneous						
Total[d]	$818	$612	$1014	$1082	$915	$1157

13	14	15	16	17	18	19	20
Bag-type Cartridge with Prefilter	Automatic Roll with Bag-type Afterfilter	Rigid-type Cartridge	Rigid-type Cartridge with Prefilter	Electrostatic Auto-Roll Afterfilter	Bag-type Cartridge	Bag-type Cartridge with Prefilter	Electrostatic Bag-type Afterfilter
2 × 3 4 ft × 6 ft	4 ft 8 in × 6 ft 8 in	2 × 3 4 ft × 6 ft	2 × 3 4 ft × 6 ft	4 ft 8 in × 8 ft 8 in	2 × 3 4 ft × 6 ft	2 × 3 4 ft × 6 ft	4 ft 8 in × 6 ft 8 in
80–85%	80–85%	85–90%	85–90%	90%	93–97%	93–97%	93–97%
5500	3750 6000	2500	3500	2500 4000	2500	3500	2,500⁶ 12,000
0.37 1.00 0.685	0.77 1.50 1.135	0.30 1.00 0.65	0.30 1.00 0.65	0.45 0.45 0.45	0.48 1.00 0.74	0.48 1.00 0.74	0.55 1.00 0.775
$378 78 <hr>$456	$1203 169 215 <hr>$1587	$367 78 <hr>$445	$367 78 <hr>$445	$2990 169 397 <hr>$3556	$437 78 <hr>$515	$437 78 <hr>$515	$2730 156 215 <hr>$3101
$337	$ 72 337	$332	$332				
3	2 3	3	3	2 2	3	3	2 3
$536 44 360 <hr>$940	$ 658 83 596 <hr>$1337	$1161 96 342 <hr>$1599	$ 829 69 342 <hr>$1240	$157 104 236 <hr>$497	$1411 96 388 <hr>$1895	$1008 69 388 <hr>$1465	$294 83 408 <hr>$785

TABLE 11.1 AIR FILTER TYPE COMPARISON (*Continued*)

Filter Type	21 HEPA *without Prefilter*	22 HEPA *with 85% Prefilter*	23 *Activated Charcoal*	24 *95% DOP Rated Filter with Prefilter*	25 *95% DOP Rated Filter with 85% Prefilter*
Size of Filter Bank					
No. high × no. wide	2 × 5	2 × 5	1 × 5	2 × 5	2 × 5
Height × width	4 ft 0 in × 10 in	4 ft 0 in × 10 ft 0 in	2 ft 0 in × 10 ft 0 in	4 ft 0 in × 10 ft 0 in	4 ft 0 in × 10 ft 0 in
Average Efficiency					
NBS atmospheric					
Dust spot	100%	100%		99%	99%
Other methods	99.97% DOP	99.97% DOP		95% DOP	95% DOP
Filter Life, h					
Prefilter					
Afterfilter	6000	24,000	8760	6000	24,000
Pressure Drop, in W G					
Initial	1.00	1.00	0.35	0.50	0.50
Final	2.00	2.00	0.35	1.00	1.00
Average	1.50	1.50	0.35	0.75	0.75
Initial Installation [a]					
Equipment	$1300	$1300	$2600	$1300	$1300
Installation	390	390	975	390	390
Electric wiring					
	$1690	$1690	$3575	$1690	$1690
Cost of One Set-Replacement Medium					
Prefilter [b]					
Afterfilter [b]	$ 975	$ 975	$ 598	$ 975	$ 975
Replacement Time Man-Hours					
Prefilter					
Afterfilter				6	6
Annual Operating Cost					
Material	$1424	$ 356	$ 598	$1424	$ 356
Labor @ $7/h	79	20	109	79	20
Electric power [c,e]	788	788	184	394	394
Miscellaneous					
Total [d]	$2291	$1164	$ 891	$1897	$ 770

NOTES:

1. Prefilter costs are not included in cols. 8, 11, 19, 21, 22, 24, 25. Add costs from cols. 1, 2, 3, 4, or 5 to arrive at total cost.

2. Costs are for a 10,000-cfm system. Cost per cfm will be slightly lower for smaller systems and higher for larger systems.

usually fewer rows deep than cooling coils and impose less resistance to airflow.

Both coils are in series in the airstream and impose combined resistance year-round, even though only one coil may be used at any given time. When the cooling medium is chilled water and the heating medium is hot water, it may be possible to remove the heating coil and repipe the cooling coil to provide both heating and cooling (though not simultaneously). Eliminating the heating coil will reduce the system resistance and allow savings in fan horsepower.

A side benefit of using the cooling coil for both heating and cooling is that the extra heat transfer surface of the cooling coil provides opportunities to lower the hot-water temperatures and use low-grade waste heat for heating. See Fig. 11.3.

High resistance to airflow caused by duct fittings on the inlet and discharge side of fans can be reduced by modifying the shape of the fitting. Where space permits, abrupt changes in sections of duct work where velocities exceed 2000 ft/min should be replaced with long taper fittings. Turning vanes should be installed in square bends.

Supply and exhaust fans must operate at a pressure sufficient to overcome the resistance to airflow through the duct path serving the outlet farthest

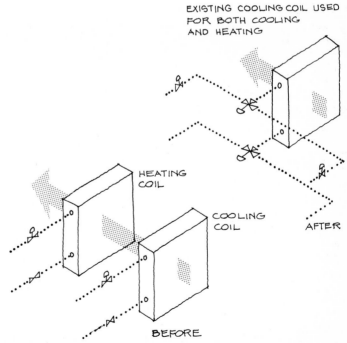

Figure 11.3 Resistance reduction with a single coil for both heating and cooling.

from the fan (index outlet). Where the index outlet is remote from other parts of the system and is served by a long run of duct, it imposes a power consumption penalty on the whole system. Replacing this particular section of duct with one of larger cross-section and lower resistance is often worthwhile. See Fig. 11.4.

Correct volumes at each grille or register are achieved by adjusting dampers in low-resistance branches until all branches are of a resistance equal to the index run. When systems are initially started, all dampers are often closed more than they need be, adding unnecessary resistance. It is also common practice to reduce fan volume by closing down dampers on the fan inlet or outlet rather than by reducing fan speed. With either of these situations, the system should be rebalanced. To start, open fully any dampers which are not used to proportion airflow between main branches (main dampers close to fan). Identify the index outlet of the branch and fully open any dampers between this outlet and the fan. Measure the volume at the index outlet and adjust branch dampers successively (starting with the next longest run) until proportional volumes are achieved at each outlet. This is a trial-and-error process as each damper adjustment will affect flow rates in branches already adjusted, but two or three successive adjustments of the whole system will give good balance.

ACTION GUIDELINES

☐ 1. Reduce duct work resistance in the longest or index circuit so that all balancing dampers can be opened and total system resistance reduced.

☐ 2. Clean filters, blower fan wheels, and fan scroll blades frequently to reduce friction.

☐ 3. Replace filters which have low efficiency and high resistance with filters offering less air resistance and greater efficiency.

☐ 4. Clean the air side of all coils in air-handling units to reduce air resis-

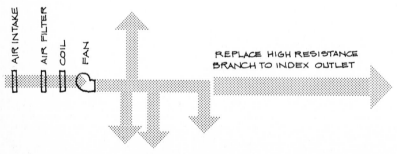

Figure 11.4 System improvement by reduction of the index duct resistance.

tance. (Cleaning coils also increases their efficiency so that less energy for operation of the heating òr cooling primary equipment is required.)

☐ 5. Eliminate preheat coils whenever possible without increasing freeze-up possibilities.

☐ 6. Remove unnecessary dampers, high-resistance elbows and fittings, and sections of duct work with high resistance to lower the total system resistance.

☐ 7. Install gauges to measure the resistance across filters and across coils to indicate when cleaning is necessary.

11.3 REDUCE VOLUME OF FLOW IN AIR DISTRIBUTION SYSTEMS

A reduction in the resistance to airflow increases air delivery which, in turn, permits a reduction in fan speed to reduce the air volume. This reduction, plus the reduction in air volume due to decreased building loads, can equal significant energy savings in fan horsepower.

In locations where excessive noise is not a problem, axial fans have certain advantages. They are better for variable flow and have a better capacity-to-power ratio.

For multivane centrifugal fans (the type most commonly used in HVAC systems), the power input (BHP) varies directly with the cube of the speed. Any reduction in speed results in a very sizable decrease in power input.

The basic laws for operation of centrifugal fans follow:

1. Volume varies directly with the speed.

2. Pressure varies directly with the square of the speed.

3. Power input (BHP) varies directly with the cube of the speed.

4. Volume varies directly with the square of the pressure.

If it is possible to reduce the volume of air in a system by 10%, the savings in power will be about 27%.

To conserve energy for fan horsepower, first, reduce heating and cooling loads; second, reduce the resistance to airflow; third, measure the increased volume which results and determine the new air volume required to meet new loads; fourth, reduce fan speed accordingly; and fifth, change the motor if necessary.

The speed of belt-driven fans can be reduced by changing the size of the motor sheave. It may be necessary to change the drive belts if adjustment cannot take up the slack.

The speed of direct-driven fans can be reduced by changing the drive motor for one with a lower speed. This may be possible only if the desired fan speed falls within the motor speed ranges available or if a compromise speed is acceptable.

Reducing the supply airflow rate through the system will also reduce the resistance to flow. Reduction in resistance will follow the laws of fluid flow, and the new system resistance can be determined by using the following formula:

$$\text{New system resistance} = \text{initial system resistance} \times \left(\frac{\text{new flow rate}}{\text{initial flow rate}} \right)^2$$

The total cost of achieving savings by reducing system resistance is the sum of the costs of all the individual steps taken. The cost of reducing fan speed will vary with each individual case, but for order of magnitude, the cost of replacing a motor sheave will be approximately $100.00.

When a fan is reduced in speed, its power requirements are also reduced:

$$\text{hp} = \left(\frac{\text{rpm 1}}{\text{rpm 2}} \right)^3$$

and it is possible that the drive motor will now far exceed the new requirements. Motors operated far below rated output are inefficient, and, of equal significance, their power factors will be drastically reduced, affecting the power factor of the entire electrical system. Depending on the electrical rate structure, this reduced power factor could entail extra charges. For guidelines on electric motors, see Secs. 10.7 and 10.8.

ACTION GUIDELINES

☐ 1. Reduce air volume by reducing the speed of rotation.

☐ 2. Where motor sheave is adjustable, open the V to reduce its effective diameter and adjust motor position or change belts to maintain proper tension.

☐ 3. Change motor sheave if no further adjustment is possible.

☑ 4. If motors offer variable speeds, set controller for reduced speed.

☐ 5. After implementing changes to the system to decrease the required fan horsepower, the existing motor may be too large. If the full load on the motor is less than 60% of the nameplate rating, consider changing the motor to a smaller one.

☐ 6. If the fan is the direct-drive type, changing the speed of rotation is expensive. However, electric motors also tend to be oversized, and if the system resistance is reduced significantly, the next-smaller size motor, running at a slower speed, may well be suitable. In this case, calculate the savings that will accrue owing to reduced power requirements to determine whether changing motors to achieve the speed reduction is economical.

☐ 7. Losses in the drive train and bearings can, if maintenance is poor, amount to as much as 20% of total power input. Examine all bearings for wear and resistance to movement. Excessive belt tension will also impose an added load on the motor. Adjust or replace slipping fan belt drives.

☐ 8. Operate exhaust systems intermittently to reduce total operating time. In many buildings the toilets and other areas require full exhaust for limited periods only. With proper programming and wiring to integrate the operation of exhaust fans and dampers with light switches, an office building with a system which exhausts 100 cfm through each of 20 toilet rooms could reduce its exhaust to an average of only 400 cfm throughout the day.

☐ 9. Where direct-drive fans are used and changing fan speed or motor is difficult or costly, blank off unused portions of exhaust hoods in kitchens and cafeterias to reduce air volume and horsepower.

☐ 10. Install a control to reduce total air volume during the heating season when the same system is used for both heating and cooling. During the cooling cycle, more air is required because of lower temperature differentials.

☐ 11. In large buildings where steam is available, consider turbine-driven fans with variable speed control.

☐ 12. In high-bay buildings, lower the levels of supply diffusers to decrease the operating time of fans and equipment.

☐ 13. Seal and caulk leaking joints in ductwork, air-handling units, and flexible duct connectors. Although air leakage from the ventilating system does not increase the resistance to airflow as such, it does waste energy in two other ways. It increases the total quantity of air handled by the fan above that required to meet the room temperature, and it increases the quantity of air that must be cooled or heated, an expenditure of energy above that which otherwise would be required to meet desired room conditions.

Energy requirements for centrifugal fans with forward-curved blades are given in Fig. 11.5, and for centrifugal fans with backward-curved blades in Fig. 11.6. To quantify energy reductions due to air volume and air resistance reductions, consider this example.

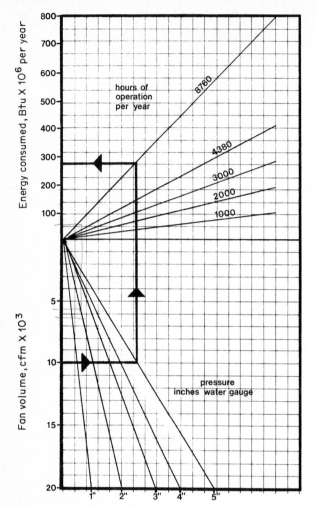

Figure 11.5 Energy requirements for centrifugal fans with forward-curved blades. (From manufacturers' fan capacity tables.)

Example

A building has a heating load of 620,000 Btu/h and a fan volume of 15,000 cfm. Energy conservation measures have been implemented to reduce the heating load to 510,000 Btu/h. If the temperature drop remains the same, the volume of air required to meet the reduced load is 12,340 cfm. In addition, system resistance has been reduced by modification to filters and heating coils.

Initial system characteristics were measured at 15,000 cfm and 4 in wg, point A in Fig. 11.2. The measurements taken after lowering system resistance were 16,620 cfm and 3.8 in wg, point B on the new system curve.

Now the fan speed is to be reduced to give the new required volume of 12,340 cfm. The new pressure at this volume is measured at 3.27 in point C.

Assuming a backward-curved fan operating 4380 h/yr, use Fig. 11.6 to determine energy savings. Enter Fig. 11.6 with the initial conditions of 15,000 cfm at 4 in wg and read that yearly energy consumed is 170×10^6 Btu. Reenter Fig. 11.6 with the final conditions of 12,340 cfm at 3.27 in wg and read yearly energy consumed at 110×10^6 Btu. The energy saved equals 170

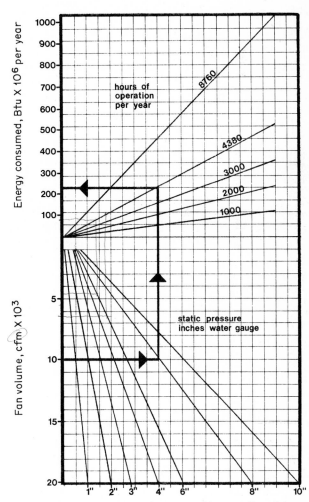

Figure 11.6 Energy requirements for centrifugal fans with backward-curved blades. (From manufacturers' fan capacity tables.)

$\times\ 10^6$ minus 110×10^6 Btu/yr or 60×10^6 Btu/yr. This can be converted to kilowatthours per year as follows:

$$\frac{60 \times 10^6}{3413} = 17{,}580 \text{ kWh/yr}$$

At \$0.03/kWh, the savings are $0.03 \times 17{,}580$ or \$527 per year.

11.4 REDUCE THERMAL LOSSES IN AIR DISTRIBUTION SYSTEMS

Though ductwork for both heating and cooling is commonly insulated, warm-air ducts alone are often installed without insulation and are typically routed from the equipment room through unoccupied spaces, shafts, and ceiling voids where their heat loss is unproductive in meeting the occupied-space heating load. Although the temperature difference between duct and ambient air is relatively small, heat loss in long duct runs can be significant. Of equal importance is the temperature drop of supply air that accompanies heat loss. In long duct runs serving many rooms in one zone, the temperature of the supply air will be lower in the last room than in the first. The tendency in this case is to heat the last room to comfort conditions, resulting in overheating in each preceding room with consequent additional waste of energy over and above the heat loss in the duct.

Air ducts may be insulated with rigid fibrous material applied with adhesive or fixed with special clips or bands. Ducts may also be insulated with flexible mats clipped or wired on (this is particularly applicable for round or oval ducts), or with spray-on foam or fibrous material as described for insulating the undersides of roofs (see Sec. 6.16). When a building is renovated, a single contract for insulating roofs and ducts is worth considering.

Insulation applied to ducts supplying only warm air need not be vapor sealed. Insulation applied to ducts supplying warm air in winter and cold air in summer must be vapor sealed to prevent condensation from forming within the insulation. Fig. 11.7 gives the value of insulation with respect to heat loss and heat gain from ductwork.

Insulate bare ductwork passing through untempered spaces or rooms. Insulate the exterior surfaces of hot-air and forced warm-air furnaces and air-handling units with insulation of sufficient R value that exterior surface temperatures do not exceed 80°F at full-load operating conditions for units located in the building. For rooftop units, the R value should not be less than 2.5 in all areas which have fewer than 7000 degree-days and not less than 3.5 where degree-days exceed 7000.

To determine the heating energy saved by insulating ducts, calculate the exposed surface area of uninsulated ductwork and determine the temperature differential between the duct and ambient air.

Enter Fig. 11.7 at insulation thickness and, following the direction of the example line, intersect the appropriate temperature differential line. Read out the heat loss in Btu per hour per square foot.

Subtract the insulated duct loss from the bare duct heat loss and multiply by the surface area of insulated duct to derive the saving in Btu per hour. Determine the hours of system operation per year and multiply by the hourly saving to derive the total yearly saving in Btu.

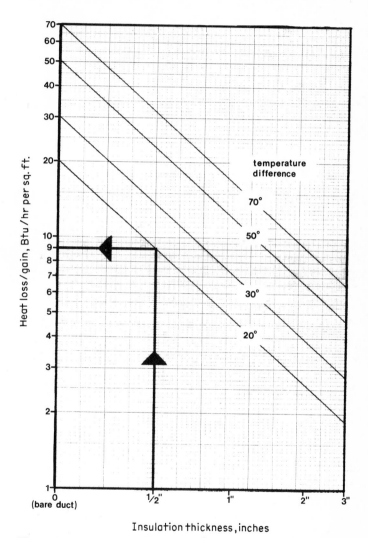

Figure 11.7 Heat loss or gain for various duct insulation thicknesses. (*From ASHRAE Handbook of Fundamentals, 1972.* Assumes rigid insulation with *k* value of 0.27 at 75°F.)

11.5 REDUCE RESISTANCE TO FLOW IN PIPING SYSTEMS

The resistance to flow of any given piping system is the sum of the resistances of all its individual parts in the index circuit. Some of the parts cannot be easily modified, but others are candidates for reduction of resistance.

Strainers are often dirty, and the filtration media are often rusty, corroded, and deformed. These should be cleaned or replaced on a regular maintenance schedule.

Heat exchangers have a high resistance to flow and are prone to fouling by scale deposits and dirt. A program of maintenance and water treatment for heat exchangers should be based on regular observations of pressure drop and temperature differentials. Over long periods of time, formation of scale deposits which occur in heat exchangers and throughout the system may result in radically increased resistance. If such scaling exists, it can be removed by chemical cleaning; there are many specialized contractors who have the expertise for this work.

Pumps must develop sufficient head to overcome the resistance to flow through the longest piping circuit (index circuit) even though this head may exceed the requirements of the other piping subcircuits. If the index circuit resistance grossly exceeds resistances of other circuits and cannot be further reduced, a small booster pump can be installed for the index circuit only to reduce the head of the main pump. See Fig. 11.8.

When existing piping systems are first operated, the installing contractors usually balance flows by trial and error. They often close balancing valves to a greater extent than is needed, imposing extra head on the pump. In addition, designed safety margins often result in oversized pumps, and the excess head is absorbed by closing down the valve on the pump discharge. To reduce resistance to flow, rebalance the system by first opening fully the

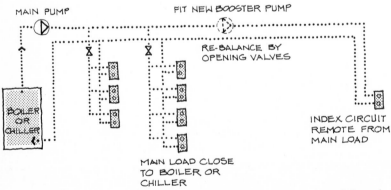

Figure 11.8 System improvement by reduction of the index piping circuit resistance.

balancing valve on the index circuit and the pump discharge valve, and remove any orifice plates from the pump circuit. Then adjust the circuit balancing valve on the next longest circuit, then the next, progressing to the shortest circuit, to achieve proportional flow rates. This process is one of trial and error; each valve adjustment will affect flow rates in circuits already adjusted. Nevertheless, two or three successive adjustments of the whole system will give good balance. When all available options to reduce system resistance have been exercised, measure or calculate the new system characteristic and determine the new operating point.

ACTION GUIDELINES

☐ 1. Open all throttling valves and remove high-resistance fittings on pipe sections to reduce frictional resistance, then adjust the speed of variable-speed motors to a lower speed.

☐ 2. Replace high-resistance elements of filters and strainers with low-resistance elements or baskets.

☐ 3. Regularly adjust all pumps to control leakage at the pump packing glands. Curtail excessive waste of water by repacking, not only to conserve water and reduce losses, but also to avoid erosion necessitating costly repairs of the shaft.

☐ 4. Check for high temperature differences in heating and cooling heat exchangers, which may be an indication of air binding, clogged strainers, or excessive scale.

☐ 5. Chilled-water and hot-water systems require near-complete elimination of air; check vents before reducing water temperature to improve cooling performance or increasing water temperature to meet room heating loads.

11.6 REDUCE VOLUME OF FLOW IN WATER DISTRIBUTION SYSTEMS

When the heating or cooling load is actually less than the original design or is further reduced, the flow rate through the hot-water or chilled-water systems may be reduced in proportion. The water pumps can be slowed down. The reduction in pump speed is derived with changes in flow rate as entrant to the manufacturer's pump curve. Any reduction in pump speed will be approximately proportional to the change of flow rate.

The speed of indirect-drive pumps can be changed by changing the size of the motor sheave. To reduce the speed of direct-drive pumps, the drive motor must be exchanged for one of lower speed. (This is possible only within the range of commercially available motor speeds, and a compromise may be necessary.)

If it is not possible or economically feasible to change the speed of a pump, the flow rate can be changed by changing the impeller size. One of smaller diameter can be substituted, or the existing impeller can be skimmed down. Seek manufacturers' recommendations for each specific application.

Reducing the water flow rate through a piping system also reduces the resistance in accordance with the laws of fluid flow. To determine the new system resitance, use the following formula:

$$\text{New system resistance} = \text{initial system resistance} \times \left(\frac{\text{new flow rate}}{\text{initial flow rate}} \right)$$

Any further reductions in system resistance from other actions should be accounted for with this same formula.

ACTION GUIDELINES

☐ 1. Reduce water flow where the terminal heating device can deliver heating or cooling adequately with a lower rate. For a chilled-water system, take care to avoid reducing flow to the point that a lower suction temperature must be used. In general, the energy consumed at that lower suction temperature to reduce chilled-water temperature (to maintain temperature or humidity control) would outweigh the energy saved by reducing the power for pumping.

☐ 2. For small systems, under 10 hp, throttle the pump discharge to supply only the amount of water required to serve the load.

☐ 3. For larger systems, change the speed of rotation of pumps by substituting motors of lower horsepower and lower speed where the drive permits, or remove the impellers and replace with impellers of smaller diameter.

☐ 4. Add a new, smaller pump in parallel with an existing pump and provide controls to sequence the operation of both so that the smaller pump can handle the required volume of water at part loads.

☐ 5. In new installations, choose modular pumps sized for one-third and two-thirds of the load where the system permits reduction in water volume. The larger pumps can be of variable volume to give complete modulation over the entire range of loads.

☐ 6. Check the power input to the pump motor against the nameplate rating. If the motor is drawing less than 60% of the rated input, analyze the potential energy and cost savings with a smaller motor and pump against the initial cost of replacement.

To determine energy consumed by pumps, refer to Figs. 11.9 and 11.10. Select the appropriate graph and enter at the initial flow rate. Following the direction of the example line, intersect with the initial system resistance and hours of operation.

Read out the yearly energy used in Btu per year. Repeat this procedure, using the new flow rate and new system resistance. Subtract the yearly energy used at the reduced flow rate from the yearly energy used at the initial flow rate. To convert Btu per year to kWh per year, divide by 3413 to obtain yearly savings.

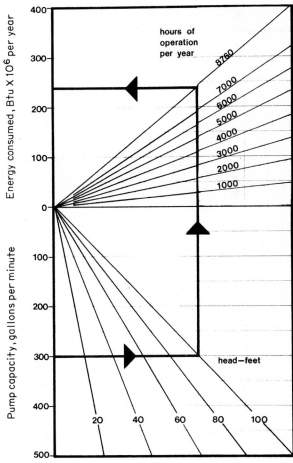

Figure 11.9 Energy requirements for pumps up to 500 gpm.

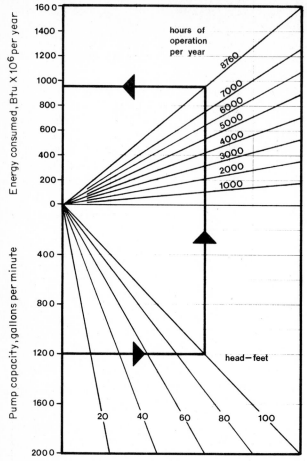

Figure 11.10 Energy requirements for pumps up to 2000 gpm.

The cost of achieving these savings will be the sum of the costs for the individual steps taken. The cost of fitting new pump sheaves will vary with each case but, for order of magnitude, will be approximately $100.00. The cost of replacing the impeller or skimming it down in diameter should be obtained from the pump manufacturer.

Figures 11.9 and 11.10 are based on the standard pump formula:

$$\text{Brake horsepower} = \frac{\text{lb/min} \times \text{feet head}}{33{,}000 \text{ ft/lb} \times \text{pump efficiency}}$$

Pump efficiency was assumed to be an average of 70%. Brake horsepower was converted to Btu by the factor 2544.43 Btu/horsepower-hour. The upper half of the graph proportions the hours of pump operation per year from 100% (8760 h), down to 11% (1000 h).

11.7 REDUCE THERMAL LOSSES
IN WATER AND STEAM DISTRIBUTION SYSTEMS

In addition to frictional resistance, energy losses from piping systems occur in the form of heat gain to chilled water and heat loss from hot water or steam systems. Insulation for piping is very important. Insulation may be a sectional, rigid type of fibrous material, plastic, or glass wool and must be capable of withstanding the surface temperature of the pipe. Plastic magne-

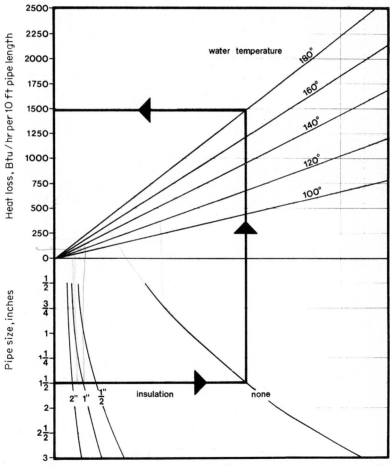

Figure 11.11 Low temperature piping heat losses for various diameters, water temperatures, and insulation thicknesses.

TABLE 11.2 PIPE INSULATION COSTS (APPROXIMATE)

| Pipe Sizes, in | Price per lin. ft Installed Insulation Thickness | |
	1 in	1½ in
½	$1.35	$2.05
¾	1.40	2.10
1	1.45	2.15
1¼	1.50	2.20
1½	1.55	2.25
2	1.60	2.35
2½	1.65	2.45
3	1.70	2.50
4	2.00	3.05
5	2.25	3.15
6	2.55	3.30
8	3.15	4.20
10	3.85	5.00
12	4.50	5.60
14	5.20	6.45
16	6.00	7.20
18	6.70	7.60
20	8.25	8.50
24	9.00	9.70

TABLE 11.3 HEAT LOSS THROUGH PIPES*

| Pipe Size, in. | Bare Pipe | Insulation Thickness, in | | | | | | |
		½	¾	1	1¼	1½	1¾	2
½	0.63	0.163	0.135	0.116	0.105	0.098	0.091	0.086
¾	0.76	0.191	0.155	0.135	0.120	0.110	0.103	0.096
1	0.93	0.211	0.179	0.153	0.136	0.125	0.115	0.108
1¼	1.14	0.263	0.210	0.178	0.158	0.143	0.132	0.122
1½	1.27	0.287	0.232	0.194	0.172	0.154	0.142	0.132
2	1.53	0.345	0.271	0.229	0.198	0.178	0.163	0.151
2¼	1.87	0.425	0.325	0.270	0.237	0.210	0.190	0.175
3	2.15	0.487	0.368	0.309	0.251	0.214	0.211	0.195
4	2.65	0.600	0.447	0.375	0.305	0.279	0.252	0.231
5	3.2	0.663	0.500	0.407	0.346	0.305	0.271	0.245
6	3.7	0.852	0.628	0.536	0.432	0.379	0.341	0.305
8	4.75	1.090	0.828	0.650	0.549	0.486	0.433	0.388
10	5.75	1.341	0.990	0.778	0.678	0.580	0.511	0.457
12	6.75	1.550	1.152	0.920	0.802	0.664	0.604	0.541

*Expressed as Btu/(h)(ft)(°Td). Based on an ambient temperature of 68°F and a K for insulation = 0.3.

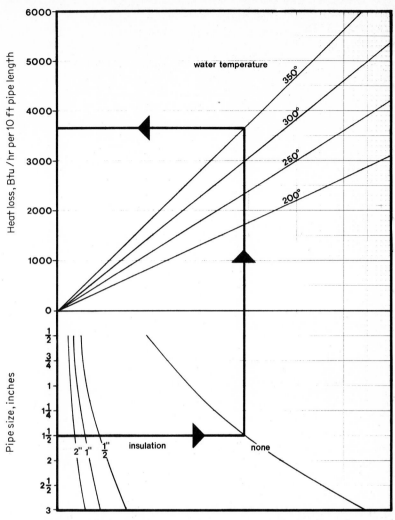

Figure 11.12 High temperature piping heat losses for various diameters, water temperatures, and insulation thicknesses.

sia is an alternative insulation that is applied wet; it is particularly suitable where there are many changes of direction or many valves and fittings which make difficult the application of sectional insulation. If existing insulation on pipework is bad, it should be replaced or repaired. See Table 11.2 for costs and Table 11.3 for heat loss through pipes. Figures 11.11 to 11.13 give the measure for various pipe size and insulation conditions.

Defective steam traps can be another serious source of energy loss. Steam traps are installed to remove condensate, air, and carbon dioxide from the steam-utilizing unit as quickly as they accumulate. With time, internal parts begin to wear, and the trap fails to open and close properly. Figure 11.14 quantifies steam loss through leaking steam traps.

ACTION GUIDELINES

☐ 1. Repair, increase, or replace the insulation on water and steam piping. Heat loss from piping should not exceed 0.25 Btu/ft² per degree difference between fluid temperature and ambient air temperature.

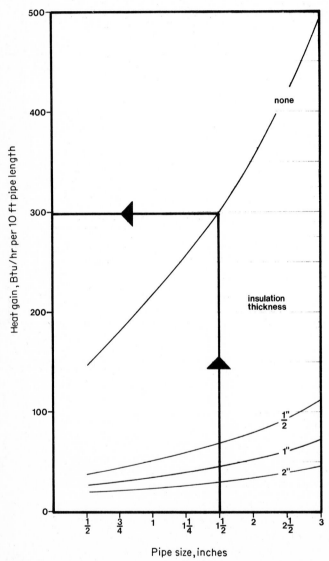

Figure 11.13 Cold piping heat gains for various diameters and insulation thicknesses.

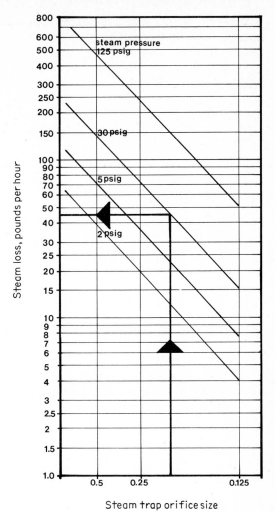

Figure 11.14 Steam losses through leaking steam traps. (From manufacturers' data.)

☐ 2. Insulate valve bodies, fittings, and other pipe appurtenances except steam traps and condensate legs within 5 ft of the trap.

☐ 3. Repair leaks in steam traps or replace them. Confirm the location of a leaking trap by testing the temperature of return lines with a surface pyrometer and by measuring temperature drop across the suspected trap. Lack of drop indicates steam blow-through. Excessive drop indicates the trap is holding back condensate.

To determine the annual savings (in Btu) obtained with pipe insulation, use Table 11.3. Subtract the value for the insulated pipe from the value for bare pipe. Multiply this unit loss first by the total number of feet in the system, then by the difference, in degrees Fahrenheit, between hot-water temperature and ambient air temperature, and finally by annual hours of operation. To determine costs, add to the total lineal feet of piping 3 lin ft for each fitting or pair of flanges to be insulated. (See Table 11.2.)

To determine the steam loss through a malfunctioning steam trap, enter Fig. 11.14 with the orifice size of the trap, intersect with steam pressure, and read the steam loss in pounds per hour. At a typical cost of $4.00/1000 lb steam, the dollar loss due to bad steam traps can be rather surprising.

Figures 11.11 to 11.13 are based on the addition of insulation with a thermal conductivity k = 0.3 (Btu)(in)/(ft²)(h)(°F td) and an ambient air temperature of 68°F.

The following formula is the basis for determining the heat emitted (heat loss) from the pipe:

$$q = \frac{T_s - T_a}{\frac{d_1}{2k}\left(\ln \frac{d_2}{d_1}\right) + \frac{d_1}{fd_2}}$$

where q = heat emission from insulated pipe, Btu/(h)(ft²) of hot pipe surface
T_s = temperature of pipe surface, °F
T_a = ambient air temperature, °F
d_1 = outside diameter of pipe, in
d_2 = outside diameter of insulation, in
k = thermal conductivity of insulating material, (Btu)(in)/(ft²)(h)(°Ftd)
f = surface coefficient, Btu/(ft²)(h)(°F)

11.8 DIRECT HOT-WATER AND STEAM-HEATING SYSTEMS

Direct hot-water or steam systems employ direct radiation, fin tube convectors, fan coil units, cabinet heaters, or fan-driven horizontal or vertical unit heaters with hot water, steam, or electric coils. Fan coil units and induction units equipped with heating coils can be used for both heating and cooling.

ACTION GUIDELINES

☐ 1. Clean the air side of all direct radiators, fin tube convectors, and coils to enhance heat transfer.

☐ 2. Keep radiators free from blockage. A 1-ft clearance in front of convectors, radiators, or registers is desirable. Heating systems, particularly hot-water or electric baseboard radiators and low-level warm-air supply registers, work more efficiently if they are not blocked by furniture. Keep all books or other impediments from blocking heat or air delivery from the top of horizontal shelves or cabinets which enclose radiators, fan coils, unit ventilators, or induction units.

☐ 3. If a radiator is set directly in front of a window where the glass extends below the top of the radiator, or in front of an uninsulated wall, insert a 1-in-thick fiberglass board panel, with reflective coating on the room side, directly between the radiator and the exterior wall, to reduce radiation losses to the outdoors.

☐ 4. Vent all hot-water radiators and convectors to ensure that water will completely fill the interior passages.

☐ 5. Check radiator steam traps to ensure that they are passing only condensate, not steam.

☐ 6. Make sure that all fans, frequently inoperative in unit heaters, fan-coil units, and unit ventilators are running normally to increase the heat transfer rate from heating coils.

☑ 7. Use electric or infrared units as spot heaters for remote areas (a reception desk in a large lobby, for example) rather than operate an inefficient central system for a small area in the building.

☐ 8. In the public spaces such as lobbies, corridors, stairwells, vestibules, and lounges consider turning off all unitary equipment.

☐ 9. Make sure that heat from overhead unit heaters is directed to the floor. Where possible, draw return air from the floor.

11.9 SINGLE-DUCT, SINGLE-ZONE SYSTEMS

These are the simplest and probably the most commonly used of the HVAC systems. They can comprise just a single supply system with air-intake filters, supply fan, and heating coil or can be quite complex with the addition of a return-air duct, return air fan, cooling coil, and various controls to optimize their performance. The system supplies air at a single predetermined temperature to one zone or the entire building. See Fig. 11.15.

Control cycles may provide fluids (hot water, steam, refrigerant, or chilled water) to the coils continuously, may modulate the flow, or may cycle it on and off. At the same time, the fan may be on continuously, modulating, or cycling on and off in accordance with room temperature demands and fluid flow. The characteristics and performance of each system vary widely and preclude a general rule for the optimal control cycle, but any measure which prevents simultaneous heating and cooling and delivers warm or cool air to spaces in accordance with thermal needs, rather than in bursts of overheated and overcooled air, is more efficient and provides more comfort to occupants.

The energy output of a single-duct system to meet a space load is determined by the air volume/temperature differential relationship. For instance, to maintain a space temperature of 65°F, the heating load could be met by a system supplying 10,000 cfm at 105°F or one supplying 6000 cfm at 125°F.

To conserve energy in single-zone, single-duct systems, determine from measurements the system output and the building load. In the heating mode, if the maximum system output exceeds design building loads, reduce the volume of system air. This will show an immediate energy saving (90% of the original fan volume requires only 65% of the original fan horsepower). Next reduce the supply temperature. This, too, will conserve energy but not as much as reducing volume flow.

In the cooling mode, if the system output exceeds the building load, the savings by reduced airflow must be compared with the savings by increased air temperature, for the increased coefficient of performance (COP) of the refrigeration equipment may yield savings which exceed the fan power reductions.

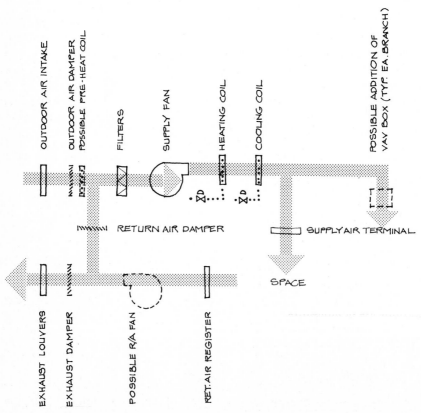

Figure 11.15 HVAC air distribution: single-duct single-zone systems.

ACTION GUIDELINES

☐ 1. Maintain filters to reduce resistance to airflow, permitting a reduction in the time that the blower and the primary heat- or cold-producing equipment runs to satisfy room loads. Where possible, change filters to low-resistance types. A V-type arrangement, as space permits, will provide more filter surface and less resistance and thus requires less frequent cleaning than a straight single-panel filter.

☐ 2. Clean coils of lint and dirt to increase heat transfer efficiency in air-handling units.

☐ 3. To reduce heat loss and heat gain, repair insulation where torn, or insulate the exterior surfaces of casings where insulation is not on the interior surface. Units located outdoors on-grade or on the roof, or in nonconditioned spaces, should be insulated with the equivalent of 3 in of fiberglass.

☐ 4. Where electric coils are used, operate the coils in stages, rather than all off and all on. Take care that coils are not on when air is not circulating, or they will burn out. Consider using the cooling coil for both heating and cooling by modifying the piping connections and removing the heating coil. The benefits are: (1) Elimination of the heating coil reduces total system resistance to airflow and allows a lower fan speed to achieve the same volume throughput of supply air with less fan horsepower, and (2) when used for heating, the cooling coils, which typically have a much greater surface area for heat transfer than heating coils, allow considerably lower water temperatures for a given heat output. The reduction in water temperature, in turn, minimizes heat losses from the system and allows more efficient operation of the boilers.

☐ 5. Maintain the hot water and the chilled water at constant flow, but reset for part load conditions of more than 2-h duration.

☐ 6. Avoid simultaneous heating and cooling except when required for humidity control in critical areas. Do not operate a large system to meet the needs of a small area for humidity control.

☐ 7. Install controls on the heating and cooling coils to modulate the supply air temperature. Allow an 8 to 10° "dead zone" between heating and cooling to prevent simultaneous heating and cooling and rapid cycling. If possible, use an economizer cycle wherever the total heat of the outdoor air is favorable.

11.10 TERMINAL REHEAT SYSTEMS

Terminal reheat systems were developed to overcome the zoning deficiencies of single-duct systems. They comprise a single-duct, single-zone system with individual heating coils in each branch duct to zones of similar loads. Reheat systems were also developed to give closer control of relative humidity in selected spaces. See Fig. 11.16.

Terminal reheat allows each zone to be individually controlled, but wastes energy in the cooling season as all the supply air must be cooled to a temperature low enough to meet the most critical load zone but must be reheated for zones of lesser loads to avoid overcooling.

As with single-duct systems, energy can be conserved by reducing the supply air volume. Many reheat systems are controlled at a fixed supply temperature of around 55°F DB/55°F WB. To conserve energy, fit controls to reschedule the supply air temperature upward according to demands of the zone with the greatest cooling load. If one zone has cooling loads grossly in excess of all others, the controlling thermostat should be located in that space.

The greatest quantity of energy can be saved by adding variable-air-volume (VAV) boxes to each of the major branch ducts. Each VAV box should be controlled by a space thermostat located in its particular zone, and its

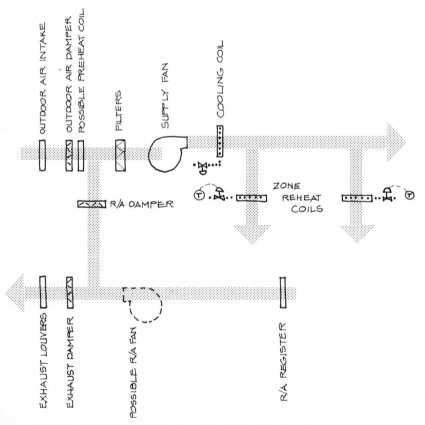

Figure 11.16 HVAC air distribution: terminal reheat systems.

associated reheat coil should be provided with controls to prevent reheat until the VAV box has reduced the zone supply air volume to 50%.

ACTION GUIDELINES

☐ 1. In buildings where a reheat system supplies zones of different occupancy, add dampers and control valves so that water can be shut off during unoccupied times.

☐ 2. Reduce the water flow and temperature of hot water to reheat coils.

☐ 3. Use waste heat from condensate, incinerators, diesel or gas engines, or solar energy for reheat, but in all cases raise the cold-duct temperature to reduce refrigeration load.

☐ 4. Terminal reheat systems were usually installed to control dehumidification. Where dehumidification can be eliminated and zone control can be satisfactorily maintained (or is not needed), operate the terminal reheat system on a temperature demand cycle only.

☐ 5. Add or adjust controls to schedule supply air temperature according to the number of reheat coils in operation. If 90% of the coils are in operation, reset supply air temperatures to a higher level until the number of reheat coils in operation falls to 10%.

☐ 6. Install an interlock between the two valves to prevent simultaneous heating and cooling.

☐ 7. If air conditioning is needed during unoccupied hours or in very lightly occupied areas, install controls to deenergize the reheat coils, raise the cold-duct temperature, and operate on demand cycle.

☐ 8. Under an alteration or expansion program, install variable volume rather than terminal reheat or other systems.

☐ 9. Deenergize or shut off terminal reheat coils, raise the chilled-water and supply air temperature of the central system, and add recooling coils in ducts in areas where lower temperatures are needed.

11.11 MULTIZONE SYSTEMS

Most multizone units currently in use have a single heating coil serving the hot deck and a single cooling coil serving the cold deck. The temperature of the supply air in each zone is adjusted by mixing the required quantities of hot and cold air from these coils. Thus, the hot-deck temperature must be sufficiently high to meet the heating demands of the coldest zone, and the cold-deck air must be sufficiently low to meet the demands of the hottest zone. All intermediate zones are supplied with a mixture of hot and cold air. This wastes energy in a manner similar to that of reheat systems. See Fig. 11.17.

New-model multizone units are now available which have individual heating and cooling coils for each zone supply duct, and the supply air is heated or cooled only to that degree required to meet the zone load. These new types of units use far less energy than units with common coils, and so

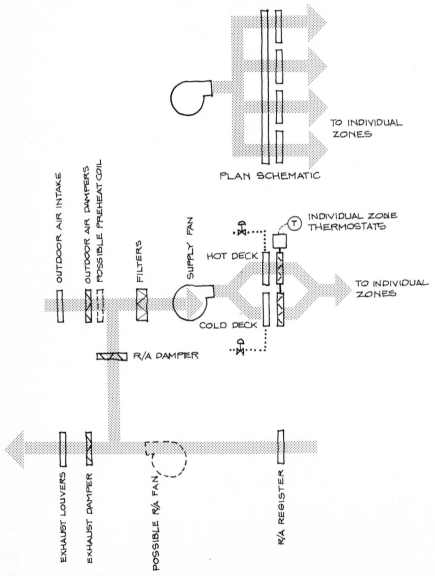

Figure 11.17   HVAC air distribution: multizone systems.

when renovations are contemplated or the existing multizone unit is at or near the end of its useful life, the replacement should be a multizone unit with individual zone coils.

Where a multizone unit serves interior zones that require cooling all during the summer and most of the winter, energy can be conserved by converting to VAV reheat. For this conversion, blank off the hot deck and add low-pressure VAV dump boxes with new reheat coils in the branch ducts after the VAV boxes. This system works well in conjunction with an economizer cycle, and reheat energy is minimized. Careful analysis, however, is required of the existing system and the zone requirements to make the correct selection of equipment.

ACTION GUIDELINES

☐ 1. Analyze multizone systems carefully; treat each zone as a single-zone system, and adjust air volumes and temperature accordingly.

☐ 2. Hot-deck and cold-deck dampers are often of poor quality and allow considerable leakage even where fully closed. To conserve energy, check these dampers and modify to avoid any leakage.

☐ 3. Install controls or adjust existing controls to give the minimum hot-deck temperature and maximum cold-deck temperature consistent with the loads of critical zones.

☐ 4. Arrange the controls so that when all hot-duct dampers are partially closed, the hot-deck temperatures will progressively reduce until one or more zone dampers are fully opened and, when all the cold-duct dampers are partially closed, the cold-duct temperature will progressively increase until one or more of the zone dampers are fully opened.

☐ 5. Install controls to shut off the fan during unoccupied periods. Install additional controls to shut off all heating control valves during unoccupied periods in the cooling season and all cooling valves during unoccupied periods in the heating season.

☐ 6. Convert the multizone system entirely or in part to variable volume by adding terminal units with pressure bypass, or add fan coil units in specific areas requiring constant air volume.

11.12 DUAL-DUCT SYSTEMS

Dual-duct systems, as the name implies, have two ducts, one trunk to supply warm air and the other to supply cool air to mixing boxes which then deliver air in one duct to the conditioned space. Warm and cool supply air are proportioned in the mixing boxes at the point of use to provide supply air to the room at whatever temperature meets the thermal demand. This type of system, though not usually described as a terminal reheat system, does, in fact, operate essentially on the same principles. Air is cooled to a fixed level, then reheated to the point of use by being mixed with warm air. This system is inherently wasteful. See Fig. 11.18.

The temperature of the hot duct should be reduced and the temperature of the cold duct should be increased to that point where the heating and cooling loads of the most critical zones can just be met. Return air is a mixture of all zones and reflects the average building temperature. In some designs of central station equipment, it is possible to stratify the return air and the outdoor air by installing splitters so that the hottest air favors the hot deck and the coldest air favors the cold deck. This will reduce both heating and cooling loads.

Dual-duct high-velocity systems operate in the same manner as low-velocity systems with the exception that supply fans for high-velocity systems run at a higher pressure and that mixing boxes must have sound attenuation. Considerable energy is required to operate fans at high pressure, and close analysis of pressure drops within the system should be made to allow reduction of fan pressures to the minimum required to operate the mixing boxes.

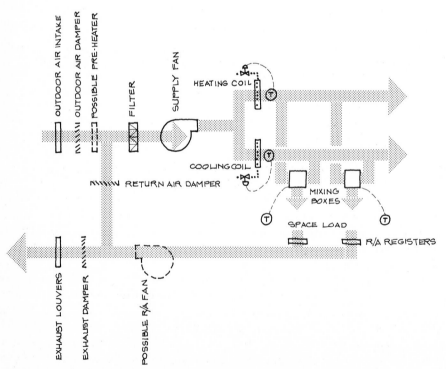

Figure 11.18 HVAC air distribution: dual-duct systems.

ACTION GUIDELINES

☐ 1. To reduce system pressure, replace existing high-pressure mixing boxes with lower-pressure mixing boxes.

☐ 2. Under conditions where there is no cooling load, install controls to close off the cold-air duct. Shut off the chillers and cold-water pumps and operate the system as a single-duct system, controlling the temperature of the warm-air duct to handle the heating loads.

☐ 3. Under conditions where there are no heating loads, install controls to close off the warm-air duct. Shut off hot water, steam, or electric heating, and operate the system with the cold-duct air only, controlling the temperature of the supply air to handle the cooling loads.

☐ 4. Replace obsolete or defective mixing boxes to eliminate leakage of hot or cold air when the respective damper is closed.

☐ 5. Provide volume control for the supply air fan and reduce air volumes by lowering fan speed when both hot-deck and cold-deck air quantities can be diminished to meet peak loads.

☐ 6. When there is more than one air-handling unit in a dual-air system, modify ductwork if possible so that each unit supplies a separate zone. This provides better opportunity to reduce hot- and cold-duct temperatures according to shifting loads.

☐ 7. Change dual-duct systems to variable-volume systems by adding VAV boxes and fan control when the payback in energy saved is sufficiently attractive.

11.13 VARIABLE-AIR-VOLUME SYSTEMS

Variable-air-volume (VAV) systems include an air-handling unit which supplies heated or cooled air at constant temperatures (which can be reset) to VAV boxes for each zone. Variable-air-volume boxes regulate the quantity rather than the temperature of the air. Varying air quantity with the appropriate supply temperature is one of the most efficient modes of operation in terms of energy usage. The fan should be run at full volume only when all VAV dampers are fully opened. See Fig. 11.19.

Fan volume is not controlled in some systems. When full volume is not needed in the conditioned spaces, supply air is dumped into a ceiling plenum, and the air is returned to the supply unit by inlet openings in the return air side of the air-handling unit. This "dump" system is necessary for direct-expansion cooling; almost 100% of the supply air passes over the coils. This VAV system is not as efficient as a fan-controlled system, but energy is saved by effectively reducing the static pressure.

The second type of VAV system controls the air volume at the outlets by closing down dampers as the zone loads decrease. This type of system varies the total supply volume handled by the fan by increasing the resistance to airflow under light load conditions.

ACTION GUIDELINES

☐ 1. To conserve energy, the fan volume can be regulated according to the demands of the system by a variable-speed motor with an SCR controller or with inlet guide vanes.

☐ 2. Provide controls to modify fan speeds up or down to maintain constant pressures in the supply duct. If not installed, provide either inlet vortex dampers or a variable-speed drive such as a multispeed motor or SCR

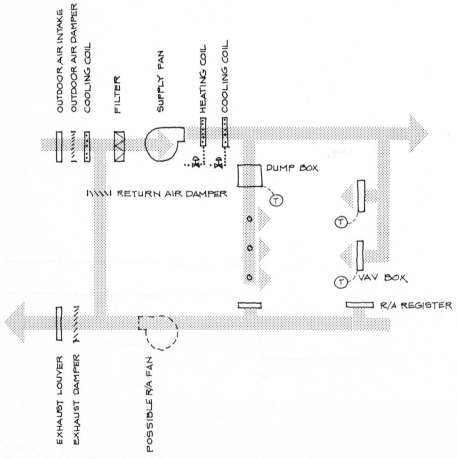

Figure 11.19 HVAC air distribution: variable-volume systems.

control so that fan volume will be reduced when dampers begin to close until one or more VAV dampers are again fully opened.

☐ 3. Provide controls to reschedule supply air temperatures that the damper of the variable-air-volume box serving the zone with the most extreme load is fully opened.

☐ 4. Install controls to reduce the hot-water temperature and raise the chilled-water temperature in accordance with shifting thermal demands.

☐ 5. With single-duct variable-volume systems with reheat, set controls to delay reheat until the volume is reduced to a minimum. Where possible, eliminate reheat coils.

11.14 INDUCTION SYSTEMS

Induction systems are commonly used for heating and cooling perimeter zones where large fluctuations of heating and cooling loads occur (see Fig. 11.20). Primary air is either heated or cooled and supplied at high pressure to the induction units located within the conditioned space, typically on outside walls. Primary air is discharged from nozzles designed to induce room air into the induction unit at approximately four times the volume of the primary air. The induced air is cooled or heated by a secondary water coil. This water coil may be supplied by a two-pipe system (either chilled water or heated water is available, but not both simultaneously), by a three-pipe system (separate supplies of hot water or chilled water are continuously available and, after passing through the unit, are mixed into a common return), or by a four-pipe system (a supply and return of hot water and chilled water are continuously available).

The primary-supply air fan operates at high pressures and requires high input in terms of horsepower. After careful analysis, the primary-air volume and pressure should be reduced to the minimum required to just operate the induction terminal units.

Induction unit nozzles may be worn through many years of cleaning and operation with a resultant increase in quantities of delivered primary air, but at lower air velocities with lower volumes of induced air. Each induction unit should be checked for needed repair or replacement.

The major causes of high-energy use with induction systems are:

1. The horsepower to maintain the high pressure to deliver primary air and induce room air flow through the unit.

2. Simultaneous heating and cooling in a "terminal reheat" type of cycle.

3. Restriction in air passages which impedes the flow of room air through the coils.

4. Dirty coils which reduce heat transfer to room air and impose an extra load on the primary-air system to compensate for the reduced induction effect.

Refer to Sec. 11.10, "Terminal Reheat Systems," for general measures also applicable to induction systems.

ACTION GUIDELINES

☐ 1. Reschedule the temperature of the heating water and the cooling water according to the load. If the building has a light cooling load, the chilled-water temperature should be raised; if the building has a light heating load, the hot-water temperature should be lowered. Avoid simultaneous heating and cooling in any one zone.

☐ 2. Typically, the temperature of the primary-supply air is maintained at

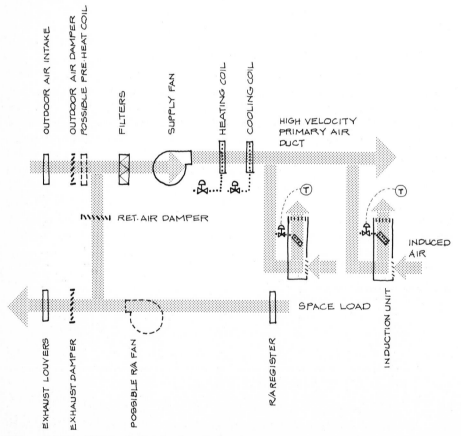

Figure 11.20 HVAC air distribution: induction systems.

constant levels for cooling and heating which necessitates reheating or recooling by the individual units to meet fluctuating load conditions. The system operates essentially as a reheat or recool system with the inevitable consequence of wasted energy. To reduce this waste, schedule the temperature of the primary-supply air at the air-handling unit according to load. Under light heating load conditions, the primary-air heating coil should be modulated to reduce the supply air temperature, and under light cooling load conditions the supply air cooling coil should be similarly modulated to increase the supply air temperature. Achieve the modulation preferably by adjusting water temperatures (rather than water volumes) to the central coils. Chillers and boilers will then operate closer to their peak efficiencies.

☐ 3. Remove lint screens in induction units and clean coils regularly.

☐ 4. For night operation during the heating season, shut down primary-air fans, raise the hot-water temperature, and operate the induction units as gravity convectors.

11.15 FAN COILS AND UNIT VENTILATORS

Fan-coil systems are composed of one or more fan-coil units set up to be two-, three-, or four-pipe systems; each unit contains a fan which blows air through heating or cooling coils into a single outlet to supply a zone or room. Unit ventilators are essentially fan coils with additional equipment for admitting partial or 100% outdoor air through the coils. Unit ventilators are larger and are typical of school HVAC systems. Usually, the heating and cooling coils of both fan coils and unit ventilators are controlled to maintain a supply air temperature to meet zone heating or cooling loads. However, in many systems the fan speed can be manually controlled (high, medium, or low) to control heating or cooling output. Often fans are operated at low speeds to reduce noise.

ACTION GUIDELINES

☐ 1. Heating and cooling coils should be cleaned, and air and water flow reduced to the minimum required to meet space conditions.

☐ 2. For fan-coil systems which have separate coils for heating and air conditioning, install a control to prevent simultaneous heating and cooling.

☐ 3. Install a 7-day timer to shut down fans, close off valves, and shut off chilled water pumps, compressors and cooling towers, or air-cooled condensers during unoccupied periods.

☐ 4. With three-pipe systems, set controls for minimum mixing of hot and cold water.

☐ 5. Where a large number of fan coils are used in a building, the coils are provided with heating and cooling media of constant temperature, and control is achieved by varying flow rates through the coils. Consider adjusting the heating and cooling systems according to load. For example, when the heating load is light and most of the heating control valves which vary the flow rate are partially closed, reduce the hot-water temperature until one or more valves are fully open.

☐ 6. In mild winter weather, and at night, shut off fans and permit the coils to operate as convectors. In severe winter weather, when the building is unoccupied but must be heated, or when excessive noise is not a problem, operate fans at high speed.

☐ 7. Clean filters and coils.

☐ 8. Close outside air dampers any time infiltration equals ventilation requirements, or block them off permanently.

☐ 9. Block off inlets where no dampers are installed if infiltration meets ventilation requirements.

☐10. Keep air outlets and inlets free of obstructions.

☐11. Where fan coil units are not located in conditioned areas, insulate casings to reduce heat loss or gain.

11.16 HEAT PUMPS AND AIR CONDITIONERS

Window and through-the-wall air conditioners have similar self-contained compressor and air-cooled condensing units. They may be equipped with electric coils for heating or, in the case of through-the-wall units, coils supplied with hot water or steam from a central source. Because both systems are relatively inefficient during the cooling cycle (owing to small, usually undersized, condensers), maintenance of coil surfaces and reduction of operating hours are very important.

Most older unitary conditioning units have a low cooling equivalent efficiency rating (EER), sometimes as low as 5 or 6 Btu/W. If these units are near the end of their useful operating life, they should be replaced with new units having an EER of 9 Btu/W or more. New units now in the development stage will have an EER of 15 Btu/W or more.

Units with electric resistance heating coils for winter use consume a large amount of energy for tempering or heating. If existing units to be replaced are equipped with electric resistance heating coils for winter use, they should be exchanged for air-to-air heat pumps (which provide for 1.75 to 3 times as much heat per kilowatt input as resistance heaters). Outdoor air blowing into the room directly through the coils of the unit or through cracks around the frame of window air conditioners can be a major source of infiltration and heat load in the winter. If they are not used for heating, remove window units and store them in the winter, or caulk the cracks and cover the units with fitted enclosures made for that purpose.

Unitary heat pumps are of either the air-to-air or water-to-air type. The older air-to-air models have the same built-in condenser and evaporator inefficiencies as room air conditioners, but the water-to-air heat pumps operate at lower condenser temperatures and are more efficient. All recommendations for condensers, compressors, evaporators, and fans apply to both types. During the cooling season, the operating efficiency of air-to-air heat pumps can be improved by shielding the condenser air intakes from direct sunlight and directing cool exhaust air from the building to the condenser air inlet.

Where possible, direct the warm exhaust air from the building to the inlet of air-to-air heat pumps to raise their coefficient of performance. Waste heat from other processes can also be used as a heat source.

ACTION GUIDELINES

☐ 1. Replace air-to-air heat pumps with water-to-air heat pumps where there is a source of heat such as ground water with temperatures above average ambient winter air temperatures or solar-heated water.

☐ 2. For incremental air-conditioning units follow guidelines for fan coil units during the heating cycle.

☐ 3. Install a 7-day timer to program operation of compressors in accordance with occupied-unoccupied periods.

☐ 4. When replacing incremental air-conditioning units which are equipped with electric heating coils, install a modal with a heat pump instead.

☐ 5. Replacement compressors should have an EER of 9 Btu/W or better during the cooling cycle.

☐ 6. Install a centralized system of on-off controls on a floor-by-floor basis using a remote-control contactor on the power circuit feeding the heat pumps. Rewire the power supply to heat pumps in such a manner that one-third, two-thirds, or all the units in any zone can be turned on and off, depending on whether the building is occupied or not and what conditions are expected.

☐ 7. When retrofitting, install one larger unit rather than multiple smaller units for greater efficiency.

☐ 8. If a heat pump is used for cooling *and* heating, sizing the unit for the cooling load may result in inadequate capacity for heating, requiring auxiliary electric resistance heating. Oversize such heat pumps for cooling to obtain adequate capacity for heating without such resistance heating. In any case auxiliary heat, if needed, can be supplied with gas or oil as well as electricity and should be considered.

11.17 EXHAUST SYSTEMS

All recommendations which have been made to reduce fan power requirements by flow rate and resistance reduction apply equally to exhaust systems and to supply systems.

ACTION GUIDELINES

☐ 1. Exhaust systems should be balanced so that exhaust airflow rates do not exceed supply airflow rates of associated systems. Ideally, exhaust should be 10% less than supply to maintain a positive building pressure.

☐ 2. Exhaust fan volumes should be modulated in step with associated VAV supply fans by installing inlet guide vanes or variable-speed controls.

☐ 3. Recirculate toilet room exhaust air through charcoal filters to reduce makeup air requirements.

☐ 4. Install controls to operate toilet exhaust fans intermittently for 10 to 20 min out of every hour and to shut them off automatically during unoccupied periods.

☐ 5. Install a motorized damper at inlet grilles with wiring from damper to light switches to reduce the air quantity when toilet rooms are unoccupied. Modulate fan volumes according to load by monitoring pressures in main exhaust stacks.

☐ 6. For new installations, or when existing fans or motors are replaced, provide variable-speed control or inlet vane control to operate in accordance with static pressure.

☐ 7. Shut off supply fans which provide makeup air for toilet rooms, and install door louvers or cut off the bottoms of doors to permit air from conditioned areas to migrate into the toilet rooms as makeup for the exhaust system. Set maximum capacity to provide 1 cfm/ft^2 of toilet area.

CHAPTER 12

BOILERS AND REFRIGERATION

12.1 PRIMARY CONVERSION LOADS: INTRODUCTION

For heating purposes, oil, coal, or gas is consumed by boiler or furnace and is converted through combustion into heat. The potential energy in the fuel is never fully realized, however, because of the difficulty in obtaining the correct mix of air and fuel for combustion and because of heat losses up the flue. Although the efficiency of a burner-furnace unit at any instant may fall just short of 90% for those fueled by oil and gas (and somewhat lower for those fueled by coal), the seasonal efficiency is generally lower by 10 to 30% owing to additional stack losses between firings, low flame temperature owing to short cycling, and inefficient partial loading. Of particular interest in this chapter are measures for increasing the efficiency that counts, the seasonal efficiency.

For cooling, energy in the form of heat operates absorption or steam turbine compression refrigeration units, and in the form of electricity operates reciprocating, centrifugal, or screw-type compressor refrigeration units. The refrigeration equipment produces either chilled water or, in the case of direct expansion units, cooled and dehumidified air. Typical seasonal coefficients of performance, or efficiency, vary from .60 for absorption chillers to 4.5 for electric-driven water chillers.

In both heating and cooling, the single goal is improved efficiency of conversion operations. The following sections show the way.

12.2 IMPROVE BOILER OR FURNACE EFFICIENCY

Although there are many types and sizes of boilers, furnaces, and burners in use today, all have certain common characteristics, and similar techniques can be used to improve their efficiency and conserve energy.

In theory, combustion of oil, gas, or coal requires a given fuel/oxygen ratio for complete burning and maximum efficiency. In practice, air (a mixture of oxygen and nitrogen) is used to provide the necessary oxygen for burning and must be supplied in excess of the theoretical requirements to ensure complete combustion. The quantity of air that just gives complete combustion is the optimum amount. Any reduction in the optimum air quantity prevents complete combustion and wastes energy; the maximum heat value of the fuel cannot be released. Any increase over the optimum air quantity for combustion will reduce not the efficiency of combustion itself, but rather the rate of heat transfer to the boiler or furnace, and will increase the stack temperatures (heat lost out the chimney). It is important, therefore, that the air supply to the combustion chamber be controlled to achieve the most favorable fuel/air ratio for any given burning rate. Measures should be taken to prevent leaks of uncontrolled air into the combustion chamber.

Burners should be cleaned and adjusted each year and should be monitored periodically during the year to ensure optimal combustion efficiency (with adjustment as necessary). In large installations combustion control and monitoring should be a daily procedure. Oxygen analyzers can be used to measure O_2, an Orsat apparatus can be used to check the CO_2 content of the flue gas, and thermometers can check stack temperatures.

After the building and distribution heating loads are reduced, nozzle sizes should be reduced and combustion chambers modified. This is an opportune time to clean and correct dirty oil nozzles and fouled gas parts.

The condition of the heat transfer surface directly affects heat transfer from the combustion chamber. The fireside of the heat transfer surface must be clean and free from soot or other deposits, and the airside and waterside must be clean and free of scale deposits. Deposits can be removed by scraping where accessible, by chemical treatment, or by a combination. In the case of steam boilers, once the waterside of the boiler is clean, correct water treatment and blowdown should be instituted to maintain optimum heat transfer conditions.

Boilers and furnaces may achieve relatively high instantaneous full-load efficiencies (80 to 87%), but because they are operated most of the time at partial load, they have lower seasonal efficiencies. Generally, measures which increase the full-load efficiency will also increase the seasonal efficiency. However, when the peak loads are of very short duration, it may be advantageous to tune the boiler for maximum efficiency at part-load conditions in order to gain greater seasonal efficiency. See Fig. 12.1.

The heating system operates between 50 and 60% full load for more hours per year than at higher or lower loads, and the optimum operating point of the heating boilers should be selected accordingly to give the highest efficiency spread over this range. For example, if the installation comprises three 8×10^6 Btu/h boilers and the maximum heating load is reduced to 18×10^6 Btu/h, then the boilers should be adjusted for maximum efficiency at 70% of the rated output. This gives three possible operating points at maximum efficiency:

One boiler at 8×10^6 Btu/h $\times 0.7 = 5.6 \times 10^6$ Btu/h or 31% full load

Two boilers at $16 \times 10^6 \times 0.7$ Btu/h $= 11.2 \times 10^6$ Btu/h or 62% full load

Three boilers at $24 \times 10^6 \times 0.7$ Btu/h $= 16.8 \times 10^6$ Btu/h or 93% full load

Indicators of maximum combustion efficiency are stack temperature and percentage CO_2 or percentage O_2.

Primary and secondary air should be allowed to enter the combustion chamber only in regulated quantities and at the correct place. Defective gaskets, cracked brickwork, broken casings, etc., will allow uncontrolled and varying quantities of air to enter the boiler and will prevent accurate adjustment of fuel/air ratios. If spurious stack temperature and/or oxygen content readings are obtained, then the boiler should be inspected for air leaks and all defects repaired before final adjustment of the fuel/air ratio.

When substantial reductions in heating load have been achieved, the firing rate of the boiler may be excessive and should be reduced. Consult the firing equipment manufacturer for specific recommendations. Reduced firing rates in gas and oil burners may require additional bricking to reduce the size and shape of combustion chambers.

ACTION GUIDELINES

☐ 1. Repair or rebuild oil burner combustion chambers to the correct size for providing optimum efficiency at 90% of the full load firing rate. Construct chambers with bricks of the refactory type (not common bricks). Incorrect matching of burner and combustion chamber and broken brickwork can result in losses of from 10 to 20%.

☐ 2. Clean and scrape firesides to remove soot and scale.

☐ 3. Clean watersides, remove built-up scale.

☐ 4. Scrape scale from steam drums.

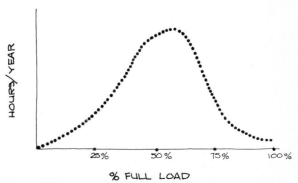

Figure 12.1 Typical boiler loading.

☐ 5. Clean airsides, remove soot, and scrape scale in forced warm-air and hot-air furnaces.

☐ 6. Insulate units which are in unheated spaces, on roofs, or in air-conditioned spaces. Repair insulation where damaged. (If boiler or furnace casing is 10 to 15% warmer than room temperature, radiation loss could be 10% or more of the capacity of the unit.)

☐ 7. Check for and seal air leaks between sections of cast-iron boilers, for air leaks into natural-draft chimneys, and for all air leaks into combustion chambers, especially around doors, frames, and inspection ports.

☐ 8. If the combustion efficiency is at a maximum but stack temperatures are still too high (over 350°F), install baffles or turbulators to improve heat transfer (consult the boiler manufacturer). Refer to Secs. 15.2 and 15.3 for discussions of excess heat recovery from breeching.

☐ 9. Maintain the lowest possible steam pressure suitable for supplying radiation or coils, and vary the pressure in accordance with the space heating or process demands; steam pressures can be reduced most of the year (and standby losses are reduced as pressures are reduced). Where high-pressure steam is required for turbines or other pieces of equipment which operate for only a portion of the year, reduce the system steam pressure for the rest of the year.

☐ 10. Maintain the lowest possible hot-water temperature which will meet space or domestic hot-water needs. In the absence of indoor-outdoor modulating controls, raise or lower operating temperature for hot-water systems to conform to indoor-outdoor conditions. For example, a boiler might operate at 120°F with an outdoor temperature at 60°F and at 160°F when the outdoor temperature is 20°F.

☐ 11. Install Airtroll fittings on hot-water systems to vent air from the hot-water piping.

☐ 12. Install and maintain proper steam regulators to reduce steam consumption. Bypasses should be used only during emergencies and during repair of regulators. On large installations, install a flowmeter and record changes so that an operator can analyze and maintain maximum efficiency.

☐ 13. Clean filters regularly in gravity and forced warm-air units to reduce the operating time of the furnace.

☐ 14. Reduce the firing rate or enlarge the return air opening, or both, if the supply air temperature of a gravity furnace is over 150°F at full load. Consider converting gravity hot-air furnaces to forced warm-air furnaces.

☐ 15. Use proved additives to improve the efficiency of combustion, to eliminate water in oil storage systems, and to reduce soot deposits in furnaces and boiler convection tubes.

☐ 16. Adjusting the firing rate of gas or oil burners at too high a rate will cause short cycling and excessive fuel consumption; too low a rate will require constant operation, and inadequate heat will be delivered to the spaces. If the boiler is oversized, adjust the firing rate to the building load, not the boiler, and adjust the firing rate through the year if seasonal loads vary widely.

☐ 17. Reset frost protection controls during unoccupied periods so that heating pumps operate only when the outdoor temperature is less than 35°F.

☐ 18. Eliminate gas pilots and install electric pilots for boilers and furnaces.

☐ 19. In very large installations, consider converting high-temperature water to high temperature–low pressure fluid systems.

☐ 20. Where air conditioning is needed during the heating season for portions of the building, use condenser water heat for reheat and for tempering makeup air for space heating, instead of operating the heating boiler.

☐ 21. Install additional condensate return lines to the boiler for feed water systems when uncontaminated condensate is being wasted. Where condensate is being wasted, consider installation of a heat exchanger to preheat domestic hot water.

☐ 22. When replacing boilers, provide for dual fuel operations as a contingency against shortages of one particular fuel.

☐ 23. Consult boiler manufacturers for feasibility of operating existing boilers without separate combustion chambers, and allow the flame to impinge directly upon boiler surfaces. This must be done only with the approval of the boiler manufacturer.

12.3 INSTALL A FLUE-GAS ANALYZER

The efficient combustion of fuel in a boiler requires an optimum fuel/air ratio providing for a percentage of total air sufficient to ensure complete combustion of the fuel without overdiluting the mixture and thereby lowering boiler burner efficiency. Excess air through a boiler can waste 10 to 30% of the fuel burned.

Optimum combustion efficiency varies continuously with changing loads and the stack draft, and can be closely controlled only through analysis of flue gases. Knowledge of flue-gas temperature and either flue gas CO_2 or O_2 content is required to allow continuous adjustment of fuel/air ratios.

Indicators are available which continuously measure CO_2 and stack temperature and give a direct reading of boiler efficiency. These indicators provide boiler operators with the requisite information for manual adjustment of fuel/air ratios and are suitable for smaller installations or for situations in which money for capital improvements is limited. A more accurate measure of efficiency, however, is obtained by analysis of oxygen content. If both are available, the cross-checking of O_2 and CO_2 concentrations is useful in judging burner performance. Owing to the increasingly widespread need for multifuel boilers, however, O_2 analysis alone is the single most useful criterion, since the O_2/total air ratio varies within narrow limits for all fuels. For larger boiler plants the installation of an automatic continuous oxygen analyzer with "trim" output that will adjust the fuel/air ratio to meet changing stack draft and load conditions should be considered.

Most boilers can be modified to accept automatic fuel/air mixture control by flue-gas analyzer, but a gas analyzer manufacturer should be consulted for each particular installation to be sure that all other boiler controls are compatible with the analyzers.

It is important to note that some environmental protection laws might place a higher priority on reducing visible stack emissions than on efficiency optimization of fuel combustion, especially where fuel oil is concerned. The effect of percentage of total air on smoke density might prove to be an overriding consideration and limit the goal of minimizing excess air. All applicable codes and environmental statutes should be checked for compliance.

The costs of adding analysis instrumentation and controls for flue gas will vary with each particular case, but for order of magnitude the following tabulated costs may be used:

O_2 analyzer with meter readout	$2000 to 3000
O_2 analyzer with chart recorder	$3000 to 5000
O_2 analyzer with trim link control	$4000 to 6000
Full metering with trim link	$11,000 to 16,000

To determine efficiency, enter Fig. 12.2 at either the O_2 content or the CO_2 content and, following the direction of the example line, intersect with stack temperature to read out the combustion efficiency against the appropriate fuel type. Percentage of total air may also be obtained from an intermediate line.

Savings are determined by calculating boiler efficiency under present operating conditions and then under improved conditions with optimization of the fuel/air ratio.

Example

A boiler burning No. 2 oil rated at 20 million Btu/h has a yearly oil consumption of 425,000 gal. A CO_2 meter is installed and reads 11% CO_2 in the flue gas and a stack temperature of 650°F. Entering the graph at the 11% CO_2 point for No. 2 oil and intersecting with a stack temperature of 650°F gives an efficiency of 78%. Air percentage of 138 indicating 38% excess air and an O_2 percentage of 6 can also be read.

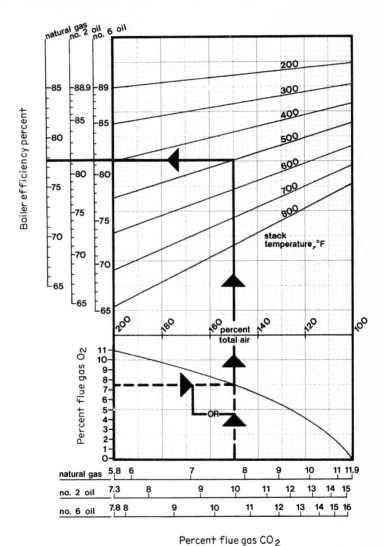

Percent flue gas CO_2

Figure 12.2 Effect of flue-gas composition and temperature on boiler efficiency.

The installation of an O_2 analyzer with an automatic fuel/air ratio control will reduce O_2 content to 3.5% and stack temperature to 530°F. Using the graph again shows the new efficiency to be 83%.

Yearly savings due to increased boiler efficiency

$$= \text{original yearly consumption} \times \frac{\text{new efficiency} - \text{original efficiency}}{\text{new efficiency}}$$

$$= 425{,}000 \times \frac{83 - 78}{83} = 25{,}600 \text{ gal/yr}$$

$$= \$10{,}240 \text{ at } 40¢/\text{gal}$$

The installation cost is approximately $16,000 and is obviously economically advantageous without further economic analysis

ACTION GUIDELINES

☐ 1. Adjust oil burner efficiencies to achieve proper stack temperatures, CO_2 emissions, and excess air settings. Adjust settings to provide a maximum of 400 to 500°F stack temperature and a minimum of 10% CO_2 at full-load conditions. Accurate testing is essential for the correct burner adjustment to attain maximum efficiency. Use appropriate instruments or controls and institute combustion testing to maintain proper fuel/air ratios.

☐ 2. Maintain continuous monitoring and keep records of boiler efficiency at various full- and part-load conditions. Install controls to prevent burner short cycling.

12.4 ISOLATE OFF-LINE BOILERS

Light heating loads on a multiple-boiler installation are often met by one boiler on line with the remaining boilers idling on standby. Idling boilers consume energy to meet standby losses which can be further aggravated by a continuous induced flow of air through them into the stack and up the chimney.

Unless a boiler is scheduled for imminent use to meet an expected increase in load, it should be secured and isolated from the heating system by closing valves and from the stack and chimney by closing dampers. If a boiler waterside is isolated, preventing airflow through the stack is important; it is possible for backflow of cold air to freeze the boiler. Large boilers can be fitted with bypass valves and regulating orifices to allow a minimum flow through the boilers to keep them warm and to avoid thermal stress when they are brought on-line again. See Fig. 12.3.

ACTION GUIDELINES

☐ 1. Install automatic draft control to shut off the breaching 30 s after burners go off and restart automatically 30 s before the next firing cycle to prevent heat loss up the chimney between firing cycles.

☐ 2. Turn off gas pilots for furnaces, boilers, and space heaters during the nonheating months and during long unoccupied periods.

☐ 3. Shut down hot-air furnaces completely when the building is not occupied and there is no danger of freezing.

☐ 4. Set operating aquastats on steam and hot-water boilers to 100°F during shutdown periods.

☐ 5. Modify piping as necessary to permit on-line boilers to maintain temperatures in idle boilers, rather than using the burners of the idling boiler.

☐ 6. Where electric coils are used in furnaces, install a control to reduce the period of time that the coils are energized.

12.5 REPLACE EXISTING BOILERS WITH MODULAR BOILERS

Heating boilers are usually designed and selected to operate at maximum efficiency when running at their rated output and have lower efficiencies at reduced output. A diagram of typical heat load distributions over a heating

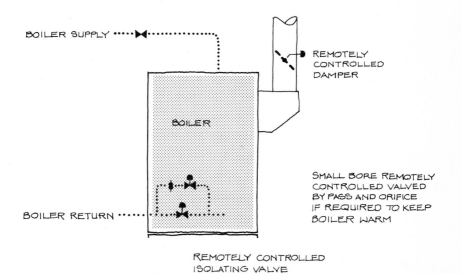

Figure 12.3 Boiler isolation controls.

season indicates that full boiler capacity is required for short periods only and that for 90% of the time the heat load is 60% of full load or less. See Fig. 12.4. Large-capacity boilers in single units must operate intermittently for the major part of the heating season, and although high-low firing capabilities may reduce cycling, the boilers can reach their design efficiency for only short periods. The result is low seasonal efficiency. See Fig. 12.5.

A modular boiler system composed of multiple boiler units, each of small capacity, will increase seasonal efficiency. Each module can be fired only when required at 100% of its capacity, and fluctuations of load can be met by firing more or fewer boilers. A small-capacity unit has low thermal inertia (giving rapid response and low heat-up and cool-down losses) and can be operated at maximum efficiency or can be turned off. Modular boilers are particularly effective in buildings that have intermittent short-time occupancy such as churches. They provide rapid warm-up for occupied periods and low stand-by losses during extended unoccupied periods.

ACTION GUIDELINES

☐1. Boiler seasonal efficiency may be improved from 68 to 75% in a typical installation when single-unit large capacity boilers are replaced by modular boilers. This represents a saving of approximately 9% of the present yearly fuel consumption.

☐2. When an existing boiler plant is close to the end of its useful life, replacement with modular boilers sized to meet the reduced heating load resulting from conservation action will be a good investment.

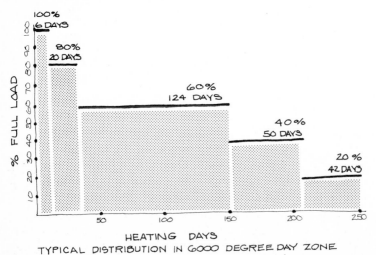

Figure 12.4 An example of the frequency of partially and fully loaded boiler conditions.

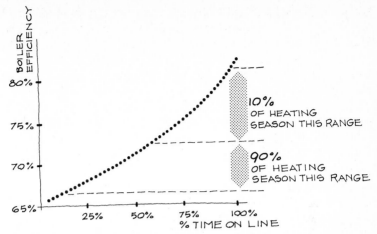

Figure 12.5 Effects on boiler efficiency due to cycling for partial loads.

☐ 3. If there is more than one boiler, operate one alone up to its maximum load before bringing other boilers on the line. Operating two or more boilers at very low capacity to carry part loads is inefficient.

☐ 4. Analyze the loads and seasonal operating efficiency and replace inefficient heating equipment with new modular-sized equipment to permit efficient operation at all part-load conditions. Select the size of multiple-boiler installations so that they can be operated at 75 to 100% of rated capacity for both summer and winter load conditions.

12.6 INCREASE BOILER AND FURNACE EFFICIENCY WITH PREHEATING AND AIR ATOMIZATION

Preheated primary and secondary combustion air will reduce the cooling effect when the air enters the combustion chamber and will increase the efficiency of the boiler. This will promote as well a more intimate mixing of fuel and air, which will also add to the boiler efficiency.

In most boiler rooms, air is heated incidentally by hot boiler and pipe surfaces and rises to collect below the ceiling. This air can be used directly as preheated combustion air by ducting it down to the firing level and directing it into the primary and secondary air inlets. Waste heat reclaimed from boiler stacks and blowdown or condensate hot wells can also be used to preheat combustion air. Heat exchange from flue gases to combustion air may be made directly using static tubular or plate exchangers or rotary exchangers. Heat exchange may also be made indirectly through runaround coils in the stack and combustion air duct.

Boiler efficiency will increase by approximately 2% for each 100°F increase in combustion air temperature. Combustion air can be preheated up to 600°F for pulverized fuels and up to 350°F for stoker-fired coal, oil, and gas. The upper temperature limit is determined by the construction and materials of the firing equipment, and manufacturers' recommendations should be obtained. See Fig. 12.6.

Waste heat from sources such as flue gases, blowdown, condensate, and hot wells, or from solar energy, may also be used to preheat oil either in the storage tanks (low-sulfur oil requires continuous heating anyway to prevent wax deposits) or at the burner nozzle. Oil must be preheated to at least the following temperatures to obtain complete atomization:

No. 4 oil	135°F
No. 5 oil	185°F
No. 6 oil	210°F

Heating beyond these temperatures will increase efficiency, but care must be taken not to overheat, or vapor locking could cause flameouts. The increase in efficiency obtained by preheating oil could be as high as 3% but depends on the particular constituents of the oil, and recommendations should be obtained from the oil supplier.

Air-atomizing burners are considerably more efficient than steam-atomizing burners and should always be considered as an alternative when replacing obsolete or defective steam-atomizing burners. They may in some circumstances be cost effective on their own merit without waiting for scheduled replacement times.

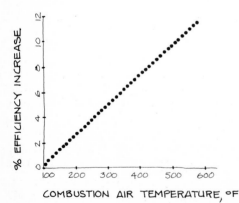

Figure 12.6 Boiler efficiency increases with preheated combustion air.

ACTION GUIDELINES

☐ 1. Use warm exhaust air from adjacent areas, or from the ceiling of the boiler rooms, to preheat combustion air.

☐ 2. Use automatic viscosity controls on fuel oil systems for purposes of attaining best atomization for efficient combustion. This control also permits flexibility of mixing and the use of any grade of fuel oil, either distillate or residual.

☐ 3. Use chemical fuel additives to reduce the flashpoint temperature of fuel oil, especially Nos. 4 and 6 oils. (Proper chemical treatment will also reduce soot deposit on No. 2 oil systems.)

☐ 4. Fit a "shell-head" adapter to the end of the gun of oil burners firing No. 2 oil to increase combustion efficiency.

☐ 5. When replacing obsolete No. 6 oil burners, consider air atomization instead of steam atomization to improve performance and efficiency.

☐ 6. Install an interlock between the combustion air unit and the burner, with proper prestart and after-firing control, to reduce the amount of cold air introduced into the building for combustion.

☐ 7. Regulate electric heating elements in steps or stages rather than with full-on, full-off staging controls. Limit the demand on the electric supply system and eliminate wasteful override.

☐ 8. When district steam is used in place of on-site boilers, utilize the condensate return to heat makeup air and domestic hot water, and for space heating.

☐ 9. Use solar energy to preheat combustion air and oil.

12.7 REDUCE BLOWDOWN LOSSES

The purpose of blowing down a steam boiler is to maintain a low concentration of dissolved and suspended solids in the boiling water and to remove sludge in the boiler to avoid priming and carryover. There are two principal types of blowdown, intermittent manual blowdown and continuous blowdown. Manual blowdown or sludge blowdown is necessary for the operation of the boiler regardless of whether continuous blowdown is installed. The frequency of manual blowdown will depend on the amount of solids in the boiler makeup water and the type of water treatment used.

Continuous blowdown is a steady drain on energy supplies because of the heat required for makeup water. Energy loss due to blowdown can, however, be minimized by the use of automatic blowdown control systems and heat recovery systems. Automatic blowdown controls periodically monitor the conductivity and pH of the boiler water and blow down the boiler only when required to maintain acceptable water quality. Further savings can be realized if the blowdown water is piped through a heat exchanger or a flash tank with a heat exchanger.

In a 100,000-lb/h, 600-lb/in² boiler, with a maximum boiler water concentration of 2500 ppm total dissolved solids, the blowoff will be 8% of the makeup water or 3500 lb/h. The total heat in the blowoff will be 1660 MBtu/h. A system using a heat exchanger with a 20°F terminal difference will recover 90% of the total heat in the blowoff, or 1494 MBtu/h. Adding a flash tank operating at 5 psig to the heat exchanger having a 20°F terminal difference will recover 93% of the total heat in the blowoff, or 1544 MBtu/h. The percentage of heat recovery will change with boiler operating conditions; the best recovery range is 78% for a 15-psig boiler to 98% for a 300-psig boiler.

Schedule boiler blowdown on an as-needed basis rather than on a fixed timetable. Smaller and more frequent blowdowns are preferable to larger and less frequent blowdowns. Be sure that boiler blowdown procedures adhere to specifications outlined by the manufacturer, the National Board of Boiler and Pressure Vessel Inspectors (Columbus, Ohio), and local codes. With few exceptions it is illegal, and in all cases undesirable, to discharge boiler blowdown directly to a sanitary sewer.

12.8 IMPROVE REFRIGERATION EFFICIENCY

All mechanical compression refrigeration systems have compressors to raise the gas refrigerant pressure and temperature, condensers to reject the heat of compression and change the state of the refrigerant from a gas to a liquid, and evaporators to absorb heat from the refrigerant (to chill water or air). The performance, or seasonal efficiency, of an existing refrigeration system can be increased by (1) changing the mode of operation and the operating conditions to conform more closely to the part-load conditions which are the rule rather than the exception in virtually all buildings, and (2) improving maintenance and service procedures.

Compressors may be:
- Reciprocating

- Centrifugal

- Positive displacement screw

- Fully or partially hermetically sealed

Compressors may be driven by:
• Electric motors

• Diesel or gas engines

• Steam turbines

Condensers may be:
• Water-cooled shell and tube

• Water-cooled evaporator type

• Air-cooled

Evaporators may be:
• Shell-and-tube water chillers

• Direct-expansion coils in HVAC duct systems or air-handling units

Unitary air-conditioning systems (such as window air conditioners, through-the-wall units, packaged heat pumps, and 3- to 150-ton self-contained packaged air conditioners) have electric-driven compressors. The smaller sizes are equipped with reciprocating or hermetically sealed compressors, and the larger sizes with centrifugal or positive screw displacement compressors. All these units incorporate compressor, condenser, evaporator filters, and supply fans in one insulated casing.

The same components are often used in "built-up" systems. Each component or group of components may be installed in a single package or in locations remote from each other but interpiped to form a single operating unit.

Larger "central-station" systems include reciprocating compressor-chillers (of up to 150 tons capacity), centrifugal compressor-chillers (from 100 to 8000 tons in size), screw-type compressor-chillers, or, as will be described below, absorption chillers. Each has separate air- or water-cooled condensers and separate air-handling units or fan coil units with filters and cooling coils which serve one area directly or multiple areas through a duct system. All drives (electric motors, gas or diesel engines, or steam turbines) may be used with all compressor-chillers with the exception of the absorption type.

The coefficient of performance of these refrigerating machines can be measured as:

$$\frac{\text{Heat absorbed in evaporator}}{\text{Heat rejected in condenser} - \text{heat absorbed in evaporator}}$$

Typical values for the COP range from 2 to 5 at full load. Air-cooled condenser equipment is in the lower range, water-cooled condenser equipment is in the upper range.

The COP is related directly to evaporating and condensing temperatures which, for a water chiller with cooling tower, are typically 40 and 100°F, respectively. If the evaporating temperature can be raised (by using chilled water at a higher temperature of, for instance, 50°F, instead of 40°F), or if condensing temperatures can be reduced by circulating more air or water, then the COP will increase and a greater cooling effect will result from the same power input. Consider each individual refrigeration machine separately to determine the extent to which its COP can be increased. In general, however, raising the chilled-water temperature 10° and reducing the condensing-water temperature 10° will result in an increase in efficiency of approximately 20 or 25%, and at part-load conditions this increase in efficiency is even greater. On a seasonal basis this will have a greater effect in conserving energy than would be indicated by consideration of full-load operating conditions only.

Absorption chillers achieve a cooling effect without the use of mechanical compression; instead they use heat directly as the driving force. They normally include a heat-activated generator-absorber, shell-and-tube condenser, and shell-and-tube evaporator. Water is used as a refrigerant with an absorbant such as lithium bromide. The COP of an absorption machine is not as favorable as that of a mechanical chiller and will normally be on the order of .67. Absorption machines are particularly sensitive to condensing temperature and will show a good improvement in COP at lower condensing temperatures, but if condensing temperatures are too low, the absorbent for most existing absorption units will crystallize. If the absorption machine is operated from waste heat, it should be operated as close as possible to its maximum output and should be modulated with other mechanical refrigeration machines, if installed, according to load.

Cooling towers lower the temperature of condenser water by direct evaporation of the water to outdoor air. The condenser water is sprayed over a series of baffles or fill and then drains by gravity into a sump. Outdoor air is drawn through the tower, over the fill, and is then discharged to the atmosphere. The intimate mixing (through counterflow of air and water) promotes evaporation of the condenser water, increasing the moisture content of the air. Each pound of water evaporated removes 1000 Btu from the condenser water system. The rate of evaporation is directly affected by the wet-bulb temperature of the incoming air and the condenser water temperature. The difference between these temperatures is known as the "approach" temperature; cooling towers are commonly sized for 10°F approach (i.e., if the design outdoor wet-bulb temperature is 75°F, the lowest temperature condenser water that can be obtained from the cooling tower at its full rating will be 85°F). Any reduction in condenser water flow rate or airflow rate through the tower, or any fouling or blocking in the fill, will reduce the tower's effectiveness and increase the approach temperature, thus increasing the condenser water temperature and in turn lowering the chiller efficiency.

Because water is constantly being evaporated in the tower, total dissolved solids in the condenser water system increase scale at the spray nozzles and on the baffles and fill. Scale formation on the spray nozzles not only reduces the quantity of water flow, but will also inhibit the fine atomized spray necessary for evaporation. Correct rates of blowdown will hold total dissolved solids in the condenser water system to a tolerable level, and correct water treatment will prevent scaling both in the tower and in the refrigeration machine.

If the cooling tower is located in strong sunshine, there is danger that rapid algae growth will clog spray nozzles and coat the fill, reducing the tower's efficiency. If the cooling tower is contaminated with algae or bacterial slime, it should be thoroughly cleaned with chlorine and flushed through to remove all deposits; this should be followed with periodic treatments with algicides. Shading the cooling tower from direct sunlight helps to inhibit the growth of algae.

In addition to increasing the efficiency of the compressor or chiller system by improving the performance of the cooling towers and air-cooled condensers, it is also possible to reduce the energy consumption of the tower itself. Air-cooled condensers discharge heat to a flow of air through a finned coil containing the hot refrigeration gas. The rate of heat rejection is directly affected by the dry-bulb temperature of the airstream and by the efficiency of the heat transfer surface. In geographic locations that experience long periods of high dry-bulb temperatures, the efficiency of an air-cooled condenser can be increased by using the cool exhaust air from the building as a source of cooling air for the condenser.

Because it forms part of the refrigerant system, the air-cooled condenser (with its connecting pipework) imposes a resistance to refrigerant flow and increases the pressure at which the compressor must operate. Condensers are frequently installed in locations remote from the refrigeration compressor, but often they can be easily relocated to reduce the length of connecting pipework. Any reduction in pipework length will decrease the friction loss and increase the efficiency of the refrigeration machines.

Each individual piece of equipment that composes a chiller system has built-in inefficiencies. If these are minimized, the total efficiency of the unit will increase. Leaks, for instance, from the refrigerant high-pressure side of the system will reduce the refrigerant charge and, hence, the refrigeration effect that can be obtained for a given power input. Leaks on the low-pressure refrigerant side (if the pressure is subatmospheric) will allow the entry of air into the refrigerant system. Air is composed of noncondensable gases and will reduce both the rate of heat transfer of the condenser and evaporator and, again, the refrigeration effect available for a given power input. Common sources of leaks include shaft seals, inlet guide vane seals, valves, and pipe fittings.

Compressor prime movers and drive trains, if poorly maintained, can absorb as much as 15% of the total energy input into the compressor. Speed-reducing gear boxes should be routinely examined for quality and quantity of lubricating oil, gear backlash and wear, thrust bearing condition, and main bearing condition. Belt drives should be examined for correct tension. Where multiple belts are used, all belts should be replaced at the same time, or different tensions will result, causing a loss of transmission efficiency.

Heat transfer is inhibited if water-cooled shell and tube condensers become fouled with scale or bacterial slime. Reducing the condenser fouling factor from 0.002 to 0.0005 will result in approximately a 10% increase in efficiency. Scale should be removed from the tube surface by chipping, by chemical means, or by a combination of both. Once the tubes are clean and free of scale, a policy of correct water treatment should be enforced. Bacterial slime can be removed from the condensers by flushing and by chemical treatment; its reappearance can be prevented by periodic shock treatments with bactericides and algicides such as chlorine. With good maintenance of the heat transfer surfaces in shell and tube condensers, a supermarket in Atlanta, Georgia, of 50,000 ft², equipped with a 200-ton centrifugal electrical chiller could reduce the annual power requirements from 600,000 to 540,000 kWh. The maintenance program would entail cleaning the surface twice yearly to reduce the average fouling factor to 0.0005. At 4 cents/kW, a reduction of 10% in annual energy usage saves $2400 per year.

ACTION GUIDELINES

☐ 1. Water-cooled shell and tube evaporators are not as prone to scale or bacterial fouling as condensers, but they should be inspected and, if necessary, cleaned. If the fouling factor of a water-cooled evaporator decreases from 0.002 to 0.0005, the efficiency of the machine will increase by approximately 14%.

☐ 2. Direct-expansion evaporator coils, installed in duct systems, quickly become fouled with dust, particularly if the filtration systems are ineffective. These coils should be inspected on a periodic basis and cleaned with steam or compressed-air jets.

☐ 3. Maintain a full charge of refrigerant; repair refrigerant leaks.

☐ 4. Maintain evaporator heat-exchange surfaces in clean condition.

☐ 5. Clean all cooling coils, air and liquid sides.

☐ 6. Clean all condenser shells and tubes.

☐ 7. Clean all air-cooled condenser coils and fins on a regular basis with compressed-air or steam jets.

☐ 8. Remove obstructions to free airflow into cooling towers and fans.

☐ 9. Remove bacterial slime and algae from cooling towers.

☐ 10. Institute and maintain a continuous water treatment program for cooling towers.

☐ 11. Clean and descale spray nozzles and descale fill in cooling towers.

☐ 12. It is often more effective to operate chillers at full load in the morning when outdoor wet-bulb temperatures are low and when low condensing water temperatures can be obtained from the cooling tower. Subcool the building, then turn the chillers off when wet-bulb temperatures rise and allow the building temperature to drift up. In large buildings, the extensive chilled-water piping system provides a degree of thermal storage.

☐ 13. Lubricate speed-reducing gear boxes.

☐ 14. Replace worn bearings.

☐ 15. Maintain proper tension on V-belt drives.

☐ 16. Clean and replace as necessary all strainers to reduce resistance to refrigerant or water flow.

☐ 17. Select a water treatment system for cooling towers that allows high cycles of concentration (suggested target greater than 10.7) and reduces blowdown quantity.

☐ 18. Maintain boiler and burner efficiencies where steam or hot water is generated for absorption cooling units.

Many of these options can be implemented by the building maintenance staff. An alternative, however, would be to let a service contract. Typical costs for an inspection and labor-type contract for servicing simple air-conditioning and fan coil units on an annual-fee basis to cover expenses of labor (for inspection, maintenance, breakdown repair) and material (such as filters, oil, grease) might be:

Air-conditioning units:

System tonnage	Price, $
2.5	168
5	202
7.5	253
10	308
15	410
20	495
25	595
30	685
40	865

Fan-coil units:

System tonnage	Price, $
½	25
¾	35
1	45
1½	70
2	95

12.9 IMPROVE COMPRESSOR-EVAPORATOR PERFORMANCE

The COP of any refrigeration machine increases, and less power is used, as the evaporator temperature is raised. The efficiency of a direct expansion refrigeration unit increases by about 1½% for each degree rise in evaporator temperature. Refer to Fig. 12.10 for the increase in COP per degree rise in chilled-water temperature.

The evaporator temperature is a function of operating mode and maintenance; all measures to reduce building loads provide opportunities to raise

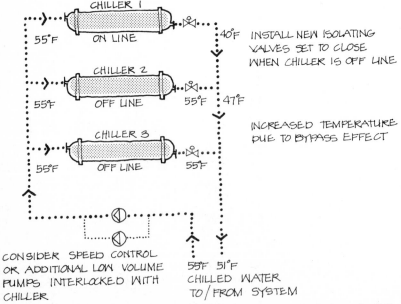

Figure 12.7 Cooling system efficiency increases with isolation of off-line chillers.

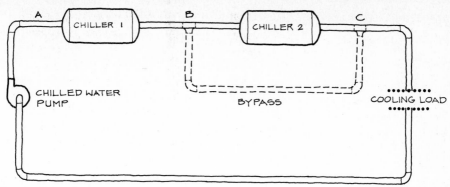

Figure 12.8 Chillers arranged in series.

this temperature at peak loads and present even greater opportunities at part-load conditions. Adjustments for load often permit an increase in the supply air temperature even with a reduction in the total amount of air circulated. This will increase the evaporating temperature and suction temperature of a direct-expansion system and conserve power for the compressor. Raising supply air temperatures with chilled-water systems permits higher chilled-water temperatures, higher evaporator temperatures, higher suction temperatures, and a reduction in power used by the chiller.

For commercial refrigeration units, for instance, very substantial savings can be realized if freezer temperatures are maintained at just the required level and are not set too low. A unit operating at $-30°F$ uses more than twice the energy required for setting at $-10°F$.

Light cooling loads on a multiple chiller installation are often met by circulating chilled water through all chillers even though only one chiller is operating. This wastes pump energy by maintaining an unnecessarily high system flow rate and, owing to the bypass through the off-line chillers, forces the remaining on-line chiller to produce chilled water at very low temperatures. The on-line chiller requires low evaporator temperatures, and the COP drops. Under light loads the off-line chillers should be isolated, and the flow rate of the chilled water should be reduced. See Fig. 12.7.

System efficiency is affected by the piping arrangement of the chillers. Generally, rather than piping chillers in parallel so that each must produce the coldest water in the system, the chillers should be piped in series as shown in Fig. 12.8. Chiller No. 2 operates at a higher suction pressure and uses fewer kilowatts per ton. One possible drawback to series arrangement is that chilled water is pumped through the off-line chiller, adding unnecessary resistance to the piping system. However, this can be avoided by adding a bypass around the off-line chiller. With series piping, chilled-water volume can be reduced (coils can take a higher temperature differential) to lower pumping resistance.

Figures 12.8 and 12.9 illustrate typical series and parallel piping arrangements for a two-chiller installation. An analysis of power required for each

TABLE 12.1   CHILLER OPERATIONS, SERIES AND PARALLEL

Load	Series			Parallel		
	Full	½	¼	*Full*	½	¼
Temperature A (sketches)	54°F	54°F	54°F	54°F	54°F	54°F
Temperature B (sketches)	49°F	49°F	51.5°F			
Temperature C (sketches)	44°F	49°F	51.5°F	44°F	44°F	49°F
Chilled water gpm	1200	1200	1200	1200	600	600
Chiller #1 input, kW	187	187	88.7	212	212	93.3
Chiller #2 input, kW	212			212		
Total kW	399			424		
Pump input, kW*	37.3	(33.2) 37.3	(33.2) 37.3	40.1	20.1	20.1
Total input, kW	436.3	(220.2) 224.3	(121.9) 126.0	464.1	232.1	113.4

*Can be reduced by modifying pumping arrangements. All power inputs are in kilowatthours. Numbers in parenthesis represent consumption in kilowatts for series arrangement with bypass installed.

arrangement was done under various load conditions. Each chiller is rated at 250 tons output at 44°F for the leaving chilled water temperature and 95°F for the leaving condenser water temperature. Table 12.1 summarizes the operation of each system under full- and part-load conditions. It can be seen from the table that the series arrangement saves 464.1 − 436.3, or 27.8 kW, per hour of operation under full-load conditions, a saving of 6%. Depending on the length of the cooling season, energy and dollar savings can be substantial over a year of operation. The COP at various chilled-water temperatures (Fig. 12.10) was determined by the relationship:

$$COP = \frac{output\ Btu}{input\ Btu}$$

The change in COP was expressed in terms of the COP at nominal conditions and then plotted as a function of chilled-water temperature.

ACTION GUIDELINES

☐ 1. Raise chilled water temperature to "follow the load." Install a limit switch in each modulating or diversion valve to measure whether the valve is fully or only partially open. Arrange control circuits so that when all coil control valves are either closed or in a partially open position (indicating light-load conditions), the chilled-water temperature supply set point is raised until one or more coil control valves returns to the fully open position.

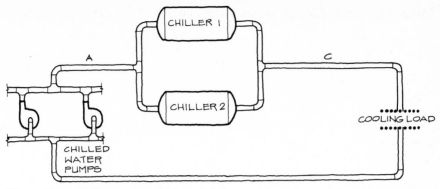

Figure 12.9 Chillers arranged in parallel.

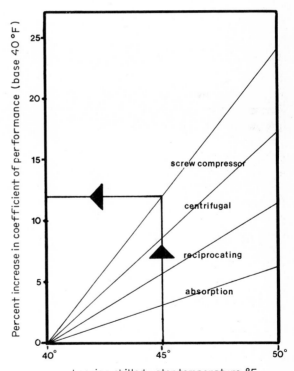

Leaving chilled water temperature,°F

Figure 12.10 The increase in chiller performance with an increase in chilled water temperature. (From manufacturers' data for different sizes and types of chillers.)

☐ 2. Provide controls to raise supply air temperature to follow the load.

☐ 3. Permit relative humidity to rise in the conditioned-air space.

☐ 4. Decrease the superheat on DX coil (consult the manufacturer first).

☐ 5. Fully load one compressor before starting the next refrigeration unit module.

☐ 6. Install a new smaller refrigeration unit for light loads when existing modules are all oversized.

☐ 7. Repipe chillers in series.

☐ 8. Isolate off-line chillers.

☐ 9. Install enthalpy controller for chilled water temperature.

12.10 INCREASE CONDENSER PERFORMANCE

The efficiency and the COP of chillers and compressors increase as condensing temperature is decreased. Each type of machine has practical limits of the lowest acceptable condensing temperature, and these limits should not be exceeded, particularly with some of the older-type absorption machines.

The condenser temperature is a function of the outdoor or ambient dry-bulb conditions (for operation of the air-cooled condensers), wet-bulb conditions (for operation of the cooling towers or evaporative condensers), and the condition and operating mode of the air-cooled condensers, cooling towers, and shell and tube water-cooled condensers.

Cooling towers, air-cooled condensers, and evaporative condensers are usually selected to provide a given condensing temperature for maximum expected outdoor conditions but will provide a lower condensing temperature when outdoor conditions are cooler. The maximum cooling load, however, usually occurs at the same time as maximum outdoor conditions. Rather than allow chiller operation to follow this load, it is often more effective to operate chillers at full load in the morning when outdoor wet-bulb temperatures are low and when low condensing-water temperatures can be obtained from the cooling tower.

If low-temperature well water is available, it can be used for condenser water rather than using cooling towers. The energy required to pump the well water is likely to be approximately the same as required for the condenser water pump and cooling tower fans, but energy savings in compressor horsepower will be worthwhile.

Where existing cooling installations are set to operate at constant condenser water temperatures by cycling cooling tower fans, the controls can be modified to operate these fans continuously when a machine is on line. This allows the condenser water temperature to drop to a predetermined low limit, at which point the cooling tower fans can be allowed to cycle on and off.

In reducing condensing temperatures and maintaining low fouling factors, automatic tube cleaners are useful. One such device is a cylindrical brush in each tube which is periodically forced from one end of the condenser tube bundle to the other by a reversal in the direction of water flow. This type of system can be very effective in keeping condenser tubes clean and, depending on the existing level of condenser tube fouling, can yield substantial energy savings.

Figure 12.11 gives percent increases in coefficient of performance for each degree Fahrenheit reduction in condensing temperature. The condensing temperature of a refrigeration machine may be significantly lowered by replacing an air-cooled condenser by a cooling tower in locations with fewer than 15,000 wet-bulb degree-hours. This can be effected on a one-to-one basis if the air-cooled condenser is large, or one cooling tower can replace a number of small air-cooled condensers. Whichever method is adopted, each

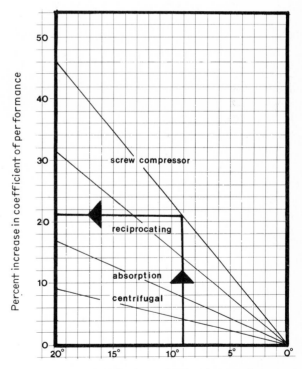

Figure 12.11   The increase in chiller performance with a reduction in condensing temperatures. (From manufacturers' data for different sizes and types of chillers.)

refrigeration machine will require the addition of a water/refrigerant heat exchanger (shell and tube condenser).

Air-cooled condensers commonly operate at temperature differences of 20°F between air and refrigerant; thus, if the outdoor air temperature is 95°F DB, the condensing temperature will be 115°F. Cooling tower performance is related to wet-bulb temperature, and cooling towers commonly operate at a temperature difference of 10°F between the air wet-bulb and condenser water. Thus, if the outdoor air is 72°F WB, the condenser water temperature would be 82°F off the tower. The 82°F water is raised to 97°F through a typical condenser for a mean condensing temperature of 90°F.

For example, if an air-cooled condenser used with a reciprocating refrigeration machine were replaced by a cooling tower in an area that experienced 95°F DB/72°F WB design conditions, then the condensing temperature would be reduced from 115 to 97°F for an 18°F difference. Such a reduction of condensing temperature will increase the COP of the refrigeration machine. Refer to Fig. 12.11 and read from the graph an increase of 28% in the COP due to the 18°F reduction in condensing temperature. In Fig. 12.11, the COP at various condensing temperatures was determined by the relationship:

$$COP = \frac{\text{output Btu}}{\text{input Btu}}$$

The change in COP was expressed in terms of the COP at nominal conditions and then plotted as a function of condensing temperature.

ACTION GUIDELINES

☐ 1. If a condenser is air cooled, increase air volume through it, add additional air-cooled condensers in parallel to increase surface area, or replace the condenser coil with a larger heat transfer surface and remove obstructions to airflow.

☐ 2. Remove high-resistance strainers, fittings, and elbows in refrigerant lines to reduce head pressure. (Obtain manufacturers' recommendations on oil return.)

☐ 3. Shade condenser from direct sunlight.

☐ 4. Consider a water spray in the airstream before the inlet side of the condenser coil.

☐ 5. Modify the exhaust ductwork to discharge cooled air from the building into the air intake of air-cooled condensers or cooling towers.

☐ 6. Install automatic tube cleaners in condensers.

☐ 7. Install sensing controls to permit cycling of cooling tower fans under light load conditions when refrigeration load is small and the ambient wet-bulb temperatures are low.

☐ 8. Investigate increasing condenser water flow rates to permit lower condensing temperatures by: (1) reducing frictional pipe losses, (2) operating on-line and standby condenser water pumps together, (3) changing pump drive on motor.

☐ 9. Use well water, ground water, seawater, or lake water, if the temperatures of the water from these sources are lower than the wet-bulb ambient temperature in the case of cooling towers, and the dry-bulb temperatures in the case of air-cooled condensers.

☐ 10. Adjust fan-blade pitch or fan speed of cooling towers for minimum power input.

☐ 11. Install baffles in cooling towers and air-cooled condensers to reduce or eliminate bypass of hot air around coils.

☐ 12. Install controls to permit refrigerant migration on large systems to obtain refrigeration effect without power input to the chiller.

☐ 13. Modify condenser water bypass around cooling towers to reduce condenser water temperature to the lowest level compatible with the refrigeration machine.

☐ 14. Replace air-cooled condensers with cooling towers.

☐ 15. Install air-cooled condensers and cooling towers in series where ambient wet- and dry-bulb conditions indicate that lower condensing temperatures can be obtained (not worthwhile unless reduction in temperatures is at least 6°F).

12.11 IMPROVE EFFICIENCY WITH EQUIPMENT MODIFICATIONS

Where existing centrifugal chillers are driven by steam turbines and the exhaust steam is condensed before being returned as feedwater to the boiler, an absorption chiller can be substituted for the condenser to produce additional chilled water from the waste heat. For example, a low-pressure turbine driving a centrifugal chiller requires 12 lb dry steam at 145 psig for each ton of refrigeration. Exhaust steam from the turbine at 15 psig and 20% moisture content will yield approximately 780 Btu/lb when cooled and condensed to 220°F, giving a total usable heat of $12 \times 780 = 9360$ Btu/ton. An absorption chiller used as a steam condenser requires a heat input of 18,000 Btu/ton and will yield $9360/18,000 = 0.52$ ton for each ton of refrigeration produced by the centrifugal chiller. (Newer units in development will require only 13,000

Btu/ton.) In this case the addition of an absorption piggyback chiller to condense waste steam will decrease the steam per ton of refrigeration by 30%.

Example

Assume a centrifugal chiller with a capacity of 600 tons and operating at an equivalent of 1400 full capacity hours per year or $1400 \times 600 = 840,000$ ton-hours per year. For turbine-driven machines only, 12 lb steam per ton $= 10.08 \times 10^6$ lb steam per year. If this is reduced by 30% by using exhaust steam in an absorption machine, the savings $= 10.08 \times 10^6 \times 0.3 = 3.024 \times 10^6$ lb steam per year. From records, 1 gal oil produces 96 lb steam. Therefore, fuel savings $= (3.024 \times 10^6)/96 = 31,500$ gal oil per year, which at 33 cents/gal $= \$10,395$ per year.

The capital cost of a 200-ton absorption chiller and all peripheral equipment is $54,000. This investment gives a 5-year payback and is worthwhile. Further economic analysis will indicate cash flow benefits. Under part load conditions, it is more efficient to use the piggyback absorption machine at full capacity and to modulate the turbine driven machine as required to meet the remainder of the load.

Piggyback absorption is particularly effective in buildings using purchased steam when condensate is not returned to the supplier. In such cases, it is worth considering the purchase of a larger machine and derating it by operating at outlet temperatures down to 180°F. Doing this will reclaim the maximum possible heat from exhaust steam and condensate. Consider also using lower-temperature condensate from the absorption machine as a heat source for low-grade systems such as domestic hot water and ventilating air pre-heaters.

Low-temperature absorption machines are also currently being developed for operating inlet temperatures of 190°F or lower while still retaining a reasonably high COP. Consider using a low-temperature absorption machine in buildings that have large quantities of low-grade waste heat (e.g., cooling water from engine/generator sets). Additional equipment modifications are included in the guidelines below.

ACTION GUIDELINES

☐ 1. Use piggyback absorption systems to produce additional chilled water when existing centrifugal chillers are driven by steam turbines.

☐ 2. When replacing absorption refrigeration equipment, compare the performance of new high-temperature absorption units with new low-temperature, low-pressure, low-energy-absorption units and select the one based on best seasonal performance. (Where high-pressure boilers are in use, analyze the advantages of turbine-driven equipment versus absorption-driven equipment.)

☐ 3. For expansion or modification, study the use of diesel engines and generators on a standby or full-time basis to reduce electric loads, and consider using waste heat from the engines for other useful work.

☐ 4. Fit capacity controls to existing refrigeration equipment where possible and, when replacing units, purchase those with four steps of capacity control.

☐ 5. When replacing existing absorption units, analyze energy requirements for new units versus electric centrifugal compressors as well as steam turbine–driven and engine-driven units, and select the system requiring the least annual energy usage.

☐ 6. Install 7-day timers to program shutdown of refrigeration equipment, chilled-water and condenser pumps, and cooling tower fans for unoccupied periods during the week.

☐ 7. Convert straight compressors and chiller systems to heat pump operation by adding additional condensers and evaporators and/or repiping as required.

CHAPTER THIRTEEN

CENTRAL CONTROL SYSTEMS

13.1 APPLICATION

A central control system provides the chief engineer of the physical plant and the energy management team with a means of constant surveillance of the building and helps in making efficient and effective use of physical plant systems and personnel. A central control system can:

- Monitor all systems for off-normal conditions.

- Monitor all fire alarm and security devices.

- Monitor operating conditions of all systems and reschedule set points to optimize energy use.

- Monitor on a continuous basis selected portions of any system and store this information in bulk memory for later retrieval and use in updating software. (This information can be used to determine the effectiveness of the energy management program and to indicate changes or modifications in approach that should be made.)

- Limit peak electrical demand values by predicting trends of loads and by shedding nonessential services according to programmed priorities.

- Optimize the operation of all systems to obtain the maximum effect for the minimum expenditure of energy.

- Optimize maintenance tasks to effect maximum equipment life for minimum manpower labor and costs.

- Provide inventory control of spare parts, materials, and tools used for maintenance.

By judicious use of these functions, the engineering staff can operate all systems from the central console and will have minute-by-minute control of each system. Any critical alarm for the physical plant can be reported automatically at the console, and the operator can then scan the system in trouble, analyze the fault, and dispatch the correct maintenance person for repairs. Maintenance alarm summaries can be made available on demand or may be printed daily to enable better work scheduling of maintenance personnel.

Sophisticated central control systems are applicable only to larger, more complex, buildings, or groups of buildings, and should be considered when net floor area exceeds 40,000 ft^2, or when energy use is high because of extended occupancy or special equipment. However, smaller buildings can also use load management and less sophisticated central control systems.

Central control systems vary from the relatively simple to the very complex. The system should be tailored to the requirements of the building and the operations to be controlled; if money for investment is limited, attention should first be given to those functions which will show the quickest and greatest return in energy saved. Provisions should be made, however, for expansion of system hardware and software capacity as investment money might later permit. An initial system, for instance, may be limited to a console and central processor unit with one single printer output and may be programmed to perform only start-stop and simple reset functions and to report alarms. Optimization of system operation and load shedding in this case would be achieved manually at the central console on the basis of decisions made by the operating staff and would involve overriding the programmed start-stop times and reset points for the various systems.

13.2 SYSTEM COMPONENTS

Proprietary central computerized control systems are marketed by each of the major temperature control manufacturers. These systems have common features and can accomplish a similar range of tasks, but each manufacturer uses coding and computer languages which are unique to the system and cannot be decoded by any other system. Once a basic system has been selected and installed, all subsequent additions must be obtained from the original manufacturer. Central computer control systems are usually modular in design, and additional hardware and software can be added if the necessary provisions have been made in the basic system.

Each manufacturer's system is made up of standard hardware, but the application is always tailored to the specific project (see Fig. 13.1). Basically, any system is composed of four major parts:

- Interface panels which are located at strategic points throughout the building, usually in equipment rooms.

- The transmission system between the central console and all interface

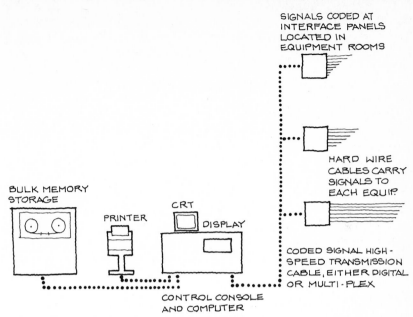

Figure 13.1 Typical central control system.

panels. This system can be single-core cable for digital transmission or multi-core cable for multiplex transmission.

- A central control console and associated computer hardware located in a control room, usually close to the boiler room and the chief engineer's office. The console, computer, and associated hardware form the point at which the operator enters all instructions and retrieves all data and from which all routine computer instructions are made.

- Software programs. Program inputs from magnetic tape, paper tape, or cards contain the basic operating instructions for the computer and are stored in core memory or in bulk memory. For large quantities of data, external bulk memory is cheaper than core memory.

13.3 INTERFACE REQUIREMENTS

To obtain maximum benefit from a central console system, it is necessary to measure, monitor, and control many different items of equipment. The central control system must be interfaced with the operating equipment to obtain this information in rational form and to exert its control functions.

Table 13.1 shows desirable interface points and the type of function required for typical building systems. An interface of two types of signals is required: binary signals (which comprise only two alternatives—on-off, open-close, etc.) and analog signals (which measure a value against a particu-

TABLE 13.1 INTERFACE POINTS

Item	Start/Stop Lead/Lag	Read Out	Reset	Alarm° 1	2
VENTILATION					
Outside air temperature (dry-bulb)		X			
Outside air temperature (wet-bulb)		X			
Air temperature on preheat coil (dry-bulb)		X		X	
Air temperature on preheat coil (wet-bulb)		X			
Air temperature on cooling coil (dry-bulb)		X			
Air temperature off cooling coil (dry-bulb)		X			
Air temperature off humidifier (dry-bulb)		X			
Air temperature off humidifier (wet-bulb)		X			
Air temperature off reheat coil (dry-bulb)		X			
Critical area temperature (dry-bulb)		X		X	
Critical area humidity		X		X	
Return air temperature (dry-bulb)		X			
Return air temperature (wet-bulb)		X			
Fresh air/exhaust air/recirculation air dampers		X	X		
Air velocity beyond filters					X
Supply fan	X	X		X	X
Humidifier spray pump	X	X			X
Supply airflow switch		X		X	
Exhaust fan	X	X		X	X
Heating coil control set point			X		
Cooling coil control set point			X		
Humidifier control set point			X		
HOT WATER HEATING					
Boiler fuel input	X	X	X		
Boiler supply temperature		X	X	X	
Boiler return temperature		X			
Boiler high-limit temperature				X	
Hot-water pumps	X	X		X	X
STEAM					
Boiler fuel input	X	X	X		
Boiler steam pressure		X	X	X	
Boiler condensate feed temperature		X			
Boiler high steam pressure limit				X	
Boiler low-water level limit				X	
Boiler feed pumps	X	X		X	X
CHILLED WATER					
Chiller power/fuel input	X	X	X		
Chiller supply temperature		X	X	X	
Chiller return temperature		X		X	

TABLE 13.1 INTERFACE POINTS (*Continued*)

Item	Start/ Stop Lead/Lag	Read Out	Reset	Alarm* 1	2
Condenser supply temperature		X	X		
Condenser return temperature		X	X		
Chilled water pumps	X	X		X	X
Chilled water flow rate		X			
Condenser water pump	X	X		X	X
Condenser water flow rate		X			
Cooling tower fan	X	X			X
Air-cooled condenser fan	X	X			X
DOMESTIC HOT WATER					
Domestic hot-water heater fuel/power input	X	X	X		
Domestic hot-water storage temperature		X	X		
Domestic hot-water high-temperature limit					X
Domestic hot-water return temperature		X			
Domestic hot-water circulating pumps	X	X			X
COLD WATER					
Cold-water pressure boost pumps	X	X		X	X
Cold-water boosted pressure		X		X	
OTHER SERVICES					
Control system air pressure		X		X	
ELECTRICAL					
Total building power input kVA and kW		X		X	
Primary isolation switch		X			
Load interrupter switch		X			
Transformer winding temperature		X		X	
Transformer secondary voltage fixture		X		X	
Emergency generator	X	X		X	X
Parking lot lights	X	X			
Selected lighting circuits	X	X			

*Alarm 1 denotes critical alarm. Alarm 2 denotes maintenance alarm.

lar scale—temperature, pressure, etc.). Binary signals are used as instructions to start and stop equipment, to open and close valves and dampers, and to retrieve data such as on-off and open-closed. Analog signals are used as instructions to raise or lower temperature set points, to adjust damper positions, to raise or lower pressure set points, and to retrieve data such as temperature, pressure, and humidity.

To select a central control system, it is first necessary to assemble a complete list of all desired interface points under the two categories of binary and analog, then to arrange these points in groups served by individual interface panels. Depending on the type of existing controls, motor starters, contactors, etc., modifications and additions may be required to allow satisfactory interface. For instance, if a motor starter does not have spare auxiliary contacts, a relay must be added to the control circuit.

13.4 SIGNAL TRANSMISSION

To make effective use of the computer's capabilities, information between the computer console and the interface panels must be transmitted at high speed in a format easily handled by the computer. The most convenient method of transmission is digital, in which information is represented by pulses arranged serially and is transmitted through a single-core conductor. Digital cable is commonly coaxial, although some systems use a twisted pair of insulated wires. Voice intercommunication is carried over a separate screened cable.

Less frequently used is a multiplex cable that is multicore for transmission of information signals which are multiplexed and transmitted in parallel. Multiplex cable is typically made up of 50 to 100 separate wires and has an overall diameter of 1 in. Installing this cable is more difficult than coaxial cable, particularly in existing buildings where empty conduit and throughways are not available. Some systems still use multiplex cable on the ground that analog signals can be transmitted in unmodified form, whereas coaxial transmission requires conversion from analog to digital form with a small loss of accuracy at the interface panel. But analog signals are sensitive to interference from "spikes" and other spurious signals induced by adjacent building wiring, and digital signals are not. Finally, digital transmission provides more options for later additions; multiplex systems are limited in the number of points that can be connected.

13.5 CONTROL CONSOLE AND HARDWARE

The control console, printers, graphic display, cathode-ray tube (CRT), and input-output keyboards are generally of comparable performance for the various systems available, and individual options are available on request to suit the prospective user's particular requirements. Different levels of access into the computer from the keyboard are available, typically from three to five levels. Access ranges from normal operation at the lowest level to reprogramming at the highest level, controlled by a key switch. Such key-switch control is vital to protect memory from accidental erasure.

The central processor is supplied with integral core memory. Core capacities vary typically from 16,000 to 64,000 words, with each word containing 16 bits of information. Core memory is expensive compared with bulk memory, and if the core capacity will be exceeded, then the addition of a bulk memory unit should be considered (but first make sure that the computer selected can be interfaced with external memory). Bulk memory can be either disk or tape, both with random access. Disk memory is preferred for control applications and typically has a capacity of 192,000 words.

13.6 SOFTWARE

Programs for common applications are available from each of the major manufacturers of controls from their libraries of application programs developed over a number of years. The cost of these programs is low compared with custom-written software. When a central control system is being selected, the library of programs available should be investigated, for the extent and range of the library are good indexes of a manufacturer's commitment and ability in the computer control field.

Programs and their applications are:

EXECUTIVE PROGRAM This program is essential to the operations of the computer itself; it controls the priority of signal processing, directs information in and out of memory, controls the operation of all associated hardware, and encodes and decodes incoming and out-going signals. An essential function of the executive program is the orderly shutdown of the computer on power failure to ensure that all information is safely stored in permanent memory before all power is lost. Beware of systems that store information in volatile memory which evaporates on power failure.

START-STOP PROGRAM This is the simplest program, yet with modifications it becomes a most effective energy saver, particularly when used in conjunction with other programs and routines. In its simplest form the program is arranged to start and stop equipment according to predetermined times, including on weekends and holidays. The program can also open and close dampers and turn lights on and off. It can, in addition, reschedule temperatures from occupied levels to unoccupied levels, as these functions are binary in nature. Manual override can be provided with automatic reversion to program control at the end of the overridden period, or for similar causes, an interface with other programs can allow the basic program to be overridden, and equipment can be turned on and off to meet other criteria. Start sequences can be arranged in cascade to avoid simultaneous starting and the cumulative effect of inrush currents that cause a high peak electrical demand.

LOAD-SHEDDING PROGRAM This program is used in conjunction with the start-stop program to override normal operation and turn off selected equipment as a predetermined electrical peak load value is approached. The program can be modified to include a rolling priority feature which turns on equipment on an assigned priority basis for an adjustable period of time, after which the equipment is turned off. The rolling priority is activated as a predetermined peak load value is approached, and equipment turned off is limited to that necessary to maintain the load below a given level. Interface with a profile routine allows a continuous profile of electrical demand and enables calculation of a predicted load in the immediate future. This predicted load initiates the rolling priority to alter the slope of the prediction curve. The current maximum peak load experienced in the previous 11 months is held in memory and is used as the criterion for load shedding. This value is constantly modified with monitored information.

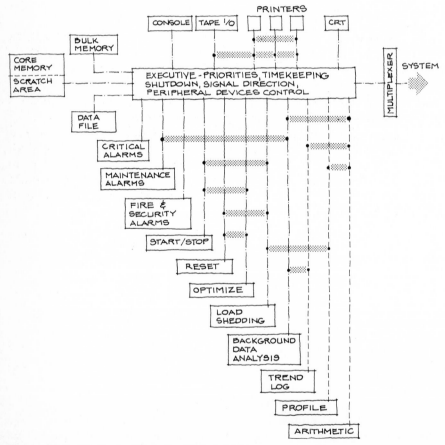

Figure 13.2 Control system program interfaces.

OPTIMIZING PROGRAMS These programs tend to be more sophisticated and require interface with an arithmetic routine to allow calculation of optimum settings. The program interfaces with the start-stop, load-shedding, and reset programs to modify operation of equipment. Optimizing programs can provide economizer cycle control of outdoor air according to enthalpy and loads and can modify chilled-water and boiler water temperatures to provide optimum equipment operation according to load.

RESET PROGRAMS These programs reset the control points of thermostats and pressure controllers and interface with the load-shedding and optimizing programs. In place of or in addition to turning off equipment for load shedding and energy conservation, loads can be trimmed; that is, if a peak load is approaching, space conditions can be reset to reduce cooling or heating loads, or outdoor air dampers can be reset.

ALARM PROGRAMS These programs report alarms generated either outside the computer (fire alarm pull boxes, boiler high pressure, etc.) or in the computer software. Upper and lower limits can be set in software for values of any monitored point and arranged to report an alarm when exceeded. Alarm programs can also be interfaced with the start-stop program to monitor hours run and report a maintenance alarm when a designated number of running hours is exceeded. Information on particular maintenance required can be stored in memory and can be printed out when the alarm is reported.

Figure 13.2 shows typical relationships between programs. The program functions described are an indication of what can be achieved with a central computer control system and are intended to stimulate investigation into all possibilities, particularly those in reference to energy conservation.

CHAPTER FOURTEEN

ECONOMIC ANALYSIS FOR BUILDINGS

14.1 THE PROBLEM

When capital investment is suggested for the purpose of energy conservation, the first inevitable questions are, "Will it pay, and how fast?" The answer is, unfortunately, not a simple equation of energy dollars saved less capital dollars spent. The answer must include other factors: expected return on investment, debt service, taxes, depreciation, changes in fuel costs, changes in labor and maintenance costs, and changes in rental income. The essence of the problem is the summation of *all* costs in the life of the purchased improvement and the presentation of this investment in terms of yearly aftertax discounted rates of investment return. This chapter gives one procedure for such life-cycle costing.

In spite of the importance of life-cycle costing, however, there is obviously at times such immediate saving with certain energy conservation expenditures that a simple calculation will do. If capital cost divided by annual savings equals 3 or less, payback will usually occur between 6 and 10 years at good rates of return; model analysis is then unnecessary.

14.2 A COMMERCIAL BUILDING INVESTMENT MODEL

The purpose of the model is to isolate costs, revenues, and changes in operating expenses associated with an energy conservation improvement and to evaluate these items separately and independently from the general operating income and expenses of a building. In using the model to evaluate the economic viability of a particular energy conservation measure, care must be taken to consider only those variables which are associated with the energy conservation measure.

The model presented below is designed to evaluate energy conservation measures which require capital expenditure. To evaluate measures which are free, simply leave blank the spaces for capital investments and debt service. In addition, the model is designed for owners subject to federal and state income taxation. If economic analysis is desired for governmental or nonprofit owners, the model can be used but must be modified for differing applications of the tax laws. In such cases, income tax, investment tax credit, and depreciation are not generally taken into consideration. In addition to these categories, property taxes, increases in rental income, and debt service are not applicable to government owners but may apply to private nonprofit organizations and institutions.

The economic viability of an energy conservation improvement is based on total cost savings, discounted to reflect the time value of money, which are expected to accrue over the useful life of the improvement. The forms given contain enough columns to evaluate improvements with useful lives of up to 10 years, though many improvements will have useful lives greater than 10 years, in some cases as long as the life of the building. In any case the useful life assigned to an energy conservation improvement should be no greater than the remaining useful life of the building in which it is installed. As a practical expedient, a building owner may choose an evaluation period shorter than an improvement's estimated useful life to simplify the computations involved. The rate of return computed using a truncated useful life of not less than 10 years can be reasonably accurate, for as time increases, the present value of energy savings and the certainty with which such savings can be predicted decrease. Assumptions of cost savings projected far into the future are risky, and the discounted present value of far-future savings is significantly less than current savings.

In preparing inputs for the cost model, it is suggested that all costs be presented in current-year dollars. This assumes that increased expenses resulting from inflation will be offset by equally inflated increased revenues. In the case of religious and nonprofit institutions, however, an inflation factor of 6 to 12% should be included.

To use the model, standard real estate investment analysis interest rate, discount, and mortgage tables will be required. A recommended source of such tables is the *Realty Bluebook* published by Professional Publishing Corporation, P.O. Box 4187, San Rafael, California, 94903 [telephone (415) 472-1964].

14.3 INVESTMENT MODEL OPERATIONS

The model consists of five basic tables. Table 14.5 summarizes the discounted rate of return and other data on a proposed energy conservation measure, with multiple columns to facilitate comparison of alternative conservation measures. Table 14.2 is used to compute aftertax cash flow from an energy conservation improvement. Table 14.4 is used to convert this stream of yearly aftertax cash flow into a discounted rate of return. Tables 14.3 and

14.1 present supporting data from which much of Table 14.2 is computed. The sample figures included within the tables are for a set of energy-conserving capital expenditures for a hypothetical office building. In completing these tables, the following order of preparation should be followed:

1. Complete all applicable items in Table 14.1 and post the computations to Tables 14.2 and 14.3 as indicated, Nonapplicable items should be left blank.

2. As applicable, compute lines 2 to 5 on Table 14.2.

3. Compute net operating profit (Table 14.2, line 7). This involves adding up the items previously computed and posted to Table 14.2. Enter in col. *a* of Table 14.3 the net operating profit by year as computed in Table 14.2.

4. Complete Table 14.3 and enter the projected income tax on line 11 of Table 14.2.

5. Complete the computations on Table 14.2. Post aftertax cash flow (Table 14.2, line 13) to Table 14.4. Complete the computations in Table 14.4. This will yield the discounted rate of return on the investment, which is the end result of the cost model.

6. Post the discounted rate of return (Table 14.4), payback period, and maximum cash required (Table 14.2, line 14) to Table 14.5.

TABLE 14.1   ENERGY CONSERVATION INVESTMENT DATA SHEET FOR COMMERCIAL BUILDING

(Building)

(Type of Investment)

1. Projected Energy Savings (Table 14.2, line 1)
   (electrical)

Year	Btu Savings per Year (000,000) (a)	Projected Cost per Million Btu (b)	Energy Savings (Col. a × b) (c)
0			
1	1813.06	$ 8.79	$15,937
2	1813.06	9.32	16,898
3	1813.06	9.87	17,895
4	1813.06	10.47	18,983
5	1813.06	11.10	20,125

Year	Btu Savings per Year (000,000) (a)	Projected Cost per Million Btu (b)	Energy Savings (Col. a × b) (c)
6	1813.06	11.76	21,322
7	1813.06	12.47	22,609
8	1813.06	13.22	23,969
9	1813.06	14.01	25,401
10	1813.06	14.85	26,924

2. Increase in Rental Income (Table 14.2, line 6)

Year	Rent Loss during Construction (a)	Rent Loss due to Decrease in Rentable Space (b)	Rent Increase (Decrease) in Remaining Space (c)	Net Increase (Decrease) in Rental Income (Col. c − a − b) (d)
0				
1				
2				
3				
4				
5				
6				
7				
8				
9				
10				

3. Financing

Purchase price $101,200

Less:
Salvage proceeds received from sale of old asset

Amount financed   75,000

Net investment (Table 14.2, line 8) $ 26,200

TABLE 14.1 ENERGY CONSERVATION INVESTMENT DATA SHEET FOR COMMERCIAL BUILDING (*Continued*)

4. Debt Service

Amortization of amount financed

Payback period in years	10
Annual interest rate	12%

DEBT SERVICE SCHEDULE

Year	Interest (Table 14.3, col. c)	Principal	Total Debt Service (Table 14.2, line 9)	Remaining Principal Balance
0				75,000
1	$9000	$4274	$13,274	70,726
2	8487	4787	13,274	65,939
3	7913	5361	13,274	60,578
4	7269	6005	13,274	54,573
5	6549	6725	13,274	47,848
6	5742	7532	13,274	40,316
7	4838	8436	13,274	31,880
8	3826	9448	13,274	22,432
9	2692	10,582	13,274	11,850
10	1422	11,859	13,274	0

5. Basis of New Asset

Purchase price	$101,200

Plus: Book value (original cost less accumulated depreciation) of old asset replaced by new asset

Less: Trade-in received on old asset replaced by new asset

Basis of new asset (use this figure for completing the depreciation schedule shown below)	$101,200

6. Depreciation

Useful life in years (not longer than remaining life of building or lease)	20
Depreciation method used	Straight line

DEPRECIATION SCHEDULE

Year	Depreciation Expense (Table 14.3, col. b)	Undepreciated Balance
0		$101,200
1	$5060	96,140
2	5060	91,080
3	5060	86,020
4	5060	80,960
5	5060	75,900
6	5060	70,840
7	5060	65,780
8	5060	60,720
9	5060	55,660
10	5060	50,600

7. Gain (Loss) on Disposal of Retired Asset

    Salvage proceeds received from disposal
      of retired asset $ 0

    Less: Original cost of retired asset

    Plus: Accumulated depreciation on retired asset

    Gain (loss) on disposal of retired asset

        (Table 14.3, col. *d*) $ 0

8. Investment Tax Credit

    Eligible: Yes_____ ·No__X__

    Useful life in years

    Amount of credit (Table 14.2, line 12):

      Useful life less than 3 years   -0-

      Useful life at least 3 years but less than
         5 years (purchase price × 0.07 × 0.333)

      Useful life at least 5 years but less than
         7 years (purchase price × 0.07 × 0.667)

      Useful life greater than 7 years
         (purchase price × 0.07)

TABLE 14.2 ENERGY CONSERVATION INVESTMENT COMPUTATION OF CASH FLOW FOR COMMERCIAL BUILDING

(Building) _____

(Type of Investment) _____

Line Number		0	1	2	3	4	5	6	7	8	9	10	Total
	Reduction (Increases) in Operating Expenses												
1	Energy savings (Table 14.1, part 1)		$19,492	$20,660	$21,892	$23,216	$24,607	$26,080	$27,658	$29,308	$31,072	$32,927	$256,912
2	Labor savings												
3	Materials savings												
4	Property taxes												
5	Other												
6	Net increase (decrease) in rental income (Table 14.1)												
7	Net operating profit (sum of lines 1 to 6)		19,492	20,660	21,892	23,216	24,607	26,080	27,658	29,308	31,072	32,927	256,912
8	Net investment (Table 14.1, part 3)	$26,200											26,200
9	Debt service (Table 14.1, part 4)		13,274	13,274	13,274	13,274	13,274	13,274	13,274	13,274	13,274	13,274	132,740

TABLE 14.2 (*Continued*)

Line Number		0	1	2	3	4	5	6	7	8	9	10	Total
						Year							
10	Pretax cash flow (lines 7 to 9)	(26,200)	6,218	7,386	8,618	9,942	11,333	12,806	14,384	16,034	17,798	19,653	97,972
11	Income tax (Table 14.3, col. f)		2,716	3,557	4,460	5,444	6,499	7,639	8,880	10,211	11,660	13,223	74,289
12	Investment tax credit (Table 14.1, part 8)												
13	Aftertax cash flow (line 10−11+12)	(26,200)	3,502	3,829	4,158	4,498	4,834	5,167	5,504	5,823	6,138	6,430	23,683
14	Cumulative aftertax cash flow (cumulative total of line 13)	(26,200)	(22,698)	(18,869)	(14,711)	(10,213)	(5,379)	(212)	5,292	11,115	17,253	23,683	

TABLE 14.3   ENERGY CONSERVATION INVESTMENT COMPUTATION OF INCOME TAX FOR COMMERCIAL BUILDING

_____

(Building)

_____

(Type of Investment)

Year	Net Operating Profit (Table 14.2, line 7) (a)	Depreciation (Table 14.1, part 6) (b)	Interest (Table 14.1, part 4) (c)	Gain (loss) on Disposal of Retired Asset (Table 14.1, part 7) (d)	Taxable Income (col. a − b − c + d) (e)	Income Tax @ % (Table 14.2, line 11) (f)
0						
1	$19,492	$5,060	$9,000		$ 5,432	$ 2,716
2	20,660	5,060	8,487		7,113	3,557
3	21,892	5,060	7,913		8,919	4,460
4	23,216	5,060	7,269		10,887	5,444
5	24,607	5,060	6,549		12,998	6,499
6	26,080	5,060	5,742		15,278	7,639
7	27,658	5,060	4,838		17,760	8,880
8	29,308	5,060	3,826		20,422	10,211
9	31,072	5,060	2,692		23,320	11,660
10	33,927	5,060	1,422		26,445	13,223

14.4 SPECIFIC INSTRUCTIONS

ENERGY SAVINGS   *(Table 14.2, line 1, and Table 14.1, part 1):* In projecting energy savings, it is necessary to estimate the future cost of energy, and the amount of energy saved. Gross Btu saved are converted into dollar cost by multiplying the projected cost per Btu for the particular source of energy used. If the projected cost per Btu for a particular source of energy is expected to increase at a rate greater than that of general inflation, the

TABLE 14.4 ENERGY CONSERVATION INVESTMENT COMPUTATION OF AFTERTAX RATE OF RETURN FOR COMMERCIAL BUILDING

(Building)

(Type of Investment)

Year	Aftertax Cash Flow (Table 14.2, line 13)	12% Discount Factor	Present Value	13% Discount Factor	Present Value
0	$(26,200)	1.000	$(26,000)	1.000	$(26,000)
1	3,502	0.892	3,124	0.884	3,096
2	3,829	0.797	3,052	0.783	2,998
3	4,158	0.711	2,956	0.693	2,881
4	4,498	0.635	2,856	0.613	2,757
5	4,834	0.567	2,741	0.542	2,620
6	5,167	0.506	2,615	0.480	2,480
7	5,504	0.452	2,488	0.425	2,339
8	5,823	0.403	2,347	0.376	2,189
9	6,138	0.360	2,210	0.332	2,038
10	6,430	0.321	2,064	0.294	1,890
Total			253		(912)

amount of this increase should be factored into the projected energy cost savings.

INCREASE IN RENTAL INCOME *(Table 14.2, line 6, and Table 14.1, part 2):* Enter on this line the projected net change, if any, in rental income resulting from an energy conservation improvement. Rent loss during the

construction or installation period of an energy conservation improvement should be included along with any projected decrease in the amount of rentable space as a result of the installation of such an improvement. Line 6 (of Table 14.2) should also include any projected change in rental income due to the nature of an energy conservation measure, such as a decrease resulting from the elimination of windows or an increase resulting from improved interior aesthetics. If appropriate, a factor should be added for increased marketability or rentability resulting when a building operates more effectively than properties of competitors under conditions of mandated energy restrictions or curtailments or during periods of fuel shortages. Many factors listed above will necessitate value judgments by persons familiar with the rental market in the area where a building is located. If the net change in rental income is negative, this should be indicated on Table 14.2 by the use of parentheses.

NET INVESTMENT *(Table 14.2, line 8, and Table 14.1, part 3):* The net investment for purposes of the private ownership cost model is the amount of cash which is expended at the time the improvement is made. It is the purchase price of the improvement less the principal amount of any note, loan, or mortgage used to finance the improvement. Salvage proceeds, if any, from the sale of replaced assets should also be deducted from the purchase price in determining the net investment.

TABLE 14.5 ENERGY CONSERVATION INVESTMENT ANALYSIS SUMMARY FOR COMMERCIAL BUILDING*

(Building)

	Type of Investment			
1. Aftertax rate of return (Table 14.4)	12.2%			
2. Payback period, years (Table 14.2, line 14)	7			
3. Maximum cash required (Table 14.2, line 14)	$26,200			

*Cost saving resulting from multiple energy conservation investments in a single building will be less than the cumulative cost savings from each investment, unless the engineering data used to compute energy savings took into consideration the diminishing-returns effect of multiple energy conservation improvements to a single structure.

DEBT SERVICE   *(Table 14.2, line 9, and Table 14.1, part 4)*: Completion of this item is applicable only if a portion of the cost of an energy conservation improvement is financed by a loan, note, mortgage, or some other form of indebtedness. When no loan is involved, the loss of income from other potential investments should be weighed against the return from investment in energy conservation. The annual debt service on the amount borrowed is the sum of principal and interest payments paid during a given calendar year. In order to compute the annual debt service for a loan, it is necessary to determine the amount of the loan, the payback period in years, the payback provisions, and the annual interest rate. With these, the annual debt service can be computed using standard mortgage amortization tables, or the amount of annual payments for interest and principal can be obtained from the bank, loan company, or other institution where financing is expected to be obtained.

BASIS OF NEW ASSET   *(Table 14.1, part 5)*: The basis of a new asset is the sum of the purchase price of the asset and the book value (original cost less accumulated depreciation) of any earlier asset replaced, less any trade-in allowance received on the earlier asset. Income tax laws do not allow the recognition of a gain or loss resulting from disposal of an earlier asset when the earlier asset is traded in on a new asset. Any gain or loss resulting from such a trade-in is simply added to the basis of the new asset being purchased. If an earlier asset being replaced by a new asset is sold rather than traded in, any gain or loss resulting from the sale of the asset is recognized at the time of the sale.

DEPRECIATION   *(Table 14.1, part 6)*: The accounting definition of depreciation is the loss of usefulness, expired utility, or diminution of service yield from a fixed asset caused by wear and tear from use, misuse, or obsolescence. Both accounting theory and tax theory assume that depreciable assets have a limited life and that the cost of such assets should be depreciated over that useful life. Depreciation represents the theoretical reduction in value of property improvements by reason of physical deterioration or economic obsolescence. In practice, properly maintained improved real estate reduces in value at a far slower rate than that allowed by income tax laws and accounting theory. This does not affect an investor's right, however, to depreciate an investment over the shortest period allowed by tax statutes.

The useful life of an improvement to a building is defined for tax purposes by specific provisions in federal tax laws. A taxpayer who uses the asset depreciation range (ADR) system must assign a useful life to the energy conservation improvements which falls within the ranges specified by the ADR provisions of federal tax statutes. If a taxpayer does not use the ADR system, the "facts and circumstances" surrounding each particular investment should be used to determine its depreciable life. In determining this actual useful life, the taxpayer's past experience is very important. But in either case, the useful life of building improvements should not be longer than the remaining useful life of the building itself, and if improvements are made by a lessee, such improvements should be amortized over a useful life no longer than the remaining life of the lease.

Generally speaking, in order to maximize the aftertax benefits of an energy conservation improvement, building owners should depreciate such improvements as fast as possible. Therefore, owners should use the shortest useful life allowed by the ADR system, if they use the ADR system to depreciate their assets, or the shortest useful life under the "facts and circumstances" method if they do not use the ADR system. If taxpayers use the ADR system for part of their capital acquisitions during a given year, they must use the ADR system for all their capital acquisitions made during that year. Taxpayers cannot use the ADR system for determining the useful life of some of their assets unless they use it for all of them.

Several rates of depreciation are available for investments in energy conservation improvements. Straight-line depreciation assumes that the diminution in value of a fixed asset occurs evenly over its useful life. When straight-line depreciation is used, a building improvement with a useful life of 40 years would be depreciated by 2.5% ($100 \div 40$) of its original cost for 40 years. There are several methods of accelerated depreciation, such as the declining-balance and sum-of-the-year's-digits methods, which provide for a greater amount of depreciation at the beginning of an asset's useful life and a lesser amount during the last years of its useful life. The theory behind accelerated depreciation is that an asset loses value more quickly during its first years than during its last years. Accelerated depreciation changes the timing of depreciation taken on a piece of property, but it does not change the total amount of depreciation which can be taken; this total on any property can be no more than its basis (see the previous section). Declining-balance depreciation is a very common form used for real estate investments. Under declining-balance depreciation, the straight-line depreciation rate (100 divided by the useful life of an asset) is multiplied by a specific percentage—125, 150, or 200%—to obtain an accelerated rate. In the example cited above, a building improvement with a 40-year useful life would have a straight-line depreciation rate of 2.5%, a 125% declining-balance rate of 2.125%, a 150% declining-balance rate of 3.75%, and a 200% declining-balance rate of 5.00%. These accelerated rates are applied to the undepreciated balance in each year to obtain the depreciation expense for that year, resulting in a steadily declining depreciation expense over the entire life of the asset. Generally speaking, the maximum depreciation rate available for energy conservation improvements to real property such as office buildings and retail facilities is a 150% declining balance. It should be noted that under certain circumstances a portion of such accelerated depreciation may be subject to special taxation when the property is sold. Furthermore, there are special taxes on accelerated depreciation deductions for certain high-income individuals.

GAIN ON DISPOSAL OF RETIRED ASSET *(Table 14.1, part 7):* The gain or loss on the sale of an asset which is replaced by an energy conservation improvement is computed by subtracting the original cost of the retired asset from the sum of the accumulated depreciation taken on such a retired asset and any salvage proceeds received from the disposal of the retired asset.

INVESTMENT TAX CREDIT *(Table 14.2, line 12, and Table 14.1, part 8):* The investment tax credit is applicable only to personal property used in a trade or business. A building's structure and improvements are considered real property and as such are not eligible for an investment tax credit. As a general rule, building components which can be picked up and moved, such as window air-conditioning units and bookcases, are considered personal property and are eligible for the investment tax credit. Structural portions of a building such as central heating and air-conditioning units, roofs, and windows are real property and are not eligible for the investment tax credit. Energy conservation improvements which are structural portions of the building are not eligible for the investment tax credit, and for such components the investment tax credit line (Table 14.2, line 12) should be left blank.

An investment tax credit is received in the year in which qualifying property is placed in service. The amount of credit depends upon the useful life of the qualifying property. If the useful life of the property is less than 3 years, no credit is available. If the useful life of the property is at least 3 years but less than 5 years, the amount of credit is 7% times one-third of the purchase price of the property. If the useful life of the property is at least 5 years but less than 7 years, the amount of credit is 7% times two-thirds of the cost of the property. If the useful life of the property is greater than 7 years, the amount of the credit is 7% of the purchase price. The amount of investment credit may not exceed a taxpayer's tax liability. If tax liability exceeds $25,000, the amount of credit is limited to $25,000 plus 50% of the tax liability in excess of that amount. It should be noted that the investment credit has been changed several times since it was originally introduced in 1962, and because it is regarded by the federal government as a means of manipulating the economy, one can reasonably assume that the provisions of the present investment tax credit will be changed as economic conditions change.

LABOR AND MATERIAL SAVINGS *(Table 14.2, lines 2 and 3):* The amount to be entered in these lines should be the projected net change in labor and material costs expected to result from the installation of a particular energy conservation improvement. In some instances, no savings will be made in these categories. If there are no projected labor or material savings, these lines should be left blank. In other circumstances, labor and material costs may actually increase as a result of energy conservation improvements. If this is the case, negative amounts, designated by parentheses, should be entered.

PROPERTY TAXES *(Table 14.2, line 4):* This line should be used to enter projected changes in local property taxes expected to result from installation of major energy conservation improvements in a building. As a practical matter, the majority of energy conservation improvements will not result in property being reassessed and will cause no increase in property taxes. In those circumstances where an energy conservation improvement will increase the assessed value of a building, however, the property tax applicable to the improvement can be estimated by multiplying the estimated increase in the value of the building by the local property tax assessment ratio times the local property tax rate. For example, a $100,000 energy conservation addition to a building in a jurisdiction which assesses property at 50% of fair market value and which has a tax rate of 4% of assessed valuation would result in an annual increase in property taxes of $2000 ($100,000 × 0.50 × 0.04).

REDUCTION IN OTHER OPERATING EXPENSES *(Table 14.2, line 5):* For certain energy conservation improvements, this may be an important line. This is the entry for hard-to-quantify cost savings such as productivity changes resulting from installation of energy conservation measures or projected increased marginal income resulting from the extension of the useful life of a building. This line should also be used to enter other miscellaneous cost savings or increases applicable to a particular energy conservation improvement which are not covered elsewhere in Table 14.2.

NET OPERATING PROFIT *(Table 14.2, line 7):* The net operating profit resulting from an energy conservation improvement is a summation of energy saved, the reduction in other operating expenses, and any changes in rental income. It is the sum of lines 1 to 6 on Table 14.2.

INCOME TAX *(Table 14.2, line 11, and Table 14.3, col. f):* Taxable income is depreciation and interest expense subtracted from the sum of net operating profit and gain on the disposal of retired assets. Income tax is computed by multiplying taxable income times the individual or corporation tax rate, which, of course, varies. The tax rate used should be a composite rate of federal, state, and local income taxes which would apply to a marginal change in the taxpayer's taxable income. Negative income tax figures can occur and are based on the assumption that the property owner has other taxable income to which the initial tax losses resulting from the energy conservation improvement can be applied.

AFTERTAX CASH FLOW *(Table 14.2, line 13):* The amount of aftertax cash flow is the key factor in determining the economic viability of an energy conservation improvement for a taxable entity. It represents the amount of cash which will be generated by an energy conservation improvement after payment of applicable income taxes. It is computed by subtracting income tax (Table 14.2, line 11) from pretax cash flow (Table 14.2, line 10). Investment tax credit, if any, is added to pretax cash flow in determining aftertax cash flow.

COMPUTATION OF RATE OF RETURN *(Table 14.4):* The present value of a dollar received currently is greater than the present value of a dollar to be received at some time in the future. A dollar to be received at some date far into the future has virtually no present value. The ultimate economic evaluation of an energy conservation investment is the aftertax cash flow expected to be generated by the investment, discounted to reflect the time value of money. A common method of evaluating the time value of returns received from investment is to determine the discount rate which, when applied to the projected stream of aftertax cash flow from the investment, yields a present value of zero. If an energy conservation investment yields a cash flow which, when discounted at the rate of 10%, has zero present value, the discounted rate of return is 10%. Such a 10% rate of return is equivalent to a fixed income investment which yields 10%/yr for the same time period for which the energy conservation improvement was evaluated and at the end of that period returns the amount of principal invested. When alternative energy conservation improvements are evaluated, the improvement with the highest aftertax discounted rate of return should be given first priority. If the discounted rate of return available from an energy conservation improvement is less than the projected rate of return from alternative investments which an individual or a corporation can make, the energy conservation improvement cannot be justified solely on economic grounds.

To compute the discounted rate of return for a particular cash flow, the discount rate for a particular interest rate should be applied. These discount rates are available in standard real estate investment analysis tables. The rate which produces a zero present value is arrived at by a trial-and-error-procedure, which with practice can be done quickly. For example, a 12% discount factor is applied to the projected aftertax cash flow from a hypothetical energy conservation improvement to yield a present value of $253. When the discount factor is increased to 13%, the present value is negative $912.

Through interpolation, it can be said that the discounted rate of return for the hypothetical investment is approximately 12.2%.

COMPUTATION OF PAYBACK PERIOD *(Table 14.5):* The payback period is the number of years required to recoup the out-of-pocket cash investment in an energy conservation investment. It is the number of years required before the cumulative aftertax cash flow (Table 14.2, line 14) from an energy conservation investment becomes a positive amount.

COMPUTATION OF MAXIMUM CASH REQUIRED *(Table 14.5):* The maximum cash required is the amount of cash, after financing and income taxes, needed to purchase and install an energy conservation improvement. It is the largest negative amount appearing in line 14 of Table 14.2. Building owners cannot install an energy conservation investment, regardless of its life-cycle rate of return, unless they have cash available for investment equal to or greater than the maximum cash required. The cost model presented in this chapter shows the net amount and negative cash flows. Within a given year, timing differences in the receipt of these positive and negative streams of cash may result in interim cash requirements which are greater than net cash required for the year as a whole.

CHAPTER FIFTEEN

ALTERNATE ENERGY SOURCES

15.1 THE POSSIBILITIES

With the exception of the nuclear bonds of the atom, the molten bowels of
the earth, and the tides of the ocean, the sun is the only source of energy for
human beings. The fossil fuels of coal, oil, and gas are the locked-in photosynthetic collection of ths sun's energy from eons past. The wood we burn
and the food we eat are products of the photosynthetic collection of the sun's
present energy. The rivers which give hydroelectric power could not exist
without the evaporation of the oceans by the sun, and the winds themselves
are driven by differential heating of the earth's surface. All these forces are of
awsome cyclic complexity, yet strewn with orders we still but dimly
perceive.

Alternative energy sources, simply defined for our purposes, means
energy alternatives to the burning of fossil fuels. For heating and cooling,
fossil fuels are burned on-site as primary energy and off-site to generate
electricity for secondary energy. At present, for building use, fossil fuels
must account as primary or secondary energy for greater than 99% of building demand. But this source, coal, oil, and gas, which we have so extrava-
gantly drawn upon for the last century or so, is not replenishable in our time
and is shockingly limited. We must finally turn directly to the sun for
renewable and nonpolluting energy. The diagram network in Fig. 15.1
nicely encompasses the potential for the sun's capture.

A primary concern for the architect and the engineer is the fit of alterna-
tive-source technology to new and existing buildings. Alternative sources
utilized off-site with an end product of electricity or district heat, a most
probable future, require no change. Alternative sources utilization at the
building site itself, however, do require a new vision of the construction art,
and architects and engineers must henceforth be open to this new vision.
The theme of this book—responsible energy management in buildings—is a
large part of this new vision.

Current promising explorations for direct utilization of alternative energies in buildings include:

- Solar thermal collection
- Solar electrical conversion
- Wind-generating plants
- Methane gas generation
- Solid waste combustion
- Nitrogen fuel cells

To this listing, within the context of this book, are added as alternative energy sources:

- Heat reclamation
- Thermal storage
- Total energy systems

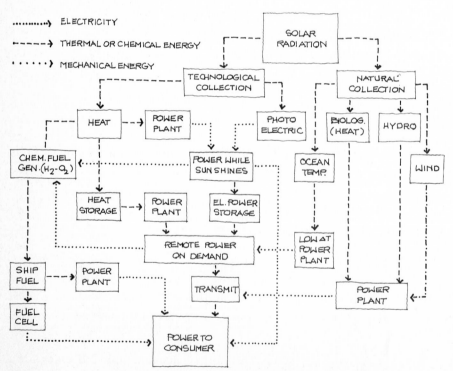

Figure 15.1 The many ways of obtaining electrical power from the sun.

These latter items are not, strictly speaking, considered to be alternative sources but are set out here, we think legitimately, as a significant classification of approaches devised to reduce building energy demand. (And they do not fit neatly into the book's classification of measures for the reduction of building load, systems load, or conversion load.)

The treatment in this chapter of sun, wind, and waste utilization is merely introductory, in deference to the several excellent bibliographic references given for these subjects. However, the treatment in this chapter of heat reclamation and total energy systems is in more detail.

15.2 HEAT RECLAMATION

Heat reclamation is the recovery and utilization of energy that is otherwise wasted. Properly captured, waste heat can be a substitute for a portion of the new energy that would normally be required for heating, cooling, and domestic hot-water systems. Heat recovery conserves fuels, reduces operating costs, and reduces peak loads. Applications of heat reclamation devices are shown in Table 15.1.

The performance of a heat recovery system will involve some of or all the following factors:

- The temperature difference between the heat source and the heat sink

- The latent heat difference between the heat source and the sink

- The mass flow multiplied by the specific heat of each source and sink

- The efficacy of the heat transfer device

- The extra energy required to operate the heat recovery device

- The fan or pump energy absorbed as heat by the heat transfer device (which either enhances or detracts from performance)

15.3 HEAT RECLAIM SOURCES

EXHAUST AIR In large buildings and in buildings requiring equipment hoods, considerable energy is lost in exhaust of conditioned air. If the controlled exhaust exceeds 4000 cfm and heating degree-days exceed 2500, and/or cooling degree-hours above 78°F DB exceed 8000, heat reclaim is definitely advised. If cooling degree-hours above 66°F WB exceed 12,000, either thermal wheels or heat pumps which can recover both sensible and latent heat are justified. If reclaim is indicated for the heating season only, an air-to-water-to-air heat pump to transfer energy from the exhaust to the ventilating intake can be considered. Another effective reclaim potential for

TABLE 15.1 APPLICATIONS OF HEAT RECLAMATION DEVICES*

Heat Reclaim Sources	Temper Ventilation Air	Preheat Domestic Hot Water	Space Heating	Terminal Reheat	Temper Makeup Air	Preheat Combustion Air	Heat Heavy Oil	Internal to External Zone Heat Transfer
Exhaust air	1a, 1b, 1c, 2, 3, 4, 5				1a, 1b, 1c, 2, 3, 4, 5	Direct		5, direct
Flue gas	1b, 3, 4		3, 4		1b, 3, 4	1b, 2, 3, 4		
Hot condensate	3, 6	3, 6	3, 5, 6	3, 6	3, 6	3, 6	6	
Refrigerant hot gas	6	6		6	6			
Hot condenser water	3, 6	3, 6	3, 5, 6	6	3, 5, 6	3, 5, 6	6	5
Hot-water drains		6						
Solid waste	7	7	7	7	7	7	7	
Engine exhaust and cooling systems	6, 8	6	6, 8	6	6, 8	6, 8	6	
Lights	5, 9		5, 9	5, 9	5, 9			5, 9
Air-cooled condensers			Direct					

*Heat Reclaim Devices:

1. Thermal wheel
   a. Latent
   b. Sensible
   c. Combination
2. Runaround coil
3. Heat pipe
4. Air-to-air heat exchanger
5. Heat pump
6. Shell/tube heat exchanger
7. Incinerator
8. Waste heat boiler
9. Heat-of-light
10. Thermal storage

exhaust air in cold climates is to use the exhaust directly for boiler or furnace combustion air or for preheating oil.

FLUE GAS The energy that goes up boiler and furnace stacks can be captured effectively with heat exchangers and can be used for space heating and preheating domestic hot water or for tempering intake air for ventilation and combustion; it can even be used for absorption refrigeration. The reclaim device must be able to withstand the corrosive effect of the flue gas, and the reduction of draft by the device must be accounted.

STEAM CONDENSATE The condensate return portions of many steam systems exhaust large quantities of heat in the form of flash steam when the hot condensate is reduced to atmospheric pressure in the condensate receiver. Waste heat can be recovered by installing a heat exchanger in the condensate return main before the receiver to reduce the condensate temperature to approximately 180°F. The heat recovered can be used to preheat water or for other purposes. See Fig. 15.2.

The quantity of heat recovered depends on the pressure and the temperature characteristics of the boiler. For example, a building system may have a

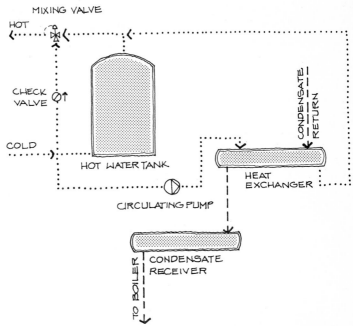

Figure 15.2 Steam-condensate–hot-water heat exchanger.

condensate return volume of 6 gpm at 260°F. A heat exchanger can be installed to reduce the condensate temperature from 260 to 180°F, and the quantity of heat recovered will be 240 × 10³ Btu/h. In this instance the entire domestic hot-water load of a typical office building might be met the major part of the year by the heat recovered in this hot-condensate heat exchanger.

REFRIGERANT HOT GAS A typical refrigeration machine with a water-cooled condenser rejects approximately 15,000 Btu/h for each 12,000 Btu/h of refrigeration. An air-cooled condenser rejects up to 17,000 Btu/h for each 12,000 Btu/h of refrigeration. Up to 5000 Btu/h of this heat rejected from either system can be recaptured. To recover the heat of compression, a heat exchanger can be installed in the hot-gas line between the compressor and the condenser of the chiller. A typical arrangement in conjunction with a domestic hot-water system is shown in Fig. 15.3. Hot-gas temperature depends on head pressure but is usually in the order of 180°F. Heat pumps, operating year-round for cooling or heating, offer a fine opportunity to use a hot-gas heat exchanger. Cold water is circulated through the heat exchanger by the circulating pump. When hot water is not being used, water is pumped back through the heat exchanger, the hot-water heater, or both. A mixing valve is provided to maintain the desired temperature.

Energy can suitably be reclaimed from refrigerant-system hot gas when there is a steady and concurrent demand for refrigeration and waste heat and when refrigeration systems operate for 750 or more hours per year. Care must be taken, however, not to reduce superheat to a point which allows liquid slugging, and the heat exchanger must be located after the hot gas bypass or

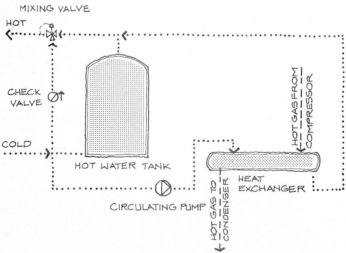

Figure 15.3 Refrigerant hot-gas–hot-water heat exchanger.

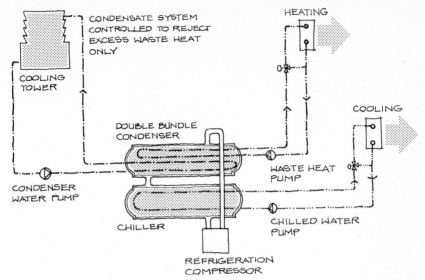

Figure 15.4 Double-bundle condenser for the reclaim of energy for heating.

other unloading devices. If the exchanger is located outdoors, drains must be provided to prevent freezing.

HOT CONDENSER WATER TO COOLING TOWER A heat exchanger or heat pipe in the hot condenser water line can temper outdoor air and preheat domestic hot water; with piping modification, condenser water can be used directly in the coils of air-handling equipment for air heating. In specification of new equipment, double-bundle or cascade condensers should be considered for flexibility in the use of refrigeration waste heat. See Fig. 15.4.

COMMERCIAL LAUNDRY, HOSPITAL, LABORATORY, AND KITCHEN WASTE WATER Hot waste water, if above 120°F and available in quantities greater than 10,000 gal/week, can readily be piped through a shell-and-tube heat exchanger for any system requiring hot water. As something of a bootstrap system, cold water for the same use can be preheated before entering the domestic hot-water heater, from 50 to 105°F, without excessive cost. Installed heat exchanger costs range from $500 per 100 gal/h for large units to $2000 per 100 gal/h for a small unit.

Consideration must be given to the characteristics of the waste water, particularly the soap or detergent content of laundry waste water and the grease content of kitchen waste water. Piping or material modifications may be necessary to enable the heat exchanger to handle water with high concentrations of these impurities. In addition, in order to maintain a steady flow

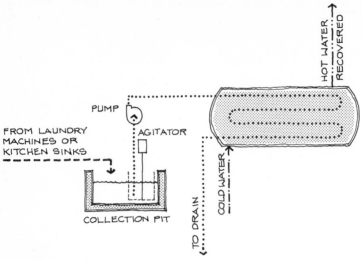

Figure 15.5 Recovery of energy from laundry or dishwashing hot-water drainage.

rate through the heat exchanger when water is being sporadically discharged, a holding tank may be required. See Fig. 15.5.

SOLID WASTE Commercial and institutional solid waste is an energy source that should be considered for on-site capture if the quantity available for incineration is greater than 1000 lb/day. The costs of required emission controls, however, should be carefully scrutinized.

ENGINE EXHAUST AND COOLING SYSTEMS The heat exchanger and the prevention of flue gas condensation impose limitations on recovery of heat from exhaust gas. The recommended minimum exhaust temperature is, for these reasons, approximately 250°F, and applications are typically limited to engines larger than 50 hp. Depending on the initial exhaust temperature, about 50 to 60% of the available exhaust heat can be removed.

LIGHTS The energy of lights suspended within a space is finally imparted to the space as heat, equal to the energy required to power the light and the ballast, too, if the light is of the electrical discharge type. Wattage times a conversion factor of 3.41 equals Btu per hour. A 100-W lamp is the equivalent of a 341-Btu/h heater. If lighting demand is such that the space is overheated, this heat can be captured to be used elsewhere or captured simply to reduce the space cooling load.

AIR-COOLED CONDENSERS In supermarkets, grocery stores, or any building with 15 hp or more of refrigerated display cases or storage boxes, and with heating degree-days exceeding 3000, the air heated by the condensers should be used directly for space heating. When air-cooled condensers are located away from spaces requiring heat, condenser heat can still be recovered through a duct system.

15.4 HEAT RECLAIM DEVICES

THERMAL WHEELS A *thermal wheel*, sometimes called a *heat wheel*, is a rotating heat exchanger with a high thermal inertia core driven by an electric motor. A thermal wheel transfers energy from one airstream to another or, for very large boiler plants, from flue gas to combustion air. The hot and cold airstreams must be immediately adjacent and parallel to permit installation of the thermal wheel, and for this purpose duct modifications may be necessary. Two types are available: one transfers sensible heat only, and the other transfers both sensible and latent heat.

Building exhaust air and outdoor air flowing in the opposite direction can each pass through half of a thermal wheel via separate but adjacent ducts. A thin cylinder containing a heat transfer medium slowly rotates between the two airstreams, and energy absorbed by the medium is transferred from one airstream to the other. For heating, incoming low-temperature outdoor air is heated and humidified by heat gained from the warm, moist exhaust air. For summer conditions, incoming high-temperature outdoor air is cooled and dehumidified as it gives up heat and moisture to the cooler exhaust air.

Thermal wheels are commercially available in sizes ranging from 3000 to 60,000 cfm. Wheels can be provided with purge sections and can be arranged for vertical or horizontal mounting. Thermal wheels of larger size can be provided with variable-speed, SCR controls. The order-of-magnitude cost for thermal wheels, excluding ductwork modifications, is between $700 and $1000 per 1000 cfm for sensible heat wheels, and these amounts plus 10% for enthalpy heat wheels.

In the heating mode, the energy contribution of the total system is equal to the heat transferred plus the heat of the driving motors. For the cooling mode, the energy contribution is equal to the heat transferred *minus* the heat of the driving motors and the heat equivalent of the additional system friction. The maximum potential heat that can be exchanged between a warm-exhaust airstream and a cold outdoor airstream is expressed with the following formula:

$$Q = (Q_e - Q_o) \times M_e \text{ or } M_o \quad \text{(whichever is least)}$$

where Q_e = heat content of exhaust air in Btu/lb
$\quad Q_o$ = heat content of outdoor air in Btu/lb
$\quad M_e$ = exhaust airflow rate in lb/h
$\quad M_o$ = outdoor airflow rate in lb/h

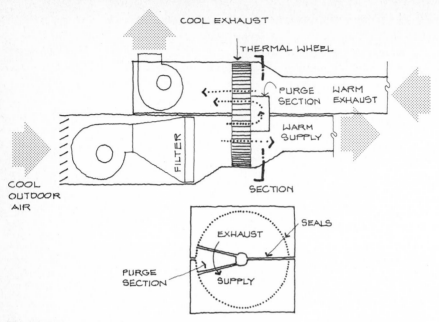

Figure 15.6 Heat exchange between exhaust and intake air with thermal wheels.

Thermal wheels typically have an efficiency of 60 to 80% for sensible heat transfer and 20 to 60% for latent heat transfer. These efficiencies should be applied to the theoretical maximum heat transfer Q to obtain the actual heat transfer rate in Btu per hour.

This formula provides the load reduction for 1 h at a specific set of conditions. The yearly load reduction is obtained by analyzing the specific conditions for each hour of operation and integrating for the total number of hours of operation. This procedure may be used for either sensible recovery or enthalpy recovery, providing the units are in Btu/lb for each respective condition.

When adding thermal wheels to ventilation systems, a roughing filter and/or insect screen should be installed to protect the core. Icing, which can occur under some outdoor conditions with enthalpy wheels, can be eliminated by preheating. When used to transfer heat from flue gases, the thermal wheel must be constructed in a way suitable for the temperatures encountered. See Fig. 15.6.

RUNAROUND-COIL SYSTEM This system is made up of two or more extended surface fin coils installed in air ducts and interconnected by piping. The heat exchanger fluid of ethylene glycol and water is circulated through the system by a pump; heat is removed from the hot airstream and is rejected into the cold airstream. A runaround-coil system may be employed in winter to recover heat from warm exhaust air for use in preheating cold outdoor air and in summer to cool hot outdoor air by rejecting heat into cooler exhaust air. The method of determining system efficiency described for thermal wheels can also be applied to runaround-coil systems.

When calculating reductions in load due to runaround coils, the effect of coil depth on the rate of heat transfer and on resistance to airflow must be recognized. A trade-off analysis should be made between the extra heat transfer obtained by deep coils against the extra resistance to airflow to determine the optimum coil size.

Where there is no possibility of freezing, water may be used as the heat transfer medium. Where frost protection is necessary and ethylene glycol or other antifreeze fluid is used as the heat transfer medium, the different specific heats and rates of heat transfer for the medium must be accounted for in the calculations.

The coil-type exchanger has the advantage of permitting heat transfer between supply and exhaust systems in widely separated portions of the building, since coils are installed in each of the intake and exhaust ducts and are connected by piping only. The liquid heat-transfer medium is pumped from coil to coil and can, with relative ease, be piped long distances through the structure, interconnecting a number of separate exhaust and fresh-air systems to form one complete system.

The initial cost of runaround coils is less than the rotary-type heat exchangers by about 40%; these coils are most economical for small systems or for heating systems only. Order-of-magnitude costs are $2/cfm for systems of up to 10,000 cfm and $1.50/cfm for larger systems. See Fig. 15.7.

HEAT PIPE SYSTEMS Heat pipes are finned tubes perpendicular to, and pinning together as it were, adjacent air ducts. The tubes are continuously exposed to the airstreams of both ducts. Each tube contains liquid refrigerant which absorbs heat by evaporation at the warm airstream end of the tube and which migrates as a gas to the cold end of the tube where it condenses to release heat into the cold airstream. The condensed liquid then runs back to the hot end of the tube to complete the cycle. For effective operation, the heat tubes must slope down from the cold end to the hot end, and accurate control of the slope must be provided. Heat pipes between hot and cold air ducts can be used only when these are immediately adjacent to each other, and some modification of existing ductwork may be necessary for installation.

In order to calculate the effectiveness and yearly load reduction of a heat pipe system, the same procedure as described for thermal wheels can be used. Heat pipe sets are available in sizes ranging from 2500 to 20,000 cfm, and order-of-magnitude cost is $2500 per 1000 cfm. See Fig. 15.8.

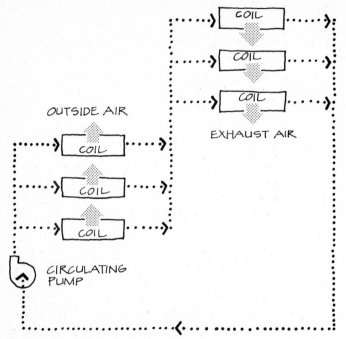

Figure 15.7 Heat exchange between exhaust and intake air with run-around coils.

AIR-TO-AIR HEAT EXCHANGERS This type of heat exchanger transfers heat directly from one airstream to another through direct contact on either side of a metal heat transfer surface. This surface may be made up of convoluted plates (common for low-temperature use in HVAC systems) or tubes (common for boiler flue gas heat transfer). The heat exchangers may be purchased as packaged units or can be custom-made. Generally, efficiencies are lower than 50%, but these exchangers are relatively inexpensive, have low resistance to airflow, require no motive power input, and are trouble-free and durable. Since the airstreams are not in direct contact and cannot mix, there is no danger of cross-contamination. See Fig. 15.9.

HEAT PUMPS AS HEAT EXCHANGERS Heat pumps are heat transfer machines and, unlike heat exchangers previously described, actually upgrade energy with efficiencies greater than unity. This feature is particularly attractive for use with low-temperature heat sources. As an example, heat can be extracted from 50°F drain water with output temperatures high enough for use for domestic hot water or for air preheating. They have the capacity to transfer latent heat as well as sensible heat.

Heat pumps are available in the following configurations, each with a particular application:

- Air to air

- Water to air

- Water to water

- Air to water

- Hybrid air and/or water to air

Sizes range from ½ to 25 tons as commercially available packaged units; larger sizes can be obtained by simple modifications to refrigeration machines. Order-of-magnitude costs are approximately $300 to $500 per ton.

As heat exchange equipment in the heating mode, heat pumps have a high COP and can be used to extract more heat from a given source than any other heat exchange device; the power input is but one-third to one-fifth of the

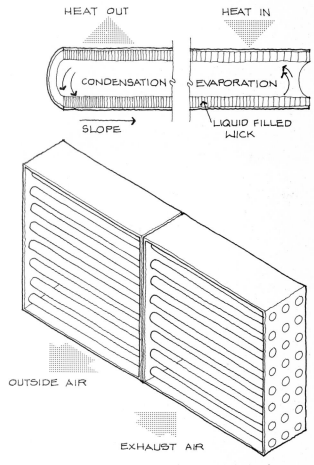

Figure 15.8 Passive heat exchange with heat pipes.

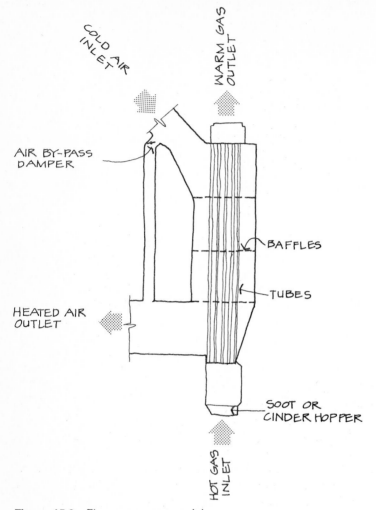

Figure 15.9 Flue-gas energy reclaim.

heat output. When used in the cooling mode, however, performance decreases, and heat pumps for this reason are best used for heating only, or for a combination of heating and cooling. Heat pumps should be selected that have the highest COP over the full range of expected operating conditions.

SHELL-AND-TUBE HEAT EXCHANGERS These heat exchangers can be used to exchange heat in the following configurations:

- Liquid to liquid
- Steam to liquid
- Gas to liquid

All three configurations are commercially available in a wide range of sizes and outputs, with reliable heat exchange data. Cost increases rapidly as the approach temperature is decreased, and this factor should be considered when making economic evaluations; the last degree of captured heat is very expensive. Particularly favorable applications are for energy capture from hot condensate, hot refrigerant gas, condenser water, hot drain lines from kitchens and laundries, and solar collector fluid loops.

INCINERATORS Solid waste burned in an incinerator can produce 6000 Btu/lb. Special heat recovery incinerators are now available in sizes down to a burning rate of 100 lb/h which generate 450,000 Btu/h. These incinerators are oil-jacketed and raise the oil temperature to approximately 450°F. The hot oil can be used in a heat exchanger as a source of either high- or low-temperature heat, but, for whatever purpose, the heat should be utilized at the highest practicable temperature. Typical applications are for absorption refrigeration, heating, and domestic hot water.

The order-of-magnitude cost for incinerators is approximately $70/(lb waste)(h). It is important to remember that local, state, and federal air pollution standards must be strictly complied with, and environmental impact statements are required.

WASTE-HEAT BOILERS These boilers can be used with high-temperature heat recovery sources, such as engine exhaust, or with the exhaust from gas turbines employed to drive refrigeration equipment and electric generators. Such an arrangement is the heart of total energy systems described in later sections of this chapter.

HEAT OF LIGHT Heat-of-light systems use special fluorescent fixtures to capture much of the heat that accompanies electric lighting for the purpose of heat reclaim and reduction of space-cooling loads. Heat-of-light systems were widely used for new construction in the 1960s and early 1970s, but with a trend to reduced general lighting levels and selective task lighting, these systems now require close examination for cost effectiveness and are generally applicable only to large buildings with large year-round interior heat gains.

Dry and wet heat-of-light systems can provide the following:

- Collection of excess interior building heat for transfer to colder exterior zones

- Reduction of interior sensible heat gain from lighting

- Increase in lamp output (cooler lamps operate at higher efficacy), shown graphically in Fig. 15.10

- Longer ballast life

- Improved occupant comfort (by reducing overhead radiant temperatures)

With dry systems, room air is extracted through the lighting fixtures, over lamps and ballasts, and is ducted to a fan or drawn into the ceiling plenum. The heated air can be supplied to perimeter zones in the heating season, can be used for control purposes, or, in the cooling season, can be recirculated back to an air-handling unit or exhausted outdoors. If recovered air is moved to perimeter zones or exhausted, a separate fan is required. When ceiling plenums are used as collection chambers, ceilings over conditioned areas should be insulated to a U value of 0.1 or lower to limit re-radiation from the ceiling surface.

With wet systems return air is extracted as with dry systems over lamps and ballasts but is then passed through built-in fixture heat exchangers to give up heat to circulated water. The heated water can be used for preheating, or the water can be cooled with a circuit through a cooling tower. One application circuits the heated water through special interior louvers at windows for perimeter heating in the winter. In the summer, the same louvers, now used to stop the sun, are circuited directly to the cooling tower for perimeter cooling independent of heat-of-light operation.

Figure 15.10 graphs the energy recapture potential through four typical heat-of-light dry fixtures at various rates of return air extraction with a lower curve and left ordinate readout, and also graphs the percentage light output increase with an upper curve and right ordinate readout.

Figure 15.11 graphs the energy recapture potential of any wet fixture in percent of total fixture power, with input of inlet water temperature and flow rate.

THERMAL STORAGE In many buildings under many operating conditions excess energy is generated which, if stored rather than exhausted, can become a valuable resource. In large buildings with interior zones that require year-round cooling, the energy typically expended to the atmosphere through cooling towers in the winter can be stored in water tanks for heating

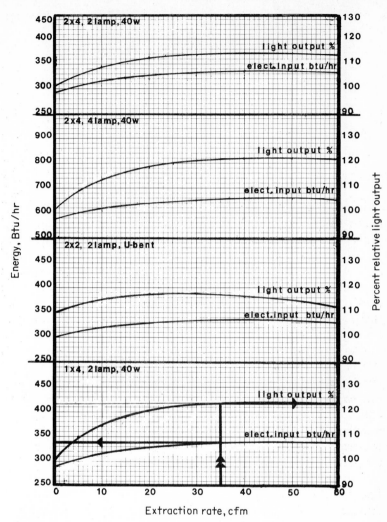

Figure 15.10 Effect of heat transfer on luminaire output and power requirements. (This graph was derived from manufacturers' data and is typical for dry heat-of-light luminaires.)

during the nighttime and unoccupied periods. In the summer the same storage system can allow the use of smaller refrigeration equipment, which, operated continuously, can build up stored chilled water during unoccupied periods to assist in peak occupied-period cooling. Such arrangements allow, in addition, higher coefficients of performance with cooler ambient nighttime operating conditions and reduce peak electrical loads for air conditioning.

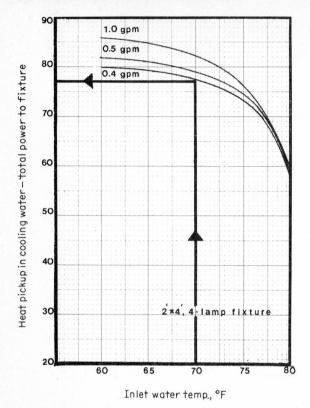

Figure 15.11 Thermal performance of wet heat-of-light luminaires. (This graph was derived from manufacturers' data and is typical for wet heat-of-light luminaires.)

15.5 TOTAL ENERGY SYSTEMS

Total energy (TE) is the name given to on-site generation of electricity if the waste heat of generation is recovered for use by the building. The lure of total energy lies in this recovered heat, since its equivalent would have to be purchased in the form of other fuels. The recovered heat can be used for heating domestic hot water, for tempering outdoor ventilation air, for cooling with absorption refrigeration for space heating, or for producing process steam.

Since all TE installations require added investment, the system is economically advantageous only when energy savings are sufficient to amortize both the higher capital costs and the added costs of service, maintenance, and replacement parts in a period short enough to be attractive to the investor. When electrical energy was plentiful and service reliable, the appeal of TE was economic, but an increasing possibility of blackout and brownout provides additional reason for considering decentralized TE systems.

TE plants consume 25 to 40% less raw source energy than that required for a combination of fossil-fueled central electrical generating plants and on-site boilers; this makes TE a highly beneficial national energy conservation measure. See Fig. 15.12.

Most total energy systems to date have been installed in shopping centers, universities, apartment houses, and hospitals, where long hours of operation provide utilization of the equipment for much of the year and faster amortization. Retail stores and office buildings which are larger in area than 50,000 ft² and which are heated and air conditioned should be considered for total energy systems. Figure 15.13 gives a detailed flow diagram of a typical total energy system.

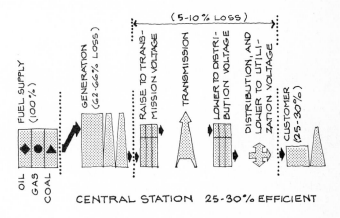

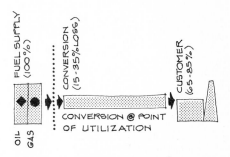

Figure 15.12 Total-energy and central-station efficiencies.

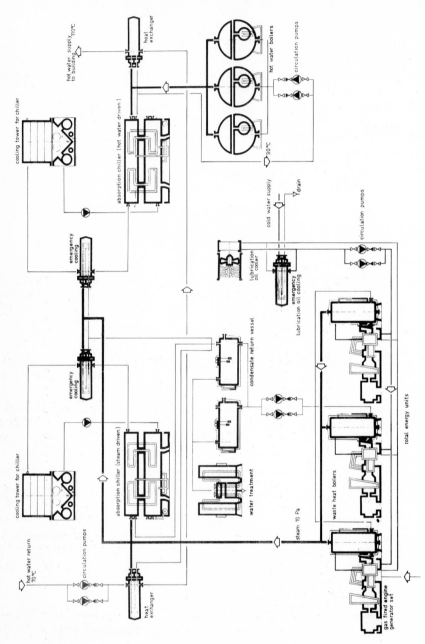

Figure 15.13 A typical total-energy system.

380

TOTAL ENERGY FEASIBILITY REQUIREMENTS

- Floor area greater than 50,000 ft²

- Occupancy greater than 38 hours per week

- A requirement for both heating and air-conditioning systems

- Gas or oil rates low enough to compete with electric rates (fuel at 20 cents/therm or lower and purchased electricity at 3.2 cents/kWh or higher)

- The availability of skilled operating personnel either as staff or by contract

- An owner's fiscal policy that permits payback periods in excess of 8 years

- Adequate space to accommodate prime movers and generating equipment

- Electrical and thermal loads that are in appropriate balance (uses for at least 50% of the waste heat)

If these requirements are satisfied, a feasibility study is in order.

ENGINEERING AND ECONOMIC ANALYSIS   The first step in the analysis is the determination of the load profile of the building, the hourly requirements for waste heat and electricity on a 48-h basis for each season of the year, stating minimum, average, and peak demands, and total consumption. In the design of the building or buildings, it is especially important to reduce the peak electrical loads to enhance the economic viability of TE systems, since peak electrical loads determine the size and cost of engines, turbines, and generators.

The load-profile curves are essential to determine the amount of waste heat that can be recovered. This amount represents the energy and dollar savings of a TE plant. A detailed example analysis is given in *Total Energy* by Dubin-Mindell-Bloom Associates.

The second step in the analysis is consideration of prime movers. Although the efficiency of reciprocating engines is considerably higher than that of gas turbines, if there is continuous use for waste heat, the total efficiencies of both prime movers with heat recovery equipment are about the same. For a project that has a high demand for waste heat in relation to the electrical load, a turbine is probably a more favorable selection than an engine. This will occur in areas with high heating and cooling loads. Where the demand for waste heat is lower in relation to the electrical load, engines will be more economical.

Large slow-speed engines cost more initially but require less maintenance and service and use less fuel than smaller, high-speed engines. Both engine types must be evaluated against turbines, which are lighter and require less maintenance but may cost more initially.

Engines and combustion turbines operate most efficiently when they are loaded to 60% or more of rated capacity. Since the electric loads vary for different conditions of occupancy, the most efficient total energy plant utilizes generating set modules such that each prime mover is always operating close to full rated capacity.

A series of prime mover-generator components or modules should be selected such that those loads which occur for the longest periods of time during the year are properly matched. Usually these periods are the normal daytime hours of maximum loading and the unoccupied hours of minimum loading. The largest modules which will still permit operation at maximum efficiency during the greatest number of operating hours should be selected. Standby requirements dictate that if any module is out of service, there will be enough capacity from modules remaining on-line to carry the maximum load. The likelihood of more than one module going out of service is not great, and the minimum standby requirement will therefore be equal to the largest module of the group of modules required to carry the maximum load.

The third step in the analysis is a consideration for the process of heat recovery and storage. For gas turbines, the waste heat available is in the exhaust gases. Heat is recovered by means of a waste-heat boiler which resembles a conventional boiler, with a duct, similar to a breeching, connected directly from the exhaust of the turbine to the boiler. After passing through the boiler, the gases are discharged to the atmosphere. Gas turbines produce about twice as much recoverable heat as reciprocating engines of the same horsepower, and a typical turbine yields 7 to 13 lb steam per hour (15 lb/in^2 steam) per kilowatt generated.

In diesel and gas engine installations, 90% of the jacket water heat and 60 to 65% of the exhaust gas heat are recoverable, for a total of about 45% of the input energy to the engine. An overall system efficiency of slightly over 65% is possible with either prime mover in total energy plants, if all the recoverable heat is used.

Heat storage systems can be used where waste requirements do not coincide simultaneously with waste heat availability. Often the heat cannot be used immediately but can be used at later periods when heat requirements exceed the heat available.

A final consideration is that conventional electric power is supplied at 60 Hz (hertz), but on-site power generation, particularly with high-speed turbines, provides the option of generating power at frequencies other than 60 Hz with little initial cost penalty. High-frequency lighting may provide additional energy conservation benefits.

Co-generation, wherein a plant generates most of its electric power on-site, utilizes the waste heat for thermal purposes, and sells excess power to the utility company when its own load is less than average peak, can be cost-effective and energy-effective for both the plant and the utility company.

In industrial plants, the generation of electricity can be accomplished with only 5500 Btu/kW (as contrasted to 9000 to 14,000 Btu/kW for utility company plants) if the plant has need for high-pressure steam. For each 20 lb steam 1 kWh can be generated at the low rate of 55,000 Btu/kW.

15.6 SOLAR ENERGY

Direct use of the sun's energy through windows, skylights, and glass doors is the most efficient way to obtain its beneficent heating, but the sun occasionally is hidden behind clouds, and it sets every night. This and the next section describe methods to collect and store the sun's energy, to be used on demand for domestic water heating and for space heating and cooling.

DOMESTIC HOT-WATER HEATING Solar water heaters have been used for more than 30 years throughout the world, especially in Florida and in Israel, Australia, and Japan. In the United States although many solar water heaters are still in operation, numerous others were abandoned when repairs were required because gas, oil, or electric heaters cost less to buy and energy was inexpensive. The high costs, now, of fossil fuels and electricity have again awakened an interest in solar water heaters.

Solar water heaters are commercially available in the United States. The components are offered separately or together in a complete package. They include a flat plate collector, storage tank (existing storage tanks can be used), piping, controls, circulating pump, and, in climates where the collector is subjected to freezing weather, a heat exchanger with a secondary pump, piping circuit, and antifreeze. Figure 15.14 shows a typical arrangement for domestic water heating alone.

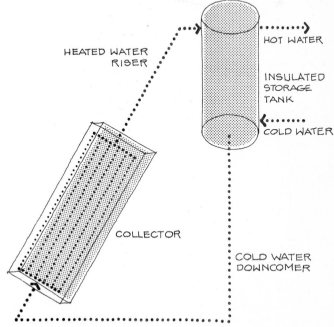

Figure 15.14 The thermosyphon solar hot-water heater.

If a collector is used solely to supply hot water, a fixed tilt of latitude plus 5° will usually be optimum—variations of 10° up or down will not seriously affect yearly performance. The collector must not be shaded more than 10% of the time; if it is, a larger collector area will be required.

For normal hot water use, approximately 1 ft² of collector and 1 gal of hot-water storage capacity per 16 gal hot water used per week are adequate. If kitchens or other processes require hot water at elevated temperature, 20 to 50% more collector area should be provided. An existing domestic hot-water heating system, even though inefficient or undersized, will usually be adequate to provide heat to supplement the solar collector system.

It is not economically feasible to attempt 100% of the domestic hot water requirement with solar heating unless the collector supplements the HVAC system, or the hot water temperature need not exceed 90°F, or the building is located in a hot climatic zone.

SPACE HEATING Prior to 1972, only some 20 solar-heated buildings existed in the United States, and of these only one was a commercial building; the rest were single-family residences. Within the past 4 years, however, there has been a proliferation of solar energy activity in the United States, and many thousands of such buildings are now in design or are under construction. While a majority of these are new buildings, the same design and construction techniques are applicable to existing buildings. From these installations, many of them instrumented as a part of continuing research activity, a considerable literature is developing and can easily be referenced for information and detail beyond this introduction.

With systems for domestic water heating the object is, of course, hot water; the collector and storage systems are therefore perforce of the water variety. For space heating, air collectors are available, and thermal storage can be accomplished with rocks or with phase-changing salts.

Contemporary HVAC systems using water typically operate at temperatures greater than 160°F. Such temperatures are attainable with available solar collectors which use water as the transfer medium, but collector temperatures are increased. For new buildings it is desirable to design the heating system for fluid temperatures of 100°F to allow low-temperature solar collection. The efficiency of the collector increases as fluid temperatures decrease. If solar collectors are considered for addition to existing heating systems, it is advisable to refit the heating system with larger coils to allow the use of lower-temperature water at about 100°F.

Solar collectors combined with existing or new heat pumps can be very effective. Low-temperature fluid delivered from the collector in cold and overcast weather can still serve as a heat source for a heat pump. The heat pump boosts the temperature to a higher level, and the useful hours of solar energy collection are substantially increased. Solar-assist heat pump installations should be considered especially in existing buildings that are already equipped with heat pumps or with electrically driven chillers which can be converted to heat pumps.

SPACE COOLING The common means for utilizing solar energy for cooling are with absorption chillers or with desiccants for dehumidification. With absorption systems, solar-heated water serves as a heat source in the absorption generator. Unfortunately, larger units in excess of 100-ton capacity require 230 to 270°F for generation, and most solar collectors on the market cannot provide these temperatures for any appreciable length of time. However, newer low-temperature absorption units are now available, in limited sizes, to operate at generator temperatures as low as 175°F at peak conditions and as low as 170°F with some loss of rated capacity.

Flat-plate collectors with selective surfaces and supplementary aluminized Mylar polyester film reflectors can perform at temperatures up to 220°F in some climatic zones for a sufficient number of hours per cooling season to operate such absorption units, but there are not many areas in the country where absorption units running from solar-collected heat alone can economically supply the majority of yearly cooling requirements of a building. Vacuum-tube collectors, and a number of focusing collectors, are now commercially available and are capable of providing the higher termperatures required to operate absorption units or the emerging Rankine cycle engines.

For cooling alone, the tilt of the collector should be equal to latitude less 10°. Where the collector is used for both cooling and heating, however, the heating efficiency suffers if the angle is so adjusted. A detailed analysis is required to determine optimum tilt angles of the collector for year-round use.

There will be very few situations where solar cooling will be economical until collector performances are improved and absorption units are produced which will operate at 160 to 180°F generator temperatures. Fortunately, both developments appear to be likely within the next 5 or 6 years.

Desiccant regeneration, used only in demonstration installations to date, appears to be cost-effective. In areas where evaporative cooling is impractical because of high relative humidity, evaporative cooling combined with desiccant dehumidification can provide adequate comfort; heat from the solar collector can supplement heat from other sources to regenerate the desiccant. Since regeneration temperatures are lower than those required for absorption cooling, the collector can be operated at cooler temperatures and is thus more cost-effective. See Figs. 15.15 and 15.16.

OTHER USES For large installations with No. 6 oil, flat-plate solar collectors can be effectively used to preheat oil before combustion to increase the efficiency of combustion. The oil storage tanks can be painted black for direct heating by the sun and can be fitted with heating coils through which the heated fluid from the collector is circulated. No additional storage system

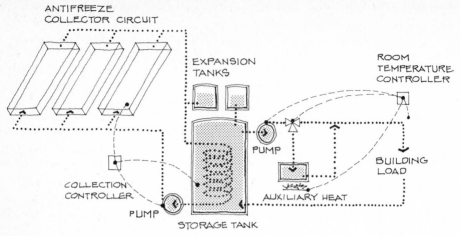

Figure 15.15 Solar collection, energy storage, and the controls for space heating with a water system.

is required. In like manner, the efficiency of oil or gas combustion increases as the temperature of combustion air is increased. Solar air or water collectors, without any additional storage facilities, can be used to temper air for combustion (or temper makeup air for ventilation). This is especially effective since the temperature of outdoor air in the winter may range from −30 to 50°F, and collector fluid only 20 or 25°F above outdoor air temperature can efficiently provide useful heat.

Though not of this sequence on thermal collection, the possibility for collection of the sun's energies as electricity should be mentioned in passing. The technology for using solar energy for the direct generation of electric power with photovoltaic cells is well known. Solar cells are used in the space program to generate electricity and in remote areas of the world to generate small quantities of electricity, primarily for signal purposes. The University of Delaware's demonstration house is equipped with such solar cells to supply electrical energy, but the hardware available for buildings is limited and too costly to consider now. There is little question, however, that with future developments, photovoltaic cells will indeed be feasible for building. Solar energy can also be used to generate high-temperature hot water or steam to run a turbine or heat-actuated engine for power generation, but these systems, too, are not commercially available, and costs are not comparable with conventional generating methods at this time.

15.7 SOLAR COLLECTION AND STORAGE

A solar collection system includes some device for absorbing the sun's energies, piping or ducting to transport collected energy, a storage unit, and a system of controls to operate and make use of the collected or stored energy.

COLLECTION   Flat-plate collectors absorb both diffuse and direct radiation. Depending upon their construction and meteorological and climatic conditions, each square foot of collector can collect from 75,000 to 250,000 Btu per heating season, and reflectors can increase this capacity from 15 to 25% at relatively low installation costs. The efficiency of the collector is inversely proportional to the collection temperature. Thus the characteristics of the building system which interface with the solar collector and storage units have a major influence on the overall economic feasibility of the complete system.

The optimum tilt for collectors used solely for space heating and domestic hot-water heating is latitude plus 15°. The tilt for collectors used for both heating and cooling is normally latitude, and for cooling only, latitude minus 10°. Orientation should be true south, with not more than 10 to 15° variation on either side of due south. A larger variation will require more collector area for the same thermal performance. See Figs. 15.17 and 15.18.

With liquid-type collectors, the higher the temperatures required, the greater are the benefits of absorber plates with selective surfaces. Absorptive surfaces (flat black, green, or even dark red) have an absorption coefficient of 0.9 to 0.95 and an emissivity of the same number. A selective surface has an absorption rate approximately the same but has an emission rate of only 10 to 15% of the radiation it receives. In many cases, only one layer of glass or plastic cover is necessary when selective surfaces are used, instead of the two that might otherwise be required.

The size of the collector varies with use, climate, and HVAC system with which it operates. The collector area required for heating ranges from 10 to 50% of the total building floor area to provide 20 to 80% of the annual requirements for heating domestic hot water and cooling energy. Collectors

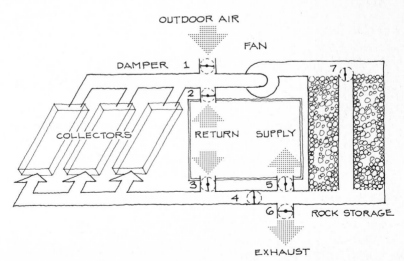

Figure 15.16   Solar collection, energy storage, and the controls for space heating with an air system with rock storage.

can be located on a building roof, on an adjacent building or garage, or on the ground. If a building is being modified, provisions made at the same time for a solar collector system can ease future installations. For example, the entire south wall can often be used as a collector, if it is designed with that possibility in mind. Collectors can also be mounted on the south wall as an eyebrow to shade the summer sun from entering indoors and to collect solar radiation at the same time.

Air-heating collectors are most suitable for space-heating applications alone (no cooling) and are particularly applicable for tempering makeup air or outdoor air for ventilation; in many cases, no storage system will be required. Most applications using air collectors will be at lower temperatures than water-type collectors, and selective surfaces with a well-designed air collector are generally not cost-effective.

STORAGE The most practical storage system with liquid collectors is hot water in metal or concrete tanks. From 2 to 5 gal of storage are required per square foot of collector. The exact amount depends upon climatic conditions, the number of consecutive cloudy days, the type of HVAC system, and the cost/benefits ratio of storage tank costs and collector performance versus the operating costs of the supplementary systems.

With the supplementary heating system that is required in virtually all cases, the costs of storage volume (and space required) can be reduced on a life-cycle cost basis by operating the supplementary backup system somewhat longer and by reducing the volume of storage. Where the entry into an

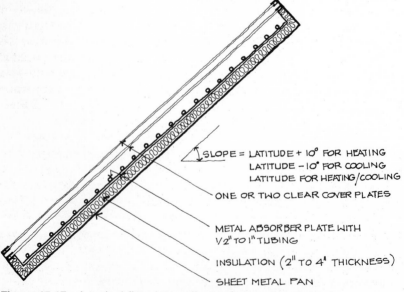

SLOPE = LATITUDE + 10° FOR HEATING
LATITUDE – 10° FOR COOLING
LATITUDE FOR HEATING/COOLING

ONE OR TWO CLEAR COVER PLATES

METAL ABSORBER PLATE WITH
1/2" TO 1" TUBING

INSULATION (2" TO 4" THICKNESS)

SHEET METAL PAN

Figure 15.17 A typical flat-plate collector for a water system.

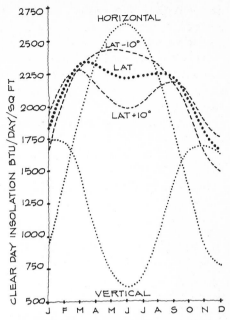

Figure 15.18 Variations in insolation due to collector tilt angles.

existing building is too small to deliver a tank, the tank can be fabricated by welding plates in the building, can be inflatable butyl rubber, or can be constructed of concrete. With hot-air systems, rocks (about 100 lb/ft² of collector) can be used for heat storage in place of water. Air should be distributed vertically, not horizontally, through rock beds, and the top of the rock bed should be charged with hot air from the collector. Air from the collector to heat the building should also be taken from the hot upper part of the rock bed storage. Latent heat storage using phase-changing salts (salt hydrates, eutectic salts, waxes, and parafins) requires less volume than water for equal thermal storage, and the potential for the development of these storage types is very good. They are just emerging commercially.

The tank and ducts or piping from the collector to the tank should be insulated to limit the thermal losses at any storage or collection temperature to 5% of the system capacity per 24 h. A minimum of two storage tanks should be considered—three is better—to permit storage of water at various temperature levels, so that high water temperature is not degraded by the collector when it cannot deliver water at tank temperatures for periods when there is little sun. With one tank only, collection is limited, and the economic feasibility is reduced since the collectors cannot be used to their maximum potential. With one tank, however, antiblending devices such as top and bottom distribution headers are quite effective in creating stratifications so that the hottest water is at the top of the tank.

CONTROLS The control system, at the least, must operate circulating pumps, divert warm water from one storage tank to another or direct to the load, or operate fans in air systems to blow air through the collectors of storage in response to the building load. A differential control is required to ensure that the fluid circulation through the collector occurs only if the collector temperature is 5°F above the storage temperature so that storage fluid temperature is not degraded. Controls are not a major cost of the installation unless elaborate monitoring, data collection, and research are undertaken.

ANTIFREEZE PROTECTION In areas where the outdoor temperature drops below 32°F, antifreeze protection is required for water collectors. The following systems may be used:

- A secondary circuit with an antifreeze solution such as glycol, triethylene glycol, propylene glycol, or light oil, with a heat exchanger and a separate pump
- Automatic draining of the collector when the outdoor temperature drops below freezing and refilling when the temperature rises above freezing
- A closed-circuit drainage system with nitrogen as a purging agent
- Bleeding a small amount of heated storage water through the collector in locations where temperatures drop below 32°F for only brief periods of time and the days are warm and clear

If the storage water bleed system shows only a small loss of thermal energy compared with useful heat collected on a seasonal basis (less than 5% loss), it is the least troublesome system. The nitrogen system is still under development, and more data are required before it can be recommended without qualification. Draining and refilling involves problems with incomplete draining and difficulties with venting when the system is refilled. Vents are subject to freezing and can become inoperative. The use of antifreeze with a secondary circuit introduces higher costs for the heat exchanger and secondary circuit, and corrosion is accelerated when antifreeze solutions reach high temperatures in the summer. Traditionally, however, most existing installations use antifreeze with a secondary circuit and heat exchanger in cold climates.

Collectors can reach high temperatures, 250 to 600°F, when exposed to direct sunlight. With no fluid flow, relief valves and expansion tanks are essential to protect the collector.

COST The best solution is the collector and system which provides the lowest life-cycle costs per million Btu delivered to the storage tank or hot-water taps. Costs vary widely with the size and type of installation and the construction conditions encountered. Flat-plate collectors cost from $5.00 to $12.00/ft²; vacuum tube and focusing collectors from $14.00 to $25.00/ft²; storage systems from $1.00 to $3.00/gal; piping and insulation from $3.00/ft² of collector and upward; and controls and wiring approximately $2.00/ft² of collector. Complete installations vary from $10.00/ft² (unusual) to $60.00/ft². Costs and collector performance are continually changing; estimates should be based on preliminary design concepts and layouts, with bids from contractors taken early, before final design is started.

GLOSSARY

Absorption Chiller   A refrigeration machine using heat as the power input to generate chilled water.

Absorption Coefficient   The fraction of the total radiant energy incident on a surface that is absorbed by the surface.

Absorptivity   The physical characteristic of a substance describing its ability to absorb radiation.

Activated Carbon   A form of carbon capable of absorbing odors and vapors.

Air Changes   Expression of ventilation rate in terms of room or building volume. Usually given as air changes per hour.

Ambient   Surrounding (i.e., ambient temperature is the temperature in the surrounding space).

Ballast   A device used in a starting circuit for fluorescent and other types of lamps.

Blowdown   The discharge of water from a boiler or cooling tower sump that contains a high proportion of total dissolved solids.

British Thermal Unit (Btu)   A heat unit equal to the amount of heat required to raise one pound of water one degree Fahrenheit.

Building Envelope   All external surfaces which are subject to climatic impact, for example, walls, windows, roof, floor.

Building Load   Combination of cooling and heating loads. *Cooling load* is the rate of heat gain to the building at a steady-state condition when indoor and outdoor temperatures are at their selected design levels, solar gain is at its maximum for the building configuration and orientation, and heat gains due to infiltration, ventilation, lights, and people are present.
    Heating load is the rate of heat loss from the building at steady-state conditions when indoor and outdoor temperatures are at their selected design levels (design criteria). The heating load always includes infiltration and may include ventilation loss and heat gain credits for lights and people.

393

Cavity Ratio   Number indicating room cavity proportions which is calculated by using length, width, and height.

Centrifugal Chiller   A refrigeration machine using mechanical energy input to drive a centrifugal compressor to generate chilled water.

Centrifugal Fan   Device for propelling air by centrifugal action. Forward-curved fans have blades which are sloped forward relative to direction of rotation. Backward-curved fans have blades which are sloped backward relative to direction of rotation. Backward-curved fans are generally more efficient at high pressures than forward-curved fans.

Coefficient of Performance   Ratio of the tons of refrigeration produced to the energy required to operate equipment.

Coefficient of Utilization   Ratio of lumens on a work plane to lumens emitted by the lamps.

Cold Deck   A cold-air chamber forming part of a ventilating unit.

Condensate   Water obtained by changing the state of water vapor (i.e., steam or moisture in air) from a gas to a liquid, usually by cooling.

Condenser   A heat exchanger which removes latent heat from a vapor, changing it to its liquid state. (In a refrigeration chiller the component which rejects heat.)

Cooling Tower   A device that cools water directly by evaporation.

Damper   A device used to vary the volume of air passing through an air outlet, inlet, or duct.

Degree-Day   The difference between the median temperature of any day and 65°F when the median temperature is less than 65°F.

Degree-Hour   The difference between the median temperature for any hour and a selected datum.

Demand Factor   The ratio of the maximum demand of a system, or part of a system, to the total connected load of a system or part of the system under consideration.

Desiccant   A substance possessing the ability to absorb moisture.

Direct Expansion   Generic term used to describe refrigeration systems in which the cooling effect is obtained directly from the refrigerant (e.g., refrigerant is evaporated directly in a cooling coil in the airstream).

Disability Glare   Spurious light from any source which impairs a viewer's ability to discern a given object.

Double-Bundle Condenser   Condenser (usually in a refrigeration machine) that contains two separate tube bundles allowing the option of rejecting heat either to the cooling tower or to another building system requiring heat input.

Dry-Bulb Temperature   The measure of the sensible temperature of air.

Economizer Cycle   A method of operating a ventilation system to reduce refrigeration load. Whenever the outdoor air conditions are more favorable (lower heat content) than return air conditions, outdoor air quantity is increased.

Efficacy of Fixtures   Ratio of usable light to energy input for a lighting fixture or system (expressed in terms of lumens per watt).

Energy Requirement   The total yearly energy used by a building to maintain the selected inside design conditions under the dynamic impact of a typical year's climate. It includes raw fossil fuel consumed in the building and all electricity used for lighting and power. Efficiencies of utilization are applied, and all energy is expressed in the common unit of Btu.

Enthalpy   For the purpose of air conditioning, enthalpy is the total heat content of air above a datum, usually, expressed in units of Btu per pound. It is the sum of sensible and latent heat and ignores internal energy changes due to pressure change.

Equivalent Sphere Illumination   That illumination which would fall upon a task covered by an imaginary transparent hemisphere which passes light of the same intensity through each unit area.

Evaporator   A heat exchanger which adds latent heat to a liquid, changing it to a gaseous state. (In a refrigeration system it is the component which absorbs heat.)

Footcandle   Energy of light at a distance of one foot from a standard (sperm oil) candle.

Heat Gain   As applied to HVAC calculations, it is that amount of heat gained by a space from all sources, including people, lights, machines, sunshine, etc. The total heat gain represents the amount of heat that must be removed from a space to maintain indoor comfort conditions.

Heat Loss   The sum cooling effect of the building structure when the outdoor temperature is lower than the desired indoor temperature. It represents the amount of heat that must be provided to a space to maintain indoor comfort conditions.

Heat Pump   A refrigeration machine possessing the capability of reversing the flow so that its output can be either heating or cooling. When used for heating, it extracts heat from a low-temperature source and increases the temperature to the point where it can be used.

Heat Transmission Coefficient   Any one of a number of coefficients used in the calculation of heat transmission by conduction, convection, or radiation through various materials and structures.

Hot Deck   A hot-air chamber forming part of a ventilating unit.

Humidity, Relative   A measurement indicating moisture content of air.

Infiltration   The process by which outdoor air leaks into a building by natural forces through cracks around doors and windows, etc. (usually undesirable).

Insolation   The amount of solar radiation on a given plane. Expressed in langleys or Btu per square foot.

Langley   Measurement of radiation intensity. One langley equals 3.68 Btu per hour per square foot.

Latent Heat   The quantity of heat required to effect a change in state.

Life-Cycle Cost   The cost of the equipment over its entire life including operating and maintenance costs.

Load Leveling   Deferment of certain loads to limit electrical power demand to a predetermined level.

Load Profile   Time distribution of building heating, cooling, and electrical loads.

Lumen   Unit of luminous flux.

Luminaire   Light fixture designed to produce a specific effect.

Makeup   Water supplied to a system to replace that lost by blowdown, leakage, evaporation, etc.

Manchester, New Hampshire, Project   A demonstration building commissioned by the General Services Administration (Isaak and Isaak, Architects) and developed by Dubin-Mindell-Bloome Associates to incorporate energy-conserving architectural features and mechanical and electrical systems.

Modular   System arrangement whereby the demand for energy (heating, cooling) is met by a series of units sized to meet a portion of the load.

Orifice Plate   Device inserted in a pipe or duct which causes a pressure drop across it. Depending on orifice size, it can be used to restrict flow or form part of a measuring device.

Orsat Apparatus   A device for measuring the combustion components of boiler or furnace flue gases.

Piggyback Operation   Arrangement of chilled-water generation equipment whereby exhaust steam from a centrifugal chiller driven by a steam turbine is used as the heat source for an absorption chiller.

Power Factor   Relationship between volt-amperes and wattage. When the power factor is unity, volt-amperes equals wattage.

Raw Source Energy   The quantity of energy input at a generating station required to produce electrical energy, including all thermal and power conversion losses.

Roof Spray   A system that reduces heat gain through a roof by cooling the outside surface with a water spray.

R Value   The resistance to heat flow expressed in units of Btu per square foot per hour per degree Fahrenheit.

Seasonal Efficiency   Ratio of useful output to energy input for a piece of equipment over an entire heating and cooling season. It can be derived by integrating part load efficiencies against time.

Sensible Heat   Heat that results in a temperature change but no change in state.

Specific Heat   Ratio of the amount of heat required to raise a unit mass of material 1 degree to that required to raise a unit mass of water 1 degree.

Software   Term used in relation to computers; it normally describes computer programs and other intangibles.

Sol-Air Temperature   The theoretical air temperature that would give a rate of heat flow through a building surface equal in magnitude to that obtained by the addition of conduction and radiation effects.

Thermal Conductance   A measure of the thermal conducting properties of a composite structure such as wall, roof, etc., including the insulating boundary layers of air.

Thermal Conductivity   A measure of the thermal conducting properties of a single material expressed in units of Btu per inch thickness (square foot) (hour) (degree Fahrenheit temperature difference).

Ton of Refrigeration   A means of expressing cooling capacity: one ton equals 12,000 British thermal units per hour of cooling.

U Value   A coefficient expressing the thermal conductance of a composite structure given in units of Btu per square foot per hour per degree Fahrenheit temperature difference.

Vapor Barrier   A moisture-impervious layer designed to prevent moisture migration.

Veiling Reflection   Reflection of light from a task, or work surface, into the viewer's eyes.

Wet-Bulb Temperature   The lowest temperature attainable by evaporating water in the air without the addition or subtraction of energy.

BIBLIOGRAPHY AND SELECTED REFERENCES

GENERAL

1. Ad Hoc Committee on Energy Efficiency in Large Buildings: "Report to the Interdepartmental Fuel and Committee of the State of New York," Albany, N.Y., March 7, 1973.

2. American Institute of Architects Research Corporation: *Energy Conservation in Building Design*, Washington, D.C., May 1, 1974.

3. American Institute of Plant Engineers: *Better Ideas for Conserving Energy*, Cincinnati, Aug. 1, 1974.

4. American Society of Heating, Refrigerating, and Air-Conditioning Engineers: *Applications Handbook*, 345 E. 47th St., New York, 1974.

5. ———: *Design and Evaluation Criteria for Energy Conservation in New Buildings*, Proposed Standard 90-P, New York, 1974.

6. ———: *Equipment Handbook*, New York, 1972.

7. ———: *Handbook of Fundamentals*, New York, 1972.

8. ———: *Systems Handbook*, New York, 1973.

9. Blackwell, H. R.: "Application Procedures for Evaluation of Veiling Reflections in Terms of ESI," *Journal of IES*, April 1973.

10. Boynton, R. M., and D. E. Boss: "The Effect of Background Luminance upon Visual Search Performance," report to the Illuminating Engineering Research Institute, University of Rochester, N.Y.

11. Brown, J. L., and C. Mueller: "Brightness Discrimination and Brightness Contrast," in *Vision and Visual Perception*, New York: Wiley, 1965.

12. Citizens' Advisory Committee on Environmental Quality: *Citizen Action Guide to Energy Conservation*, Washington, D.C., Government Printing Office.

13. Commercial Refrigeration Manufacturers' Association: *Retail Food Store Energy Conservation*, Washington, D.C., 1975

14. Daylighting Committee of the Illuminating Engineering Society: "Recommended Practice of Daylighting," New York, 1962.

15. Department of Commerce, National Bureau of Standards: *Technical Options for Energy Conservation in Buildings*, Washington, D.C.: Government Printing Office, July 1973.

16. Department of Commerce, National Oceanic and Atmospheric Administration, National Weather Service: *Operations of the National Weather Service*, Silver Springs, Md.: October 1970.

17. Departments of the Air Force, Army, and Navy: *Engineering Weather Data*, Washington: Government Printing Office, 1967.

18. Dorsey, Robert T.: "An Extension of the Energy Crisis: More Light on Lighting," *American Institute of Architects Journal*, June 1972.

19. Dubin, Fred S.: "Check-off List on Energy Conservation with Regard to Power," *Electrical Consultant*, 89:4.

20. ———: "Energy Conservation in Building Operations," address to Regional Directors of the Building Management Division of the General Services Administration, April 3, 1973.

21. ———: "Energy Conservation Needs New Architecture and Engineering," *Public Power*, March/April 1972.

22. ———: "Energy Conservation through Building Design and a Wiser Use of Electricity," address to Annual Conference of American Public Power Association, San Francisco, June 26, 1972.

23. ———: "If You Want To Save Energy," *American Institute of Architects Journal*, December 1972.

24. ———: and Margot Villecco: "Energy for Architects," *Architecture Plus*, July 1973.

25. Dubin-Mindell-Bloome Associates, P.C.: *A Study of Air Conditioning, Heating, and Ventilating Design for Veterans Administration Hospital Kitchens and Laundries*, Research Staff, Office of Construction, Veterans Administration, Washington, D.C., February 1970.

26. ———: *A Study of Design Criteria and Systems for Air Conditioning Existing Veterans Administration Hospitals*, Research Staff, Office of Construction, Veterans Administration, Washington, D.C., March 1971.

27. ———: *Energy Conservation and Office Building Design: Research, Analysis, and Recommendations: Office Building, Manchester, New Hampshire*, May 15, 1973.

28. ———: *Energy Conservation Design Guidelines for Office Buildings*, General Services Administration, Public Buildings Service, Washington, D.C., January 1974.

29. ———: *Report to Connecticut General Insurance Corporation on Energy Conservation Opportunities*, December 1973.

30. "Energy System Limits Waste," *Design Engineering Journal*, February 1974.

31. Fanger, P. O.: *Thermal Comfort: Analysis and Applications in Environmental Engineering*, New York: McGraw-Hill, 1972.

32. Federal Energy Administration: *Tips for Energy Savers*, Washington, D.C.: Government Printing Office, 1976.

33. Federal Energy Office: *Project Independence Background Paper*, Washington, D.C.: Feb. 11–12, 1974.

34. Gatts, Robert R., Robert G. Massey, and John C. Robertson: *Energy Conservation, Program Guide for Industry and Commerce*, Department of Commerce, National Bureau of Standards, Washington, D.C.: Government Printing Office, 1974.

35. Givoni, B.: *Man and Architecture*, London: Elsevier, 1969.

36. Griffith, J. W.: "Analysis of Reflected Glare and Visual Effect from Windows," paper presented to the National Technical Conference of the Illuminating Engineering Society, Detroit, Sept. 8–13, 1963.

37. "GSA Sets Energy Saving Steps," *Design Engineering Journal*, May 1974.

38. Heating, Piping and Air Conditioning Seminar: "Energy Conservation for Existing Buildings," proceedings of New York, December 1974.

39. "Heat Recovery Cuts Energy Use," *Design Engineering Journal*, February 1974.

40. Henderson, R. L., J. F. McNelis, and H. G. Williams: "A Survey and Analysis of Important Visual Tasks in Offices," Annual Conference of the Illuminating Engineering Society, 1974.

41. *IES Lighting Handbook*, 5th ed., New York: Illuminating Engineering Society, 1972.

42. "Innovative Lighting Fixture Aids," *Actual Specifying Engineer*, April 1972.

43. Jennings, Burgess H., Necati Ozisik, and Lester F. Schutrum: "New Solar Research Data on Windows and Draperies," *American Society of Heating, Refrigerating, and Air-Conditioning Engineers Journal* reprint.

44. Kaufman, John E.: "Energy Utilization: Optimizing the Uses of Energy for Lighting," *Lighting Design & Application*, October 1973.

45. Dubin, Fred: "Life Support Systems for a Dying Planet," *Progressive Architecture*, October 1971.

46. Lighting Survey Committee: "How to Make a Lighting Survey," report to the Illuminating Engineering Society, November 1963.

47. McClure, Charles J. R.: "Optimizing Building Energy Use," *American Society of Heating, Refrigerating, and Air-Conditioning Engineers Journal*, September 1971.

48. Murphy, E. E., and J. G. Dorsett: "Demand Controllers Disconnect Some Loads during Peak Periods," *Transmission & Distribution*, July 1971.

49. Northeast Utilities Company: *Guidelines for Implementing Northeast Utilities Systems' Energy Management Program*, Hartford, Conn., 1973.

50. Office of Emergency Preparedness: *The Potential for Energy Conservation*, Washington, D.C.: Government Printing Office, October 1972.

51. Olivieri, Joe B.: "Heat Recovery Systems—Part I," *Air Conditioning, Heating & Refrigeration News*, Jan. 1, 1972.

52. ————: "Heat Recovery Systems—Part II," *Air Conditioning, Heating & Refrigeration News*, Jan. 17, 1972.

53. Phipps, H. Harry: "Energy Systems Analysis—Why and How," *Building Systems Design*, May 1972.

54. Plant Engineering: "Getting the Most from Your Electrical Power System: Selected Articles." Barrington, Ill.: Plant Engineering Library.

55. Price, Seymour: *Air Conditioning for Building Engineers and Managers*, New York: The Industrial Press, 1970.

56. "The Role of Refrigeration in Energy Conservation and Environmental Protection," seminar of the Air-Conditioning and Refrigeration Institute: New Orleans, Jan. 26, 1972.

57. Ross & Baruzzini, Inc.: *Energy Conservation Applied to Office Lighting*, Contract No. 14-01-001-1845, Washington: Federal Energy Administration, April 1, 1975.

58. ————: *Lighting Systems Study*, Washington, D.C.: General Services Administration, Public Building Service, March 1974.

59. Saunders, J. E.: "The Role of the Level and Diversity of Horizontal Illumination in an Appraisal of a Simple Office Task," *Lighting Research and Technology*, 1969:1.

60. Shlaer, S.: "The Relation between Visual Acuity and Illumination," *Journal of General Physiology*, 1937:21.

61. Stewart, John L.: "Report on TC 4.5-Fenestration," *American Society of Heating, Refrigerating, and Air-Conditioning Engineers Journal*, November 1972.

62. Strock, Clifford, and Richard L. Koral (eds.): *Handbook of Air-Conditioning, Heating, and Ventilating*, New York: The Industrial Press, 1965.

63. Syska & Hennessy, Engineers, and Tishman Research Corporation: *A Study of the Effects of Air Changes and Outdoor Air on Interior Environment, Energy Conservation, and Construction and Operating Costs: Phase II*, Washington, D.C.: General Services Administration, December 1973.

64. "Three Hundred Hints to Save Energy," *Congressional Record*, Oct. 30, 1973.

SOLAR ENERGY

65. American Society of Heating, Refrigerating, and Air-Conditioning Engineers: "Low Temperature Applications of Solar Energy," New York, 1967.

66. Barber, E. M., and D. Watson: "Criteria for the Preliminary Design of Solar Heated Buildings," Sunworks, Inc., Guilford, Conn., 1974.

67. Colorado State University and University of Wisconsin: *Design and Construction of a Residential Solar Heating and Cooling System,* Report No. NSR/RANN/SE/GE-40451/PR/74/2, National Science Foundation, Washington, D.C.

68. Daniels, Farrington: *Direct Use of the Sun's Energy,* New York: Ballantine, 1974.

69. Department of Commerce, National Bureau of Standards: *Performance Criteria for the Solar Heating/Cooling of Residences,* Washington, D.C.: Government Printing Office.

70. Dubin-Mindell-Bloome Associates: "Specifications of Solar Collectors for the Federal Office Building, Manchester, New Hampshire," Boston: General Services Administration.

71. Duffie, J. A., and W. A. Beckman: *Solar Energy Thermal Processes,* New York: Wiley, 1974.

72. "Report of the Solar Heating and Cooling Committee of BRAB," National Academy of Engineers, Washington, D.C.: National Science Foundation, 1974.

73. Shurcliffe, W. A.: "Solar Heating Buildings, a Brief Survey," *Solar Energy Digest,* P.O. Box 17776, San Diego, Calif. 92117.

74. "Solar Energy Industry Report," 1001 Connecticut Ave., N.W., Washington, D.C.: 20036.

75. Yellot, John I.: "Utilization of Sun and Sky Radiation for Heating and Cooling of Buildings," *ASHRAE Journal,* December 1973.

TOTAL ENERGY SYSTEMS

76. Ahner, D. J.: "Environmental Performance," based on a paper presented at the General Electric Gas Turbine State of the Art Engineering Seminar, June 1971.

77. Barrangon, Maurice: "Conservation of Resources, How Total Energy Saves Fuel," *GATE Information Digest,* March 1971.

78. Becker, Herbert P.: "The Concept of Total Energy—Lecture No. 1," Fall Lecture Series, *ASHRAE,* October 1967.

79. Bjerklie, John W., P.E.: "Small Gas Turbines for Total Energy Systems," *Actual Specifying Engineer,* August 1971.

80. Dubin, Fred S., P.E.: (editorials) "Life Support Systems for a Dying Planet"; "Available Now: Systems that Save Energy"; "Can Building Codes Help Protect the Environment?" *Progressive Architecture,* October 1971.

81. ————: "The New Architecture and Engineering," presented at Workshop on Total Energy Conservation in Public Buildings, State University of New York, Albany, January 1972.

82. ————: "Total Energy Systems for Mass Housing," paper delivered at the Urban Technology Conference 2, San Francisco, July 1972.

83. Dubin, Fred S., P. E.: "TR-2—Total Energy for Schools and Colleges," report prepared for the Educational Facilities Laboratories of the Ford Foundation.

84. Dubin-Mindell-Bloome Associates: "Total Energy," prepared for Education Facilities Laboratories, Ford Foundation, May 1970.

85. Environmental Protection Agency: "Standards of Performance for New Stationary Sources," *Federal Register*, vol. 36, no. 247, pt. 11, December 1971.

86. Henderson, John O.: "Turbines or Engines for Total Energy?" Caterpillar Tractor Company, *Gas Age*, May 1966.

87. Holdren, John, and Philip Herrera: *Energy*, Sierra Club, 1971.

88. Huber, Ernest G., P. E.: "Total Energy Concept Applied to Small Diesel Engines and Gas Turbines," File No. 3207, *Air Force Civil Engineer*, February 1970.

89. Kennedy, J. J.: "Watch the Basic Details When Installing Gas Prime Movers," File No. 3202, *Plant Engineering*, February 1970.

90. "Total Energy Evaluated," *Energy International*, February 1972.

METRIC UNITS AND CONVERSION FACTORS

SI UNITS

BASIC SI UNITS

Measurement	Unit	Abbreviation
Length	m	meter
Mass	kg	kilogram
Time	s	second
Temperature	K	kelvin
Electric current	A	ampere
Luminous intensity	cd	candela

DERIVED SI UNITS

Measurement	Unit	Abbreviation
Area	square meter	m^2
Volume	cubic meter	m^3
Density	kilogram per cubic meter	kg/m^3
Volumetric flow rate	cubic meter per second	m^3/s
Force	newton (N)	kg/m^2
Pressure	pascal (Pa)	N/m^2
Energy	joule (J)	$N \cdot m$
Power	watt (W)	J/s
Heat flux density	watt per square meter	W/m^2
Luminance	candela per square meter	cd/m^2

UNITS IN USE WITH SI UNITS

Measurement	Unit	Symbol	Definition
Time	minute	min	1 min = 60 s
	hour	h	1 h = 3600 s
	day	d	1 d = 86,400 s
	week, month, etc.	—	
Plane angle	degree	°	$1° = (\pi/180)$ rad
Temperature	degree Celsius	C	(See Temperature table.)
Volume	liter	L	$1 \text{ L} = 10^{-3} \text{ m}^3$

PREFIXES FOR SI UNITS

Prefix	Symbol	Multiple
tera	T	10^{12}
giga	G	10^{9}
mega	M	10^{6}
kilo	k	10^{3}
milli	m	10^{-3}
micro	μ	10^{-6}
nano	n	10^{-9}
pico	p	10^{-12}

CONVERSION FACTORS

LENGTH

1 ft	= 0.3048 m
1 mi	= 1.6093 m
1 in	= 25.4 mm
1 yd	= 0.9144 m

AREA

1 ft²	= 0.092903 m²
1 mi²	= 2.58999 km²
1 in²	= 0.000645 m²

VELOCITY

1 ft/min	= 0.00508 m/s
1 mi/h	= 0.44704 m/s

VOLUME

1 ft³	= 28.3168 L
1 gal	= 3.78544 L
1 ft³	= 7.48 gal
1 yd³	= 0.7645 m³
1 gal/ft²	= 0.02454 L/m²

VOLUMETRIC RATE

1 cfm	= 0.47195 L/s
1 gal/min	= 0.06309 L/s
1 gal/(min) (ft²)	= 0.6791 L/(s) (m²)
1 cfm/ft² (air)	= 0.1968 L/(s) (m²)

MASS

1 lb	= 0.453492 kg
1 oz	= 28.3495 g

MASS FLOW RATE

1 lb/h	= 0.000126 kg/s
1 lb/(h) (ft²)	= 0.001356 kg/(s) (m²)

TEMPERATURE

°F	= (°C × 1.8) + 32
°C	= (°F − 32) × 5/9
K	= °C + 273
°R	= °F + 462

ENERGY

1 Btu	= 1.05505 kJ
1 Therm	= 105.506 MJ
1 cal	= 4.1868 J
1 kWh	= 3.6 MJ
1 langley	= 41.86 kJ/m²

POWER

1 Btu/h	= 0.29307 W
1 ton (refrig.)	= 3.41685 kW
1 kcal/h	= 1.163 W
1 hp	= 0.74570 kW

ENERGY FLUX

1 Btu/(h) (ft²)	= 3.15469 × 10⁴ W/m²
1 langley/h	= 11.6277 W/m²
1 cal/(cm²) (min)	= 697.4 W/m²
1 Btu/(h) (ft²) (°F)	= 5.67826 W/(m²) (°C)
1 Btu/(h) (ft) (°F)	= 1.70307 W/(m) (°C)

INDEX